A

TEXT-BOOK OF BOTANY.

BY

J. M. LOWSON, M.A., B.Sc., F.L.S.

FORMERLY LECTURER AT UNIVERSITY TUTORIAL COLLEGE.

Third Edition.

LONDON: W. B. CLIVE,

𝓤𝓷𝓲𝓿𝓮𝓻𝓼𝓲𝓽𝔂 𝓣𝓾𝓽𝓸𝓻𝓲𝓪𝓵 𝓟𝓻𝓮𝓼𝓼 𝓛ᵈ.

(*University Correspondence College Press*),

157 DRURY LANE, W.C.

1903.

PREFACE TO THE FIRST EDITION.

WHILE the intention of the author in writing this text-book of Botany has been to provide a work that will be an efficient introduction to the subject, it includes all the types set for the London University Intermediate Science and Preliminary Scientific Examinations, for which it is specially adapted. Throughout, in both text and illustration, the author has aimed at presenting the chief facts in the structure and life-histories of the types selected as simply and as clearly as possible. At the same time an endeavour has been made, wherever possible, to direct the student's attention to the leading principles underlying these facts.

It has been considered advisable, in order to secure compactness of treatment and to avoid confusing digressions in later chapters, to devote the first part of the text-book to the general facts of structure and physiology. Needless to say, it is not expected that the student will be able to master the contents of the opening chapters at the first reading.

All the illustrations have been specially drawn and special precautions have been taken to make them clear and helpful. Diagrams of subjects from nature cannot in most cases materially differ in different text-books, but some of the figures in this work will be found to be quite new. It is hoped that the plan adopted for indicating the various parts of the figure will materially assist the student in his reading.

The author desires to express his indebtedness to the following standard works to which reference has frequently been made:

Vines' *Student's Text-Book of Botany*, Scott's *Structural Botany*, Willis' *Flowering Plants and Ferns*, Green's *Manual of Botany*, Goebel's *Outlines of Classification*, Strasburger's *Text-Book of Botany*. The classification of tissues, pp. 41-48, is in accordance with the excellent arrangement given by Professor Vines.

The sincere thanks of the author are due to Mr. A. M. Davies, B.Sc., F.G.S., for his kindness in reading the proof sheets, and for many valuable suggestions.

UNIVERSITY TUTORIAL COLLEGE,
October, 1898.

PREFACE TO THIRD EDITION.

In Parts I.—III. various alterations and additions have been made. The more important of these are on pp. 155-157, 197-199, 220-221, which have been partly rewritten, and on p. 304 where a note on "Double Fertilization" has been added.

The life-history of *Claviceps purpurea*, the Ergot of Rye, has been introduced in Part IV., and a description of twenty additional Natural Orders, with over fifty new diagrams, is given in the Appendix. The Index also has been enlarged.

It is hoped that these changes will render the text-book still more suitable for general readers, and for students preparing for special examinations.

September, 1903.

CONTENTS.

INTRODUCTORY 1

PART I.—GENERAL.

CHAPTER
I. EXTERNAL MORPHOLOGY AND PHYSIOLOGY 5
II. GENERAL HISTOLOGY
    A. THE CELL 17
    B. THE TISSUES 40
    C. TISSUE SYSTEMS 49

PART II.—THE ANGIOSPERM.

III. SEED AND EMBRYO 55
IV. THE STEM OF THE ANGIOSPERM
    A. EXTERNAL CHARACTERS 61
    B. INTERNAL STRUCTURE 75
        1. *THE DICOTYLEDON* 75
        2. *THE MONOCOTYLEDON* . . . 90
        3. *GENERAL* 93
V. THE ROOT OF THE ANGIOSPERM
    A. EXTERNAL CHARACTERS 95
    B. INTERNAL STRUCTURE 98
VI. THE LEAF OF THE ANGIOSPERM
    A. EXTERNAL CHARACTERS 109
    B. INTERNAL STRUCTURE 125
VII. NUTRITION AND GROWTH 130
VIII. THE PLANT AND ITS ENVIRONMENT . . 152
IX. STRUCTURE OF THE FLOWER 158
X. THE INFLORESCENCE 185
XI. REPRODUCTION AND LIFE-HISTORY OF THE ANGIOSPERM 192
XII. SEED AND FRUIT
    A. SEED 203
    B. FRUIT 205
    C. DISPERSAL OF SEEDS AND FRUITS . . 216

viii CONTENTS.

CHAPTER	PAGE
XIII. SYNOPSIS OF NATURAL ORDERS	218
QUESTIONS ON THE ANGIOSPERM	245

PART III.—VASCULAR CRYPTOGAMS AND FLOWERING PLANTS.

XIV. STRUCTURE AND LIFE-HISTORY OF THE FERN 248
XV. EQUISETUM AND SELAGINELLA
    A. EQUISETUM 268
    B. SELAGINELLA 270
XVI. THE GYMNOSPERMS—STRUCTURE AND LIFE-HISTORY OF PINUS 280
XVII. HOMOLOGIES IN ANGIOSPERMS . . . 299
XVIII. RELATIONSHIP BETWEEN V. CRYPTOGAM AND FLOWERING PLANT 305
    QUESTIONS ON THE V. CRYPTOGAMS AND FLOWERING PLANTS 310

PART IV.—THE LOWER CRYPTOGAMS.

XIX. LIVERWORTS AND MOSSES
    A. MARCHANTIA 312
    B. FUNARIA 320
XX. THE ALGÆ 330
    SPIROGYRA, 331; ULOTHRIX, 334; VAUCHERIA, 338; FUCUS, 341.
XXI. THE FUNGI 348
    PYTHIUM, 350; EUROTIUM, 354; CLAVICEPS (ERGOT), 358; SACCHAROMYCES (YEAST), 364; AGARICUS, 368; BACTERIA, 371.
QUESTIONS ON THE MUSCINEÆ AND THALLOPHYTA . 375

APPENDIX.
    A. GENERAL ADVICE 377
    B. MICROSCOPICAL WORK 380
    C. DESCRIPTIVE BOTANY 384
    D. ADDITIONAL NATURAL ORDERS . . . 388
INDEX 442

INTRODUCTORY.

§ 1. **Botany** is the science which deals with the phenomena of plant-life. It considers the form and structure of plants, their functions and life-histories; it studies the various processes of their growth and development; and, by carefully noting the resemblances and differences discernible between them, it attempts to draw up an ordered scheme or classification, expressing as clearly as possible their affinities or relationship to one another.

§ 2. **Subdivisions of the Science.** Plants, like animals, may be studied in different ways or from different points of view. These constitute different subdivisions or departments of Botanical Science. Of these the most important, and the only ones which closely concern us here, are **Morphology** and **Physiology.** We may indicate the scope of these by considering a particular plant—let us say the Sunflower. Naturally we should first of all give our attention to external features. We should find that the plant consists of certain well-defined parts or **members** —roots, stems, leaves, flowers, etc.; that, speaking generally, these have approximately the same forms in all sunflowers of the same species, while they differ in many respects from those of other plants; that the stem branches, and the branches have a definite position in relation to the leaves. By comparing the relative positions of the members in the sunflower with those in other plants it would be possible to draw up a classification of plant-members. A study of this kind, dealing with the external forms and relative positions of plant-members, is called **External Morphology.**

We might then wish to get some knowledge of the internal parts of these various members. For this purpose we should take various sections of the stem, root, leaf, etc., or examine these in various other ways. Such a study, dealing with internal structure, is called **Internal Morphology**. It may be carried on in two ways. First, we might content ourselves with the naked-eye appearance of the internal parts, and examine only the coarser features. This is **Anatomy**. Secondly, we might undertake a closer study with the help of the microscope, and make out the finer features of structure—the cells and tissues of the plant-body. This is **Histology**. Evidently, then, *morphology* has regard only to the form and structure of plants, and pays no attention to the vital processes which are carried on. But we might carry our morphological studies much further than this. Instead of examining the sunflower at any one particular stage in its growth, we might study the form and structure exhibited through all the stages, and observe the changes which take place. Beginning with the seed, we could make out the parts of the embryo plant contained in the seed; we could notice the gradual growth of the root and stem, and the formation of leaves; we could study the origin and growth of branches; finally, we could make a study of the flower and investigate the morphological changes which lead up to the formation of another seed. This is a study of **Development**. It seeks to discover the earliest beginnings of the organism and to trace the changes and modifications which occur in passing from one stage to another. Development might be defined as a history of the morphology of an organism. Another morphological study is **Classification**. It is based on Comparative Morphology and Development. Here the forms and structures of plants in all stages of their life-histories are studied and compared, and according to the resemblances and differences perceived the plants are arranged or classified.

The physiologist would approach our sunflower plant in quite a different frame of mind. Neglecting for the moment its form and structure, he would set himself such questions as:—How does the plant obtain nourishment? What is the nature of its food-material? How are these food-materials assimilated? What is growth? What processes are carried on in connexion with growth? How is the plant affected by its environment? What influence have light, heat, etc., on plant growth? How is reproduction carried on?—etc. To give an answer to these and all such questions belongs to **Physiology.** Thus Physiology concerns itself with the various life-processes—that is, with the functions necessary for the welfare of the individual and the perpetuation of the species. It regards form and structure as of merely secondary importance, and to be discussed only in so far as they aid in the elucidation of function. In this connexion the various parts of the organism are considered as **organs** adapted to the performance of various functions.

§ 3. **General Classification of the Plant Kingdom.** There is no need to enlarge on the infinite variety of form presented in the Vegetable Kingdom. The beginner in Botany is sufficiently impressed with it. The refuge from hopeless confusion is found in Classification. At the outset a certain advantage will be obtained if the student obtains, as it were, a bird's-eye view of the country he is about to enter. For this reason, and also because it will serve as a convenient table of reference, we venture at this early stage to give a *general* classification indicating clearly the position of the various plant-types we shall consider in the following pages. In very much the same way as we might ask the student to think of a bird, a fish, an insect, and a mussel, and notice how very different they are from each other, so we would ask him to consider and contrast four plant-types—namely, a buttercup, a fern, a moss, and a seaweed. In a general way, he will recognise that they present considerable differences from each other—the buttercup alone has flowers—the fern has a stout underground stem, roots and leaves—the moss is a much more delicate plant, with stem and leaves but no true roots—the seaweed bears no members resembling the stem and leaves of the other types. Now, these four plants may be taken as types of the four chief groups of the Plant Kingdom.

These and the more important subdivisions are represented in the following scheme :—

A. **Thallophyta.**
  (i) **Algæ**—mostly aquatic plants, including the seaweeds and various fresh-water forms ; *e.g.* Fucus, Sp'rogyra, Ulothrix, Vaucheria.
  (ii) **Fungi**, including moulds, toadstools, etc. ; *e.g.* Agaricus, Pythium, Eurotium, Saccharomyces.

B. **Muscineæ** or **Bryophyta**, including the mosses and liverworts, *e.g.* Marchantia, Funaria.

C. **Pteridophyta**, or **Vascular Cryptogams**, *e.g.* ferns and selaginellas (also horse-tails and clubmosses).

D. **Phanerogams, Spermaphyta** or **Flowering Plants.**
  (i) **Gymnosperms**, *e.g.* Pinus, the pine, the commonest species of which is *Pinus sylvestris*, the Scots fir (also the larches, spruces, yews, cedars, cypresses, etc.).
  (ii) **Angiosperms**—the highest or typical Flowering Plants.
    (a) **Monocotyledons**, *e.g.* grass, lily, narcissus, orchid, etc.
    (b) **Dicotyledons**, *e.g.* sunflower, buttercup, rose, etc.

The Flowering Plants were called **Phanerogams** because, bearing flowers, and forming seed, their method of reproduction was regarded as clear or evident (Gr. φανερος, *evident* ; γαμος, *marriage*) ; the other groups, Thallophyta, Bryophyta, and Pteridophyta, were grouped together as **Cryptogams,** because their reproductive processes, it was thought, were hidden or concealed (Gr. κρυπτος, *hidden ;* γαμος). These terms are still retained, although they have lost their original significance. The reproduction of the Cryptogams has been fully elucidated, and, as a matter of fact, is more evident than that of the Phanerogams.

PART I.—GENERAL.

CHAPTER I.

EXTERNAL MORPHOLOGY AND PHYSIOLOGY.

§ 1. **Unicellular and Multicellular Plants.** The lowest plants are of microscopic size and have a very simple structure. For example, in a simple Alga called *Pleurococcus* (fig. 1), each individual consists of a little round vesicle filled with a slimy granular substance called *protoplasm*, in which is embedded a denser protoplasmic body, the nucleus, along with others, the chloroplasts, through the substance of which is diffused a green colouring matter called chlorophyll.

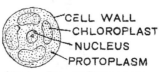

Fig. 1.—CELL OF PLEUROCOCCUS.

Such a structure is called a **cell**. The lowest forms of plant-life are one-celled or unicellular. In all except these very simple forms the plant-body is multi-cellular, *i.e.* it consists of numbers of cells aggregated together and closely united with each other.

§ 2. **Differentiation.** In unicellular plants all the vital functions are carried on by the single cell; but in multi-cellular forms, as a rule, different parts of the organism take on different functions, and each part has a form and structure adapted to the performance of its particular function. Members which have special forms adapting them to the performance of particular functions are said to be *specialized*. Thus the organism consists of parts or

6 BOTANY.

members differing from each other. Seeing that these various functions are carried on for the good of the organism as a whole, it is evident that in such an organism there is **Division of Labour.** The distribution of functions which characterises this division of labour is called **Physiological Differentiation**; and the marking off of distinct members serving as the organs of particular functions, which is correlated with it, constitutes **Morphological Differentiation.** It is evident that morphological and physiological differentiation go hand in hand. In the lower forms, the functions carried on are very simple and general, and there is comparatively little division of labour, so that morphological differentiation is only slightly marked. As we ascend from lower to higher forms, however, we find that the arrangements become more and more numerous and complex, and the division of labour correspondingly extensive. Hence it is in the highest plants that we meet with the most pronounced and far-reaching differentiation of members. As a matter of fact, speaking generally, we distinguish between lower and higher forms by the degree of differentiation and division of labour exhibited in each case.

§ 3. **The Thallus.** Amongst the Thallophyta (p. 4) the plant - body is very simple. It may be unicellular; when multicellular, it usually consists of a flattened membranous expansion, or of a mass of branched or unbranched filaments (fig. 2). Various members are in many cases more or less distinctly differentiated. As a rule, however, there is no clearly marked separation into distinct members corresponding to the root, stem, and leaf of higher plants. In the higher forms only of Thallophyta do we find

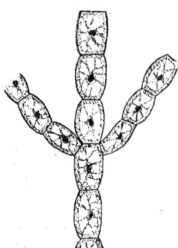

Fig. 2.—PART OF A BRANCHED FILAMENTOUS THALLUS.

indications of such a differentiation. For the most part, the lateral outgrowths reproduce the structure of the part from which they spring. A vegetative structure of this kind is called a **thallus**. It is specially characteristic of the Algæ and Fungi—although not confined to these—hence the name Thallophyta for the division in which they are placed.

§ 4. **Shoot and Root.** In plants above the Thallophytes, the plant-body usually exhibits a differentiation into distinct members, which becomes more marked and more complex as we pass from lower to higher types. In these plants a descending portion, which serves to fix the plant in the soil and absorb nourishment, is usually clearly distinguished from a part whose tendency is to pass upwards towards the light. These parts are known as the *root* and the *shoot*. Plants exhibiting this differentiation, including Bryophyta, Vascular Cryptogams, and Flowering Plants, are called **Cormophytes** to distinguish them from the Thallophytes. The shoot is nearly always further differentiated into **stem** and **leaf** (leafy shoot). The differentiation of root probably arose in adaptation to terrestrial conditions of life—the earlier and more primitive plants being aquatic Thallophytes.

§ 5. **Vegetative Shoot and Reproductive Shoot.** In many Vascular Cryptogams the same shoot carries on the nutritive or *vegetative* functions and bears the reproductive organs; but in others there is a differentiation of two kinds of shoots—one purely vegetative, the other reproductive. This differentiation is carried still further in the Flowering Plants, where the reproductive shoots (floral region of the plant) are clearly marked off in most cases from the vegetative shoots (the foliage region). Our conceptions of stem and leaf structures are derived from a consideration of the foliage or vegetative region of the plant. Nevertheless, the *flower*, although in appearance altogether different, consists, like the foliage or vegetative shoot, of stem and leaf structures, only these have been very highly specialized in adaptation to the special functions which they carry on.

§ 6. Higher Differentiation or Specialization of Members.

In the Flowering Plants, the members take on very various and often highly specialized or complex functions. In each case the special form and structure of the member are adapted to the performance of its special function. This is why in the flowering plants we meet with such a countless variety in the form of the various members. In both vegetative and floral regions, even of the same plant, stems and leaves may have many different forms, according to the special functions which they perform. When the specialization is of a very extreme character, there is often a difficulty in realizing that the structures under consideration are really stem or leaf structures—as *e.g.* in the pitcher-plant, where the leaves take the form of pitchers. Such highly specialized forms are spoken of as "*modifications.*" These so-called "modified" forms of stems, leaves, and roots are extremely numerous and interesting, and we shall have to deal with them in detail when we come to consider fully the morphology of these members. At present it is desirable that the student should recognize that these highly specialized forms have arisen simply as adaptations to a special environment and for the performance of special functions.

§ 7. Plant-Members.

Thus in the higher plants three chief categories of members—members of the first rank—are recognized, viz.: **roots, stems,** and **leaves.** According to the various well-marked "specializations" met with, these may be subdivided into members of secondary rank; thus leaves may be subdivided into foliage leaves, floral leaves, and so on. And these again may be subdivided. In each category or group the members are **morphologically similar,** but members belonging to different groups are **dissimilar.** Thus stem and leaf, stem and root, are examples of dissimilar members.

§ 8. Appendages or Outgrowths.

In addition to members which are included in one or other of the above categories, other members of subordinate rank are found in plants. They are of the nature of appendages or outgrowths of the members of the first order. Their forms are innumerable, for they include all the different kinds of hairs, prickles, etc. They may be developed on all parts of the plant, but chiefly on stem and leaf structures. Many reproductive bodies originate as special outgrowths or appendages on stem or leaf structures.

§ 9. **Symmetry of Plant-Members.** Plant-members usually exhibit more or less well-marked symmetry. They may be divided in certain directions so as to give similar halves. The forms and degrees of symmetry are various, the two chief ones being:—

> (a) **Radial Symmetry.**—Where the member can be divided into similar halves by a number of planes (two or more) passing through some particular axis.
>
> (b) **Bilateral Symmetry.**—Where the member can be so divided by only one or, at most, two planes.

For example, most stems and roots have radial symmetry; they are *usually* perfectly symmetrical round their longitudinal axis; so also many flowers, and a few cylindrical leaves (*centric* leaves, *e.g.* the onion).

There are two kinds of **bilateral symmetry**:—(i) the member may be divisible in two planes at right angles. In this case the halves formed by division along one plane resemble each other, but differ from the halves formed by division along the other plane. Thus the walnut may be divided along the line which separates the two valves of the fruit, or at right angles to this. So also the leaf of the iris. It is a vertical leaf and shows similar right and left surfaces. It may be divided longitudinally either parallel to these surfaces or at right angles to them. This form of symmetry is known as the **isobilateral**. (ii) There may be only one plane of symmetry. Here the symmetry is **zygomorphic** and the member is monosymmetrical. Examples of this are common. It is seen in many flowers, *e.g.* the pea or violet. When the zygomorphic symmetry is such that distinct lower and upper surfaces can be distinguished, the members are said to be *dorsiventral*. This is the case in the common or *bifacial* type of leaf.

§ 10. **Branching of Members.** The various members of a plant may bear other members either like or unlike themselves, *i.e.* similar or dissimilar. Thus roots may bear lateral secondary roots, *i.e.* similar members; stems may bear secondary stems and leaves, *i.e.* both similar and dissimilar members. The development of similar members is called *branching*.

There are two chief ways in which branches may be produced—two chief types of branching :—(*A*) **dichotomous branching** (fig. 3, A); (*B*) **lateral branching** (fig. 3, B, C). In dichotomous branching the growing apex of a stem or a root is divided into two and each part grows out into a branch. The branching in this case consists of a series of bifurcations. True dichotomy is comparatively rare, at least in the higher plants. It probably does not occur at all in the Flowering Plants; but examples of dichotomous branching are found in the Vascular Cryptogams and Bryophyta, and are common amongst Thallophytes. In lateral branching the branches arise as lateral outgrowths a short distance behind the extreme apex of the growing region of the parent member. This is the characteristic mode of branching in the Flowering Plants. If the parent member continues to grow, and develops numerous lateral branches, one after the other, the lateral branching is said to be **indefinite** or **racemose** (fig. 3, B). If, as is nearly always the case, these numerous lateral branches are produced in regular order, and in such a way that the youngest lie nearest the apex, they are said to be developed in *acropetal succession*. If, however, the parent member ceases to grow after producing one or a very few branches, and the growth is continued by these branches repeating the process, the lateral branching is said to be

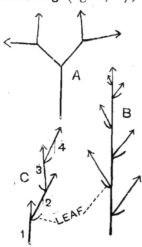

Fig. 3.—Forms of Branching. A, Dichotomous; B, Racemose; C, Cymose.

definite or cymose. Thus in fig. 3, c, axis 1 produces axis 2 and ceases to grow; 2 produces 3; 3 produces 4; and so on.
A. *Dichotomous.*
B. *Lateral.*
  (a) *Indefinite* or *Racemose* (Monopodial).
  (b) *Definite* or *Cymose.*

Here we have a general table of branching which will be supplemented and illustrated in connection with the detailed description of the morphology of roots, stems, leaves, and inflorescences.

§ 11. **Nutrition and Growth.** We have already seen that the lowest forms of plants are unicellular. In typical cases the cell (§ 1) consists of a nucleated speck of protoplasm coloured green by chlorophyll and enclosed in a delicate membrane or cell-wall. The protoplasm is the *living* substance of the cell. The cell-wall merely supports and protects the protoplasm. If we examine such an organism carefully we find that it grows in size and reproduces itself. Now it is evident that there must be some source from which it draws food-materials, and certain vital processes or functions must be carried on in order that these materials may be assimilated or built up into its own substance. These processes are the same in kind as are carried on by all green plants, but in these unicellular plants they are all performed by a single cell, and may therefore be studied in their simplest forms.

The first consideration is that food-materials, of whatever kind they may be, cannot, owing to the presence of a definite cell-wall, enter the cell in solid form, but must do so in solution. Here we have what is sometimes regarded as a fundamental distinction between a plant and an animal. All plants take in their food-material in solution.

Our green unicellular plant grows either in water or on some moist substratum. This water, with various substances in solution, passes by diffusion through the cell-wall and is absorbed into the interior of the cell. The dissolved substances are of a very simple character. The most important of them are certain mineral substances, such as nitrates, sulphates, phosphates, and carbon dioxide (CO_2). Here we have another important point of distinction between an ordinary green plant and an animal. Green plants take in their food-material in the form of simple *inorganic* substances, animals cannot live on simple inorganic substances; they must feed on complex *organic* compounds, such as carbohydrates, fats and oils, and proteids.

The question now arises—how are these simple compounds built up into living protoplasm, which is an extremely complex unstable substance containing carbon, oxygen, hydrogen, nitrogen, sulphur and phosphorus, but whose exact composition has not yet been determined? The building up of a complex unstable substance from simple stable substances necessitates the expenditure of a certain amount of energy. In animal nutrition this energy is derived from the food-materials themselves. The proteids, fats, and carbohydrates absorbed contain a large store of potential energy. Little or no energy can, however, be obtained from the very simple inorganic substances assimilated by green plants. What, then, is the source of energy? The answer will be clear if we consider the conditions of assimilation. It has been determined that from the water and carbon dioxide absorbed certain organic compounds of the nature of carbohydrates are formed. This process is called **carbon assimilation** (or **photosynthesis**). It requires the presence of light and of chlorophyll. The conclusion we come to is that sunlight is the source of energy, and that chlorophyll is a substance which enables the plant to make use of this radiant energy, just as the manufacture of cloth from raw fibre requires not only some form of energy, but also a special mechanical apparatus. More complex compounds are then elaborated, containing nitrogen derived from the absorbed nitrates. Finally these complex compounds are made use of by the protoplasm in building up its substance. If the student carefully reflects on the process so far described he will recognize that in both plants and animals living protoplasm makes use of complex compounds in building up its substance, but, whereas animals derive these compounds already manufactured (by feeding on plants or other animals), the plant must elaborate them for itself from the simple inorganic solutions which it absorbs. Hence the necessity (in most plants—see § 12) for the presence of chlorophyll and for exposure to light. The plant, as it were, starts its chemical processes at a lower level than the animal.

Thus the protoplasm in our unicellular plant obtains nourishment and increases in amount. But what of the cell-wall? It also must increase in surface-extent. How is this effected? The cell-wall consists of a substance called cellulose, much simpler in composition than protoplasm. During growth new molecules of cellulose are formed from the protoplasm and are deposited in the cell-wall. Now a complex substance to produce a simpler substance must undergo decomposition. The growth of the cell-wall, then, implies decomposition of protoplasmic substance. This is characteristic of all growth. Not only are there synthetic or "building up" processes which may result in the elaboration of protoplasm, but there are also "breaking down" processes, processes of decomposition. The former are called **anabolic**, and constitute **anabolism**; the latter **katabolic**, and constitute **katabolism**. The sum-total of the chemical processes going on, anabolic and katabolic, constitutes **metabolism**. This decomposition of the protoplasm is a process of oxidation; in connexion with it there is an absorption of oxygen; in other words, there is a process of **respiration** similar to what goes on in animals. The decomposition results in the formation of certain substances, of which some are directly used in building up the substance of the plant (*plastic substances*), others are only indirectly useful (*secretions*), and others finally have no evident use (*excretions*). The decomposition also sets free a certain amount of energy. In plants this is chiefly expended in the processes of growth, although a certain amount may be dissipated in other ways.

Here again we may institute a comparison between the plant and animal. In animals these katabolic processes are active. In a full-grown animal the loss of matter (by excretion) and energy in katabolism nearly equals the gain made in anabolism. This is correlated with the much greater activity exhibited by the animal organism. In plants, on the other hand, anabolism is greatly in excess of katabolism, so that in plants there is continual increase in the amount of plant-substance. This is correlated with the passivity and lack of movement characteristic of plants. The animal is essentially active and katabolic, while the plant is essentially passive and anabolic.

Although we have described the general processes of metabolism with special reference to a unicellular plant, they hold good for all green plants. Thus a seaweed absorbs simple inorganic solutions over its whole surface. These pass from cell to cell through the whole plant, and are elaborated as above described. In higher plants, special members are developed to perform the work of absorption. The root serves to fix the plant and absorbs solutions of mineral salts from the soil. These are conveyed in ways to be afterwards described to the leaves. The foliage leaf is the *chief* organ for the absorption of carbon dioxide which is derived from the atmosphere. In the cells of the leaf elaboration of these various food materials is carried on. The complex organic compounds thus formed are distributed over the plant and are made use of by the living protoplasm. In addition to this the leaves are active respiring organs and also give off large quantities of water-vapour (process of **transpiration**).

§ 12. **Plants without Chlorophyll.** Some plants have no chlorophyll, *e.g.* the fungi and a few flowering plants. From what has been said above it will be evident that these plants cannot assimilate simple inorganic food-materials. They can only make use of food-material consisting of complex organic compounds. In this respect they resemble animals, but the compounds are simpler than those required by animals. These complex compounds may be obtained either from living organisms or from decaying organic substance. In the former case the plants are **parasites**. They send down absorbing organs into the tissues of some living plant or animal (called the *host*), and thus obtain their nutritive material. In the latter case they are called **saprophytes**.

EXTERNAL MORPHOLOGY AND PHYSIOLOGY. 15

§ 13. **Reproduction.** Two methods of reproduction are met with in plants:—(a) **asexual** or **agamogenetic**; (b) **sexual** or **gamogenetic**. The asexual method consists essentially in the separation from the parent organism of a part which grows directly into a new organism. The part separated off may be simply a more or less specialized portion of the vegetative region, e.g. the potato-tuber. This is *vegetative reproduction*. If it is a highly specialized single reproductive cell **(spore)** it is *asexual spore-reproduction*. The sexual method consists in the separation of two sexual cells **(gametes)**, each incapable *by itself* of producing a new organism, but which fuse together and produce a new cell **(the zygote**, or sexually produced spore) with altogether new properties, which is able to develop into a new plant. In many of the lower plants (Thallophyta) the gametes are similar. In the higher types they are clearly differentiated into male and female—the former (*e.g.* spermatozoid) corresponding to the spermatozoon of the animal, the latter to the ovum. A *spore* may be defined as a highly specialized reproductive cell which is capable of giving rise directly to a **new** organism. It may be produced sexually or asexually.

§ 14. **Relation to Environment.** The intimate relation which exists between a plant and its environment is shown by the fact that plant-members always have a form and structure adapted to the particular conditions in which the plant lives. These forms and structures can only be explained by a reference to these conditions. A plant which is not adapted to or in harmony with its surroundings dies. The living protoplasm is constantly subjected to the stimulating influence of external agencies, such as light, heat, gravity, etc., and it is constantly responding in particular ways to these influences. We shall have to consider some of the results of this on growth in a later chapter **(VIII.)**.

§ 15. **Homology and Analogy.** In the course of this chapter we have suggested two distinct principles of the highest importance to the biological student. We may now clearly explain and state them. We have seen that a study of morphology leads to the recognition of similar members, *i.e.* morphologically similar. Thus stems, whatever their special form, are similar members. So also leaves. Members morphologically similar are said to be homologous, or the *homologues* of each other. Homologous members or structures are recognised by the relations of their position and development. **Homology** is the term by which we express a resemblance existing between various members as regards their position and development. We have seen, however, that homologous members may take on many different forms according to the functions which they carry on. Thus foliage leaves, floral leaves (sepals, petals, etc., of a flower), although homologous, have distinctly different forms. On the other hand, some tendrils are morphologically leaves or leaflets, *e.g.* the tendrils of the pea; while others, *e.g.* vine, are morphologically stems. Here we have two structures apparently similar, namely tendrils. Yet they are not homologous. Their resemblance to each other is physiological, not morphological. They have similar functions, and have therefore, by way of adaptation, assumed the same form. Members which present resemblances of this kind are said to be analogous, or the *analogues* of each other. **Analogy** is the term we use when we wish to express this resemblance. It is, of course, unnecessary to add that many members are both homologous and analogous, *e.g.* the ordinary foliage leaves of one plant with those of another.

The student will now be able to understand the two principles referred to. They are :—(*a*) members morphologically similar (*i.e.* homologous) may be differently modified for the performance of different functions; (*b*) members not morphologically similar may be similarly modified for the performance of the same function. These principles should be kept in mind, and illustrations of them, in the following chapters, carefully noted.

CHAPTER II.

GENERAL HISTOLOGY.

A. THE CELL.

§ 1 Cellular Structure of Plants. We have already explained (p. 5) that the substance of the plant-body is not homogeneous, but, in all except the lowest forms, consists of aggregations or unions of microscopic structures, called cells, each *living* cell consisting of a tiny mass of a viscid substance, called protoplasm, bounded by a distinct membrane, the cell-wall. These cells can be readily seen by teasing out the substance of a very ripe, mealy apple in water, and examining it under the microscope; also in thin microscopic sections of stems, roots, and other parts of plants (see, *e.g.*, figs. 19, 51). The protoplasm is the essential or living substance of a cell, and the seat of all the vital processes. The cell-wall is formed by the protoplasm, and, during the life of the cell, is added to and moulded in various ways according to the special functions it has to carry on. The cell-walls, therefore, are to be regarded as constituting a skeleton or framework, giving support to the living substance, and firmness and strength to the whole organism. There is really a continuity of protoplasm between all the living cells of a plant. The protoplasm of one cell is connected with that of others by means of extremely delicate threads passing through the cell-wall.

In the lowest forms of multicellular plants, the organism consists of an aggregation of similar cells, all carrying on very much the same functions; but in higher forms, correlated with the physiological differentiation which has taken place, there is what is called **Histological Differentiation**. In other words, many different kinds of cells, more or less definitely arranged in groups, can be recognised, the form and structure of the cells in each group depending on the functions entrusted to them. This differentiation becomes more and more marked as we pass from lower to higher types. In the Thallophyta and Bryophyta, all parts of the organism consist of typical *living* cells, although these cells may present many different forms. For this reason, these two divisions are distinguished as "**cellular plants**." A considerably higher differentiation is exhibited by the Pteridophytes and Flowering Plants. In these groups, in addition to typical living cells having an infinite variety of form, long, slender, and often tubular structures of an altogether different character can be recognized. These run through the masses of cells, sometimes irregularly, more often in definite strands or bundles (see fig. 52). They are, to a large extent, structures adapted for the rapid transmission of nutritive fluids, that is, *vascular* structures. Although very different in appearance from typical cells, a study of their development shows that all these tube-like structures are really formed by the union and alteration of young cells. The Pteridophytes and Flowering Plants are distinguished by the presence of such vascular structures in their tissues, and are therefore spoken of as "**Vascular Plants**." Hence also the term *Vascular* Cryptogams for the Pteridophytes. Thus, however extensive the differentiation may be, we may say that all parts of plants are made up of cells or of structures, or *elements*, derived from cells.

§ 2. **Protoplast or Energid.** In section, the cells of a plant present a general resemblance to the cells of a honey-comb. This was the origin of the term "cell." In many ways it is misleading and inaccurate. In plants each fully formed cell or tube has its own proper wall, and by suitable methods can be isolated for individual examination. The walls of the cells in a honey-comb, on the other hand, are common walls. Again, the protoplasmic contents are the essential part of a cell. As a matter of fact in certain stages of the life-history, certain of the protoplasmic bodies have no protective membrane—*e.g.* the ovum or egg-cell and the spermatozoid. Here the term "cell" is not at all appropriate, and the case is not improved by the use of such terms as "*naked cell*," "*primordial cell*," etc. The term "cell," however, has become so fixed in the nomenclature that it is impossible to avoid using it. At the same time many botanists have now come to apply the terms "**protoplast**" or "**energid**" to any living uni-nucleated protoplasmic unit, whether enclosed in a cell-wall or not.

§ 3. **The Young Cell.** Young cells are always found at points where growth is going on (**growing-points**)—*e.g.* at the apex of a stem. They are called initial or **meristematic** cells. They show repeated division into two, and it is in this way that new cells are produced in the plant. These young cells have very simple definite forms. They show no trace of the differentiation which in the higher types is so marked in the older parts of the plant. At the growing apex of a stem or root they are always more or less rounded or polygonal in form (fig. 4, A). In some meristematic regions (cambial layers, see § 20) they are elongated and flattened (fig. 4, B). In all cases, however, the cell-walls are extremely thin, and the protoplasm completely fills the cell-cavity. The general protoplasm of the cell (as in *all* living cells) is called the **cytoplasm**. It is always more or less granular, and embedded in it are several denser and more highly specialized *protoplasmic* bodies. The largest of these is the **nucleus**, and with it may be associated one or two smaller bodies, the **centrospheres** (fig. 5). The others usually present are known as **plastids**.

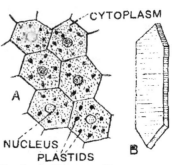

Fig. 4.—A, Young or Meristematic Cells; B, A Cambial Cell.

§ 4. **The Cell-Wall.** The cell-wall in young cells, and in many fully developed cells, consists of a substance called **cellulose**, with which is associated a smaller or larger percentage of mineral matter. Cellulose is a carbohydrate, *i.e.* it is one of a large group of organic substances, consisting of the chemical elements carbon, oxygen, and hydrogen, in which the hydrogen and oxygen are in the same proportions as in water (water = H_2O). In a general way the carbohydrates may be regarded as compounds of carbon and water in varying proportions. The formula of cellulose may be given as n ($C_6H_{10}O_5$), the value of n being doubtful. It is readily recognized by the use of reagents. With Schulze's solution it turns blue; with iodine and strong sulphuric acid it swells up and turns blue. The molecules of cellulose are probably arranged in groups forming particles, each invested by a film of the watery sap which permeates the cell-wall. It is because of this that substances in solution in water can diffuse through the cellulose wall.

§ 5. **The Protoplasmic Substance.** Protoplasm is essentially clear and jelly-like, but often presents a granular appearance, owing to the formation of various bodies within its substance. It is an extremely unstable compound of highly complex composition, and when dead is found to consist mainly of **proteid** substances, which are highly complex compounds of carbon, hydrogen, oxygen, nitrogen, and sulphur, but whose exact composition has not yet been determined. Thus protoplasm contains these five chemical elements in very complex proportions. Phosphorus, also, is present in the protoplasmic substance of the nucleus, but it does not appear to be an essential constituent of protoplasm in general. Like the cell-wall, all protoplasmic substance is permeated with water. The vital power of protoplasm seems to depend on the constant presence of water. From this alone the student can readily infer the essential importance of water to the life of the plant.

§ 6. **The Nucleus** is a body of the highest importance in the life of the cell. It appears to be the originating centre of all the vital processes—the initiator and director of all the cell's activities. It seems, however, to exercise a special control over reproductive processes. This will be evident when we come to consider the processes of cell-division and sexual reproduction. The nucleus (fig. 5) is bounded by a delicate **nuclear membrane** formed from the surrounding cytoplasm. Inside there is a semifluid ground-substance, the **nucleo-hyaloplasm**, in which is embedded a network of fine fibrils usually spoken of as the **chromatin network**. In the ground-substance also, lying in the meshes of the chromatin network, are to be found one or more small granular bodies, the **nucleoli**. Nuclei are usually spherical or oval in form. They are never formed "de novo," *i.e.* by the simple aggregation and differentiation of the protoplasmic substance, but always by division of pre-existing nuclei. The division of the nucleus is either *direct* or *indirect*. In the former method there is a simple splitting or bipartition, which is not accompanied by the division of the cell. This *direct* method is known as **fragmentation**; it is found chiefly in old cells which become multinucleate. In the *indirect* method a complicated series of changes is gone through which constitute what is known as **karyokinesis** or *mitosis*—to be described presently (§ 18). It is followed by cell-division.

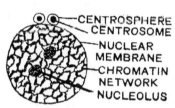

Fig. 5.—NUCLEUS AND CENTROSPHERES.

§ 7. The cytoplasm, apart from any granules embedded in it, seems to consist of a single proteid substance called *cytoplastin*. Nuclear substance consists of various proteids, some of which contain phosphorus. The chromatin network and the nucleoli are the parts which take on stains most readily. With iodine they stain a deep brown colour. This is a general action of iodine on the denser forms of proteid.

§ 8. **The centrosphere** (fig. 5) is a small spherical cytoplasmic body observed in some cells just outside the nucleus. The denser central portion is called the **centrosome**. The presence of centrospheres has as yet been recognised only in animal cells, in the cells of Algæ and Muscineæ, and in the reproductive and dividing cells of higher plants (see § 18 and fig. 105).

§ 9. **The Plastids**, like the nucleus and centrosphere, are highly specialized and differentiated portions of protoplasmic substance. Similarly, they are not formed "de novo," but always multiply by division (direct). The substance of the plastid has a spongy texture. There is a clear, semifluid ground-substance, in which is embedded a network of strands or fibrils of a denser proteid.

§ 10. **Processes of Histological Differentiation.** In the very young or embryo plant all the cells are meristematic, but in the older plant the meristematic cells are confined to certain points or regions which are distinguished as "growing-points"—as, *e.g.*, the apical cell or cells of a thallus, the apex of stem or root. This localization arises from the early commencement of histological differentiation. The cells which are formed are gradually altered or modified in various ways to adapt them to the performance of different functions, and in this way are produced all the varieties of cell and tissue met with in the fully developed organs of a higher plant. This is a point we must now insist on. *All these different kinds of cells or cell-structures are derived by modification of various kinds from the young cells produced at growing-points.* Before we can satisfactorily describe these various forms and kinds of cells, we must consider the nature of the changes or modifications which give rise to them. In other words, we must give some account of the *processes of histological differentiation.* These changes affect both the cell-wall and the cell-contents. They may be roughly tabulated thus:—

(1) Growth in surface-extent of the cell-wall.
(2) Thickening of the cell-wall.
(3) Chemical alteration and impregnation of the cell-wall.
(4) Changes in the contents.
(5) Cell-fusion.

§ 11. (1) **Growth of Cell-Wall in Surface-Extent.** In other words, the cell grows in size. This growth may be uniform or localized. If uniform, the young cell simply develops into a larger cell of the same form. If localized, the resulting cells assume new forms. If, for example, the young cell grows more especially at three or four particular

GENERAL HISTOLOGY.

points, the resulting cell will show a number of radiating arms or outgrowths (the *stellate* form—fig. 27, A). If the growth is confined to two opposite points the cell becomes very long and *pointed*. This elongated pointed kind of cell is very common. It is called the **prosenchymatous** form (fig. 6), and is to be distinguished from the **parenchymatous** form in which the cell is usually not much longer than it is broad, and is not pointed. This parenchymatous form shows considerable variety, *e.g.* it may be rounded, oval, polygonal, prismatic, tabular or flattened, stellate, etc. Localized growth in surface-extent, then, gives rise to different forms of cells.

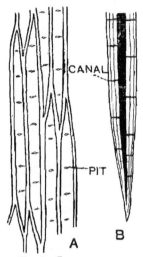

Fig. 6.—SCLERENCHYMA (PROSENCHYMATOUS).
A, Not strongly thickened; B, End of a thickened bast-fibre.

§ 12. (2) **Thickening of the Cell-Wall.** This also may be uniform (general) or localized. In the former case the cell-wall is equally thickened all round. Strictly speaking it is rarely met with. Localized thickening is the rule. In this case only certain parts of the wall are thickened. The character of the thickening varies very much. In some cases it takes place in rings, so that we get **annular** bands formed on the inner surface of the cell-wall (fig. 7, A). In other cases the thickening is along a **spiral** line (fig. 7, B). We may imagine that this is due to the annular thickenings of the former case becoming interrupted and running into each other; as a matter of fact we find cases where the

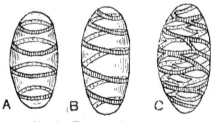

Fig. 7.—TYPES OF THICKENING.
A, Annular; B, Spiral; C, Reticulate.

thickening is partly annular, partly spiral. If now we imagine the turns of the spiral to be arranged very closely, so that at intervals they fuse, we get the next type of thickening, the **reticulate** (fig. 7, c). Here the thickening forms a network or reticulum on the inner surface of the cell-wall. The transition from this to the **pitted** or *dotted* type of thickening (fig. 8) is easy; we have only to imagine that the strands of the network become very thick and the meshes correspondingly reduced. In this case the whole of the wall, with the exception of numerous small circumscribed areas, undergoes thickening. When examined under the microscope these unthickened areas appear like perforations, apertures, or dots, according to their size, hence the terms pitted or dotted.

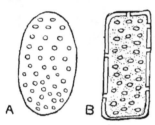

Fig. 8.—PITTED CELLS.
A, Surface view; B, Cell cut longitudinally, showing pits in section and surface view.

The student should carefully notice the transition, readily recognizable, from the simple annular type to the more perfect pitted type. The thickening substance is laid down on both sides of the original cell-wall. This, however, can generally be distinguished in the middle of the thickened wall, and is known as the **middle lamella** (fig. 9). It is evident, then, that just as localized growth in surface-extent results in the development of different *forms* of cells, so localized thickening leads to the formation of different *patterns* or markings on the cell-wall. These markings are most characteristically and most distinctly developed on wood-elements (wood-vessels and tracheides), but are not by any means confined to these.

Fig. 9.—THICKENED SCLERENCHYMA.
(Transverse section.)

Sometimes the thickening of the cell-wall is so extensive that the cavity is almost obliterated. This is frequently the case in elements forming a tissue called sclerenchyma (figs. 9, 6 B). If pits are present they become transformed into canals running through the thickened wall. A curious form of pit, the **bordered pit** (fig. 10), is developed on the walls of many wood-vessels and tracheides. Here a circular area of the cell-wall remains unthickened, and the edge of the thickening matter all round this arches over it in a dome-like manner. The apex of the dome, however, is open, so that there is an aperture leading from the small cavity covered over by the dome into the cavity of the tracheide or vessel. A similar structure is developed *at exactly the same point* in the adjacent tracheide or vessel. A glance at fig. 10, A, C, will now show that in the wall separating the cavities of the two vessels or tracheides there is a lenticular space across which stretches the original unthickened cell-membrane. In a surface view of the structure (fig. 10, B) we see the small aperture at the apex of the dome as a small circle, surrounded by a larger circle representing the edge

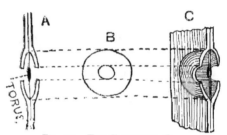

Fig. 10.—THE BORDERED PIT.
A, Longitudinal section ; B, Surface view ; C, Semi-profile.

of the unthickened membrane where the thickening matter begins to arch over the lenticular cavity. The student will get a good idea of this structure if he takes two small wooden dishes shaped like watchglasses, makes a perforation in the bottom of each, and places them rim to rim with a thin sheet of paper between. The thin paper between the dishes will represent the unthickened cell-wall. It should be noticed that the unthickened membrane in a fully developed bordered pit is usually pushed over to one side. It shows a slight thickening or swelling in the middle, known as the *torus*. These bordered pits are frequently met with on the walls of the wood-elements of Angiosperms and Vascular Cryptogams, but they are most typically developed on the wood-elements (tracheides) of Gymnosperms.

Sometimes the pits on a wood-element are very much elongated transversely. In this case the thickened bars between the elongated pits look like the rungs of a ladder; hence the name **scalariform**, applied to this type of thickening. The pits in this scalariform type are frequently bordered (figs. 11 B, 12).

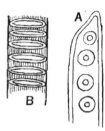

Fig. 11.—BORDERED PITS IN SURFACE VIEW.
A, Circular; B, Elongated (scalariform).

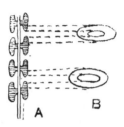

Fig. 12.—OVAL BORDERED PITS.
A, Longitudinal section B. Surface view.

Note. The growth and thickening of the cell-wall is due to the formation of new particles of cellulose by the protoplasm, and their incorporation in the cell-wall. According to some, these particles are laid down in layers on the inner surface of the wall. This is the **apposition theory.** Here, increase in surface-extent is supposed to be due to stretching of the wall. Others maintain that the new particles are intercalated, or pushed in between those already present. This is the **intussusception theory.** The balance of evidence seems to favour apposition in the case of thickening, intussusception in the case of growth in surface-extent.

§ 13. (3) **Impregnation and Chemical Alteration of the Cell-Wall.** The wall of a young cell is thin and consists of cellulose, and when thickened it may still show very much the same chemical characters. But, in many elements, the wall during growth becomes impregnated or altered in various ways. It may be cuticularized, suberized, lignified, converted into mucilage, or impregnated to a greater or less extent with mineral matter. **Cuticularization** is due to the modification of cellulose into a substance called *cutin*. This modification is most commonly seen in the outer layers of the external walls of epidermal cells. The cuticularized layers extend over the epidermis, forming a membrane known as **cuticle** (see fig. 33, A). The properties of the cell-wall are completely altered by this change. It not only gives firmness to the walls, but also renders them almost impermeable to water. **Suberization** is a very similar modification, due to formation of *suberin*.

GENERAL HISTOLOGY.

In its chemical characters suberin closely resembles cutin, and may be only a mixture of cutin and a little lignin (see below). This change is met with in cork cells, and, like cuticularization, it renders the cell-walls impermeable to water. Cuticularized or corky walls are stained yellow by iodine, yellow or brown by Schulze's solution. They are not acted on by sulphuric acid. **Lignification** is due to the formation (by modification of cellulose) of *lignin*. It is met with only in thickened cell-walls, as in the cells or elements of wood and sclerenchyma. While giving great strength and rigidity to the cell-wall, this change does not interfere with its elasticity or permeability. Lignified walls are stained a bright yellow by aniline sulphate (or chloride), yellow, or sometimes brown, by iodine, yellow by Schulze's solution; under the action of iodine and sulphuric acid they become brown and swell up. **Mucilaginous** walls, when dry, are hard and horny. The property which peculiarly distinguishes them is their great capacity for absorbing water. When moistened or soaked in water they swell up and become soft and sticky. In extreme cases the absorption of water may lead to the disorganisation of the mucilaginous cell-wall, and produce drops of gum. This is the origin of the exudation of gum seen on the stems of cherry and other trees, also in many budscales. Mucilage, in many of its other reactions, resembles cellulose.

Of the mineral substances deposited in the cell-wall silica, calcium carbonate and calcium oxalate are the commonest. Silica is often found so completely impregnating the cellulose wall that if the tissue is burned a complete siliceous skeleton of the cells is left behind, *e.g.* in the epidermal tissues of grasses. Isolated crystals of calcium oxalate or carbonate are occasionally found in cell-walls. The most characteristic form, however, in which calcium carbonate is associated with the cell-wall is that known as the **cystolith** (fig. 13). These cystoliths are found in many plants, *e.g.* the epidermal cells of plants belonging to the nettle order and of the indiarubber plant. During their development a small cellulose protuberance arises on the cell-wall internally. As the protuberance grows, it becomes impregnated with calcium carbonate. When fully developed the cystolith forms a pear- or cigar-shaped mass, attached by a short stalk to the cell-wall. It has an organic basis of cellulose. The distinguishing test for these mineral substances is dilute acetic acid. Calcium oxalate does not dissolve in this acid: but calcium carbonate does, with an evolution of gas (CO_2).

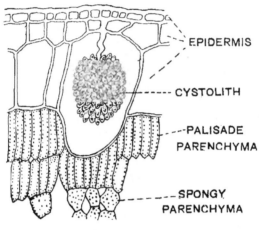

Fig. 13.—PART OF TRANSVERSE SECTION OF THE LEAF OF *Ficus elastica* (INDIARUBBER PLANT), SHOWING A CYSTOLITH.

§ 14. (4) **Changes in the Contents.** In the young cell, as we have seen, the protoplasm, etc., completely fills the cell-cavity. As the cell grows the increase in the amount of protoplasm is not sufficient to keep pace with the extension of the cell-wall. The result is that small spaces or cavities, called **vacuoles** (fig. 14), make their appearance in the protoplasm, and these become filled with a watery fluid called **cell-sap**. In the very young cell the relatively small amount of cell-sap simply permeates the protoplasmic substance and the cell-wall. These small vacuoles gradually increase in size, and finally all run together to form one large central vacuole (fig. 15). The protoplasm is now reduced to a **parietal layer**, which lines the cell-wall internally, and a number

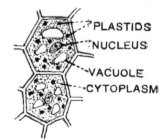

Fig. 14.—YOUNG CELLS, SHOWING FORMATION OF VACUOLES.

of delicate **protoplasmic strands** or threads which run across the vacuole to the middle of the cell. The parietal layer forms a sort of bag containing the cell-sap. Hence, when first seen, it was called the "**primordial utricle**," a name which is still in use. The nucleus in such a cell is usually embedded in the little central mass of protoplasm formed by the union of the protoplasmic strands; but it lies in the primordial utricle when, as occasionally happens, the protoplasmic strands are absent. This condition of the cell is met with in many fully developed plant-tissues, e.g. the general succulent tissue of plants.

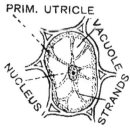

Fig. 15.—A TYPICAL, LIVING, FULLY DEVELOPED CELL.

In young cells, the **plastids** are colourless or green. They multiply by *direct* division (fig. 16), and during the growth of the cell they may undergo many changes as regards colour. It is only in the older living cells of the plant that they attain their full functional activity. Three kinds are differentiated. In the cells of underground portions of plants (*e.g.* tubers of potato), or deep-seated tissues (*e.g.* medullary rays of trees)—more generally, in tissues not exposed to light— they are colourless and called **leucoplasts** or **amyloplasts**. They have various forms— spheroidal, discoidal, rodlike, etc. Their special function is the formation of starch from soluble carbohydrates (sugar). In parts exposed to light, such as leaves and the superficial tissues of herbaceous stems, most of the plastids develop the green colouring matter called **chlorophyll**. They are therefore called **chloroplasts, chlorophyll corpuscles**, or chlorophyll grains. The chlorophyll seems to be in the form of fibrils or granules embedded in the protoplasmic substance

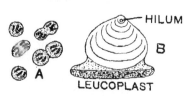

Fig. 16.—PLASTIDS (HIGHLY MAGNIFIED).
A, Chloroplasts of a Moss (*Funaria*), showing division; B, Leucoplast with developing Starch-Grain.

of the plastid. Their function is twofold. Like the leucoplasts they can form starch from soluble carbohydrate, but besides this, by means of the chlorophyll which they contain, they have the power of building up carbohydrate substance from water and carbon dioxide in the presence of light (see p. 12). In all the higher plants they are spherical or spheroidal, often more or less flattened. Sometimes the green colour is masked or concealed by the presence of other colouring matters dissolved in the cell-sap, *e.g.* leaves of the copper beech. When exposed to light leucoplasts develop into chloroplasts, while the latter lose their chlorophyll and become yellow if withdrawn from light. Frequently, however, the plastids in aerial parts contain colouring matters other than green. Such are called **chromoplasts**. They are found in the petals of many flowers—most yellow and many red flowers —and in fruits. The colours of most blue flowers and many red flowers, on the other hand, are due to colouring matters dissolved in the cell-sap. Chromoplasts may be formed directly from leucoplasts, but usually from chloroplasts. Thus the young floral leaves may be green, and the colours of autumn leaves are due to the presence of chromoplasts derived from chloroplasts by the decomposition of the chlorophyll towards the approach of winter. Chloroplasts and chromoplasts are generally spoken of together as **chromatophores** (colour-bearers).

Many **non-living substances** may be produced during the metabolism of the cell, and are found either in the protoplasm or in the cell-sap. These substances are either soluble or insoluble. In the former case they are dissolved in the cell-sap; in the latter they appear in solid form, usually in the protoplasm. They may be arranged in three groups. First, there are those substances which are at some time or other made use of by the protoplasm as foodmaterial. The most important of these are starch-grains, proteid grains (fig. 17), oil or fat (insoluble), various sugars and nitrogenous compounds of the nature of amides (soluble). These nutritive compounds are of the nature of **plastic** substances (see p. 13). Then there are **secretions** (p. 13). The more important are the organic acids, various colouring matters, and many ferments (these are soluble). Finally, there is a group of waste products or substances of no (evident) use to the plant. These are called **excretions**. Amongst these there are such forms as the alkaloids which are nitrogenous, e.g. morphia, nicotin, etc.; they constitute the active principles of medicinal plants: also ethereal oils (fig. 18), resins, and various mineral substances. Of course, only some of these are formed in any particular cell which may be examined. Some of the more important of these non-living substances will be considered more fully in § 17.

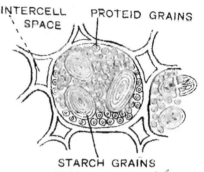

Fig. 17.—Cells with Grains of Starch and Proteid.

Finally, it has to be noticed that the contents, protoplasmic or otherwise, disappear altogether from many cells towards the close of their development. Evidently the tissues consisting of such "cells" can only discharge a purely mechanical or physical function as vascular or supporting tissues. Here the cell-walls are of importance, not the living substance. The protoplasm ends its work when it has sufficiently moulded the walls in accordance with the functions they have to carry on, and disappears. The occurrence of tissues which have lost their living substance and are therefore dead must be carefully noticed. We have examples of these in wood-vessels and sclerenchyma (figs. 6, 9).

§ 15. (5) **Cell-Fusion.** Frequently, in the development of masses or rows of cells, the cell-walls break down and disappear. The extent to which this takes place varies considerably. Sometimes a whole mass of cells disappears owing to complete absorption and disappearance of their walls, and a large irregular space or cavity is formed. This is the origin of most of the large irregular cavities met with in plants, such, for example, as the spaces found in the middle of many stems. Sometimes more definite passages are formed by a similar absorption of longitudinal series of cells. Cavities or passages formed in this way, by complete absorption of cells, are said to be developed **lysigenously** (fig. 18). On the other hand, definite tubes or **vessels** *bounded by distinct walls* are produced from rows of cells, if the absorption affects only those walls which lie between the original cells, so that their cavities become continuous. If irregular rows of cells fuse in this way, the vessels formed branch and anastomose (*i.e.* the branches run into each other) to form a network, as in the formation of laticiferous vessels (fig. 19). A straight, tubular vessel is formed if cells in a single definite longitudinal series fuse together, as in wood-vessels (fig. 52).

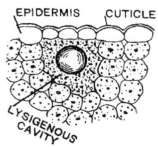

Fig. 18.—Lysigenous Cavity with Drop of Ethereal Oil.

§ 16. **Intercellular Cavities.** Young cells are all closely in contact with each other. There are no spaces between them. But, in cells growing in different directions and taking on different forms, the cell-walls must be subjected to a considerable amount of strain. The cell-walls yield to this by separating from each other at certain points, so that small cavities, called *intercellular spaces*, appear between the cells. In most cases these are small, and, *in sections*, are seen more especially at the angles of the cells (figs. 15, 17). They are not, however, isolated from each other, but communicate so as to form a continuous system. They are of great importance in the plant, as they serve for the ready passage of various gases and

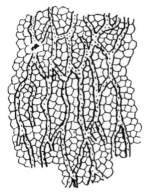

Fig. 19.—Laticiferous Vessels running through thin-walled Parenchyma.

vapours, the most important of which are oxygen, carbon dioxide, and water vapour. Frequently larger cavities or passages are formed by the *separation* of masses of cells from each other, owing to a splitting apart of the cell-walls in this way. This is the **schizogenous** method of development of such spaces and passages. Most **resin-passages**, *e.g.* ivy, Scots fir, are formed in this way.

§ 17. **Non-living Cell-Contents.** We must now give a fuller account of some of the more important non-living substances found in cells, and already mentioned on p. 30.

(1) **The Cell-Sap** is a watery fluid, containing many substances either in solution or in suspension. The water is derived from the soil in the process of root-absorption. It contains various inorganic salts dissolved in it. Amides and sugars are the most important plastic substances (see p. 13) contained in the cell-sap. The chief sugars are grape-sugar ($C_6H_{12}O_6$) and cane-sugar ($C_{12}H_{22}O_{11}$). Another carbohydrate sometimes found is inulin, a form of starch. Inulin is especially abundant in the cells of many Compositæ (*e.g.* roots of Sunflower and Dahlia). Though dissolved in the cell-sap, it is precipitated by alcohol in the form of very characteristic crystalline masses (fig. 20), showing concentric and radiating structure (*i.e.* marked by concentric and radiating lines). Colouring matters, organic acids (*e.g.* malic, citric, tartaric, and oxalic), and ferments are the most important of the secretions present. As already mentioned, excreted products such as tannin and various alkaloids are also present. The cell-sap is to be regarded as a nutritive fluid and as a receptacle for waste products.

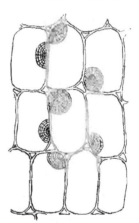

Fig. 20.—Cells with Inulin-Crystals.

It can diffuse from cell to cell, and permeates the substance of protoplasm and cell-wall, so that the protoplasm absorbs from it the substances necessary for its nourishment.

(2) **Starch** is a carbohydrate and an isomer of cellulose, *i.e.* it contains the same chemical elements in the same proportions, but has a different chemical constitution, so that its properties are different. Its formula may be given as n ($C_6H_{10}O_5$). It is found in plant-cells in the form of grains (figs. 17, 21), most abundantly in parts which serve as store-places of reserve material. Rarely the grains are formed in the general protoplasm of the cell; in this case they are small and devoid of any structure. Their formation is nearly always the work of leucoplasts, chloroplasts, or chromoplasts. They are developed inside the plastids, and chloroplasts are often much distended by them. In the case of leucoplasts they are frequently large, and appear to lie outside, owing to their formation beginning near the margins of the plastid bodies (fig. 16). When the grains are examined under the microscope they are found to exhibit stratification, a number of layers being arranged round a definite point called the **hilum**. Sometimes the layers are arranged quite regularly and concentrically (fig. 21, D); but often the arrangement is excentric, and the hilum lies near one end (fig. 21, A).

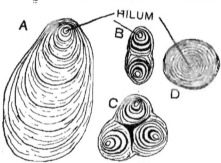

Fig. 21.—Starch-Grains.
A, Excentric; D, Concentric; B, C, Compound.

Concentric grains are formed in the centre of the plastids; excentric grains to one side of them. The reason, of course, is that, in the former case, the substance of the plastid invests the grain equally, and regular starch-layers are deposited; while, in the latter case, the starch-layers are thickest on the side next the main body of the plastid. The starch-grain contains water. The stratified appearance is due, not only to the fact that the starch is deposited in layers, but also to these layers containing varying amounts of water. The hilum is the most watery part of the grain. The grains assume many different forms, but the form is characteristic in each particular plant. Those of the potato are oval or ovate and excentric ; of the wheat, spherical or lenticular and concentric ; of the rice, polygonal. Sometimes a plastid begins to form two or three grains at the same time. As these grow they are enclosed in common layers and form compound grains (fig. 21, B, C). Spuriously compound grains are produced by separate grains simply becoming aggregated together. Starch-grains when fully formed may be found floating free in the cell-sap. Starch is stained a dark blue, sometimes violet, by iodine, and is therefore readily detected.

(3) **Proteid Grains** (figs. 17, 22). These are solid grains of proteid substance formed as reserve food-material in connexion with nutrition. They are formed in the general protoplasm. They may be found in any living cell, but are especially large and abundant in the cells of many seeds. In many cells they are extremely minute and form granular aggregations. They are spoken of generally as **aleurone grains**. If some of the larger forms met with in seeds (*e.g.* castor-oil or brazil-nut) be examined, they are found to contain a denser proteid body called the **proteid crystalloid** (fig. 22). Sometimes a number of crystalloids are present in an aleurone grain. At one end of the grain also can be seen a clear mineral body called the **globoid**, consisting of a double phosphate of calcium and magnesium. The grains, especially the crystalloids, are stained yellow or brown by iodine (cf. the nucleolus and chromatin fibrils of the nucleus). By taking on stain in this way, and also by the fact that it can be made to swell up under the action of various reagents, the crystalloid is readily distinguished from a mineral crystal. The crystalloids are not present in all proteid or aleurone grains, and may also occur by themselves, as in the outer cells of the potato-tuber.

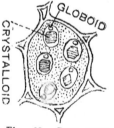

Fig. 22.—CELL WITH LARGE ALEURONE GRAINS.

(4) **Fats and Oils.** These substances occur in the form of drops or globules in the general protoplasm of cells and also in the vacuoles. As plastic substances they occur most commonly in the cells of seeds. The oil-drops found in seeds are usually insoluble in alcohol; sometimes soluble, as in the castor-oil. Various ethereal oils, however, of the nature of excretions are frequently met with in the cells of vegetative parts of plants, as in the cells of some leaves and in glandular hairs.

(5) **Resin** appears in various forms in many cells, sometimes mixed with other bodies like gum or mucilage. Frequently the resin is poured into special resin-passages.

(6) **Mineral Crystals.** Both the oxalate and the carbonate of lime are precipitated in the form of crystals or crystalline masses. They may be regarded as excretions of superfluous mineral matter. The oxalate is by far the commoner. They may occur in the form of small crystals (fig. 23), or in rounded and more or less angular crystalline aggregates called **sphæraphides** (fig. 24, B). A very characteristic form of calcium oxalate is the elongated acicular or needle-like form. Groups of these acicular crystals occur in the cells of many Monocotyledons (e.g. arum) and some Dicotyledons (e.g. species of dock). They are called **raphides** (fig. 24, A).

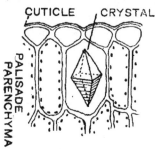

Fig. 23.—PART OF TRANSVERSE SECTION OF A LEAF (REGION OF THE UPPER SURFACE), SHOWING A CRYSTAL IN ONE OF THE CELLS.

§ 18. **Cell-Formation.** Having described plant-cells, their structure, modifications, and contents, we have now to consider the origin or formation of new cells, for it is evident that wherever growth or reproduction is taking place there must be development of new cells. In all cases new cells are formed from pre-existing cells. In the vegetative parts of plants they are nearly always produced by very characteristic **cell-division.** Each meristematic cell, after reaching a certain size, divides into two daughter-cells, which may repeat the process. The division of the cell is preceded by the karyokinetic division (p. 21) of the nucleus. Sometimes in vegetative parts, but more frequently in connexion with reproductive processes, new cells are developed by a process known as **free cell-formation** in which also there is karyokinetic nuclear division. In the development of reproductive cells, two other processes are met with, in which there is no preceding nuclear division — namely **rejuvenescence** and **conjugation**. We shall now briefly consider these methods of cell-formation.

(1) **Ordinary Cell-Division.** Here we have to describe the series of changes which take place in the indirect division of the nucleus and constitute **karyokinesis** or **mitosis** (see fig. 25). At the commencement of the process of cell-division the two centrospheres (formed by division of an original one) pass to opposite sides of the nucleus. The nuclear membrane becomes indistinct. The chromatin network becomes thicker, opens out, and finally breaks up into a number of curved, more or less U- or V-shaped rods, the **chromosomes**.

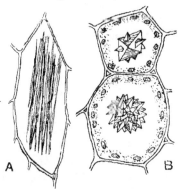

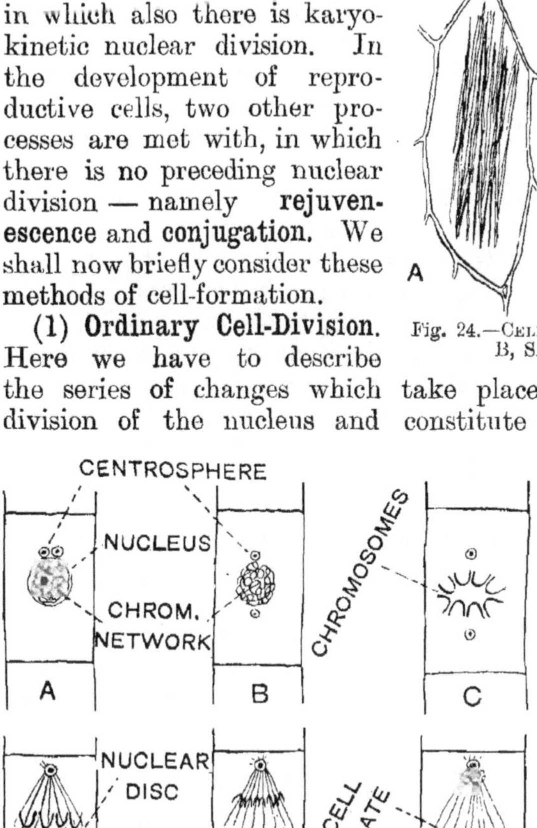

Fig. 24.—Cells with A, Raphides; B, Sphæraphides.

Fig. 25.—Diagrams illustrating Stages in Karyokinesis.

These become aggregated towards the centre of the cell, with their arms or free ends directed outwards towards the centrospheres. They give rise to a star-like structure, the **nuclear disc**, and hence this is referred to as the "*star*" or "*aster*" stage in the process. At this stage a number of fine fibrils or threads can be recognised passing from the centrosphere on each side to the chromosomes. This structure is called the **nuclear spindle**; its threads are probably formed from the surrounding cytoplasm of the cell. In the next stage each chromosome divides *longitudinally* into two thinner Us or Vs, so that the number of chromosomes is doubled. Then the chromosomes begin to move along the threads of the nuclear spindle outwards to the centrospheres, one half towards one centrosphere, the other half to the other centrosphere. In this movement the apices of the chromosomes are directed towards the centrospheres. The centrospheres lying at two opposite poles apparently act as directive centres in the process. It should be noticed that the two halves of each *original* chromosome pass to opposite poles. At the close of this stage we can recognise two stars; hence called the "*diaster* stage." At each pole the chromosomes become aggregated, and eventually two **daughter-nuclei** are formed.

Towards the close of the above processes thickenings, due to deposition of small particles of cellulose, begin to appear on the threads of the nuclear spindle across the equatorial region. These thickenings increase in size and fuse together to form a thin plate or membrane which gradually extends outwards till it reaches the wall of the cell. This is the new cell-wall which divides the original cell into two daughter-cells, each with its own nucleus.

During these processes the nucleoli are disorganised. Nucleoli reappear in the daughter-nuclei.

(2) **Free Cell-Formation.** This type of cell-formation differs from the above in that division of the cell does not *immediately* follow mitotic division of the nucleus. By repeated mitotic division of the successive daughter-nuclei, a large number of nuclei are produced lying free in the protoplasm of the cell. Towards the close of this process the protoplasm begins to aggregate round these nuclei to

form naked protoplasts (p. 18). Finally, cell-walls are formed between them. These new cells are formed *inside* the mother-cell, and the young cell-walls are entirely new structures. In ordinary cell-division the dividing plate is the only *new* part of the cell-wall formed. Typical free cell-formation is seen in the development of the endosperm of seeds. Occasionally the rapid division of the nucleus is not followed by the actual division of the protoplasm and the formation of cell-walls. We simply get a number of nuclei lying in the protoplasm, or rather, perhaps, an aggregation of protoplasts with their nuclei. Such a structure is called a **cœnocyte**. We have examples in laticiferous "cells." The cœnocyte may seem to be indistinguishable from a multi-nucleate cell. As points of difference, it should be noted that a cell becomes multi-nucleate only in an old condition and the division of the nucleus is direct, not mitotic.

(3) **Rejuvenescence and Conjugation.** Many reproductive bodies, either asexual or of the nature of gametes (p. 15), are simply naked protoplasts, either motionless, or moving by means of vibratile protrusions of the protoplasm called *cilia*. They are developed from typical cells with a cell-wall. In the process the protoplasm or part of the protoplasm aggregates in the middle of the cell, the cell-wall bursts, and the protoplasmic unit is set free. These bodies are new "cells" (protoplasts), having altogether different activities and different properties from the cells which produced them. The protoplasm has, as it were, taken on a new lease of life. Hence the term *rejuvenescence* for this form of cell-formation. It will be noticed that there is no nuclear division, and no increase in the number of cells. We can observe the other method, *conjugation*, if we follow out the destiny of the gametes. Sooner or later they fuse in pairs (corresponding to male and female: see p. 15). Not only the protoplasm, but also their nuclei, fuse together, and the result is a new cell (zygote) with altogether different potentialities. This is the process of conjugation. It will be seen that there is a fusion of nuclei and a reduction in the number of cells. The term conjugation is strictly applied to the fusion of *similar* or undifferentiated

gametes, and the zygote is called a **zygospore**. The term **fertilization** is given to the process in the higher plants where a *male* gamete makes its way, or is carried, to a distinct *female* gamete (ovum or oösphere); and the zygote is called the **oöspore**.

B. THE TISSUES.

§ 19. **A tissue** may be defined as an aggregation of similar cells or elements, united from the first, governed by the same laws of growth and development, and having therefore a similar structure adapted to the performance of the same function. The significance of the differentiation of tissues has already been indicated. The tissues of a plant may be arranged in two chief groups:—(*a*) **Meristematic Tissues**; (*b*) **Permanent Tissues**. The former are the tissues found at growing-points. They consist of meristematic cells, *i.e.* cells which possess the power of dividing. In the latter group are included all tissues derived from the former by various processes of differentiation. They consist of cells or elements which have lost the meristematic property, and taken on some fixed or permanent structure, adapted to the performance of some special function.

§ 20. **Meristematic Tissues.** As already indicated (p. 22), the regions of meristem become restricted. These merismatic (=meristematic) regions may be apical (**apical meristems**), as at the apices of stems and roots. They provide for the further growth of these members in length. But frequently we find meristematic layers situated between masses of permanent tissue (**intercalary meristems**). When an intercalary meristem provides for secondary increase in thickness of a member it is called a cambial layer or **cambium**. Meristems are also distinguished as **primary** or **secondary**. A meristem which has persisted throughout the growth of a member (*e.g.* stem or root), and which therefore was present at the first origin of the member, is a primary meristem. So also any meristematic layers which have been *directly* derived from it as in *parts* of some cambial layers. A secondary meristem is produced

GENERAL HISTOLOGY. 41

when living cells of permanent tissue take on meristematic activity. This is the case with most cambial layers.

The structural characters of meristematic cells (fig. 4) have already been described (p. 19). We may summarize the characters of the tissue thus :—A meristem is an active growing tissue. Its cells are in an active state of division. The cells have all approximately the same form and structure. In apical meristems they are usually more or less polygonal; in cambial layers they are usually flattened and more or less elongated. In all cases, however, the cell-walls are thin, and consist of cellulose; the protoplasm completely fills the cell-cavity; the nucleus is large and well-defined; there are no intercellular spaces.

§ 21. **Permanent Tissues.** Some of these are clearly marked off from each other; but others are connected by so many transitional or intermediate forms that it is impossible to classify them sharply. As a matter of fact very different classifications have from time to time been proposed. The following classification is based partly on the forms of the component cells (parenchymatous or prosenchymatous), partly according to the characters of their walls or contents (cuticularized, lignified, etc.).

§ 22. (1) **Thin-walled Parenchyma** (see fig. 19). This is one of the commonest kinds of tissue found in plants. It constitutes the greater part of the soft succulent tissue, *e.g.* the tissue of algæ and mosses, the cortex and pith of stems, the mesophyll of leaves. The cells are parenchymatous in form. They may be rounded or oval, with numerous intercellular spaces (spongy parenchyma, fig. 13), slightly elongated and arranged perpendicularly to some surface (palisade parenchyma, fig. 23), elongated and prismatic stellate, etc., etc. The thin cell-walls consist of cellulose. Primordial utricle, protoplasmic strands, nucleus, vacuole, and cell-sap are usually present. Various substances may be formed in the cells, such as starch, proteid grains, oil, resin, etc., etc. Small intercellular spaces are nearly always present. Sometimes, as in the pith of some trees, the cells entirely lose their contents.

It is a tissue essentially engaged in the processes of assimilation and nutrition. The cells containing chlorophyll can elaborate organic substances; other cells serve for the storage of these substances; and, generally, plastic substances in solution in the cell-sap are carried by slow diffusion over the whole plant.

Occasionally a similar tissue, but consisting of more or less prosenchymatous cells, is met with. This may be distinguished as **thin-walled prosenchyma.**

§ 23. (2) **Thick-walled Parenchyma.** In this tissue also the cells are parenchymatous and retain their contents, but the cell-walls are more or less thickened. The thickened walls may consist of cellulose only, as in the tissue called **collenchyma** (fig. 26). In collenchyma the cellulose thickening is laid down more especially at the angles of the cells; it is a tissue found underneath the epidermis of many stems and leaf-stalks. Sometimes the walls are not only thickened (equally), but also lignified, as in many of the thick-walled elongated parenchymatous cells of the wood (**wood-parenchyma**).

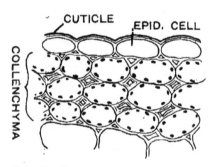

Fig. 26.—EPIDERMIS AND COLLENCHYMA OF STEM OF SUNFLOWER. (Transverse section.)

The cells of collenchyma contain chloroplasts; the wood-cells usually contain storage products. But in addition to this nutritive or assimilatory function, this tissue has in addition a mechanical function, giving strength to the parts in which it occurs.

Occasionally a thick-walled tissue consisting of prosenchymatous cells retaining their contents is met with. This may be called **thick-walled prosenchyma.** The cells may be unlignified (*e.g.* some bast-fibres), or lignified. The cells may be spoken of as fibrous cells or, if lignified, fibrous sclerotic cells.

§ 24. (3) **Sclerenchyma** (figs. 6, 9). Amongst the thin- and thick-walled tissues just described there are many transitional forms. Similarly the thick-walled lignified forms lead up gradually to the tissue called sclerenchyma. In this tissue the contents have been completely lost, and the walls of the elements are thickened and lignified. Its function in the plant is purely mechanical. The thick-walled tissues already mentioned and sclerenchyma constitute the strengthening tissues of plants. Frequently the thickening of the walls is so great that the cavities are almost obliterated (fig. 9). Sclerenchyma usually and typically consists of prosenchymatous elements (**fibrous sclerenchyma**). These elements are often referred to as **sclerenchymatous fibres**. This form of sclerenchyma is typically developed in bands or bundles (**stereid bundles**). The hard bast of most fibro-vascular bundles is a good example of this (see fig. 52). Occasionally, however, sclerenchyma consists of rounded or slightly elongated parenchymatous elements. Such **sclerotic cells** (fig. 27), as they are called, are found in some fruits (*e.g.* the "stone cells" of the pear), and in the cortex and secondary phloem of some woody stems, *e.g.* the oak. The sclerenchymatous elements as a rule have simple pits on their walls. If the wall becomes very strongly thickened, these pits are converted into elongated, often branching, canals (fig. 27, B). Practically it is convenient to regard all thickened forms of prosenchyma as sclerenchyma.

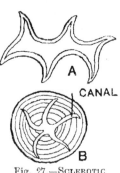

Fig. 27.—SCLEROTIC CELLS.

§ 25. (4) **Cuticularized or Suberized Tissue.** In this tissue the cell-walls are cuticularized or suberized (p. 26), partially or completely. It is met with in various parts of plants. The cells are parenchymatous in form, usually either flattened and tabular or more or less brick-shaped. The best examples of this kind of tissue are seen in the **epidermis**, **cork**, and in the **endodermis** or bundle sheath. In the epidermis, only the outermost layers of the external walls are cuticularized, forming the **cuticle** (p. 26 and fig. 26). Cork consists of more or less brick-shaped cells without intercellular spaces. Frequently the walls are comparatively thin, but examples of thick-walled cork cells are abundant. The walls are suberized throughout, and the older cork cells lose their contents. Endodermal cells may be thin, or, as frequently happens, certain of them are more or less thickened.

The function of the tissue is protective—more especially to prevent evaporation or diffusion of watery fluids.

§ 26. (5) **Tracheal Tissue.** This is the essential tissue of the *xylem* or *wood*. In it we recognise two kinds of elements:—(*a*) **Wood-vessels** (vasa or tracheæ, fig. 52); (*b*) **tracheides** (fig. 11, A). In both of these the walls are thickened and lignified, and the protoplasmic contents have disappeared. In both, annular, spiral, pitted, or scalariform patterns may be developed on the walls. The tracheide, however, is a prosenchymatous element developed from *a single cell;* whereas the vessel is a long, tubular structure derived by *cell-fusion* from a longitudinal row of cells. In the Angiosperms the vessels are the characteristic structures of the wood, although tracheides also are found especially in the secondary wood of Dicotyledons. In Gymnosperms and Vascular Cryptogams, there are, with rare exceptions, tracheides only. Tracheal tissue is usually found in bundles.

Owing to the thickening and lignification of the walls this tissue performs a mechanical or supporting function, but it is especially adapted to discharge a vascular function. It is essentially a vascular tissue. It serves for the rapid transport of watery solutions from the root, where they are absorbed, to the leaves and other organs, where they are elaborated.

A typical sclerenchymatous fibre is distinguished from a typical tracheide in that, having only a strengthening function, it is more completely thickened, and shows no large or definite pattern, like the tracheide. Transitional forms, however, are of common occurrence.

§ 27. (6) **Sieve-tube Tissue** (figs. 28, 29). This is the essential tissue of the *phloem* or *bast* (soft bast) of vascular bundles. **Sieve-tubes** are typically developed in the Angiosperms. In this group they are long, slender structures composed of elongated cells placed end on end. The walls are thin and consist of cellulose. The end walls are specially thickened and modified to form **sieve-plates**, the structures characteristic of sieve-tubes. In the thickening of these end-walls small areas remain thin, forming pits. The thin membranes closing these pits are ultimately absorbed (p. 31), so that the end-wall is actually perforated in a sieve-like manner, and the contiguous cells are placed in communication.

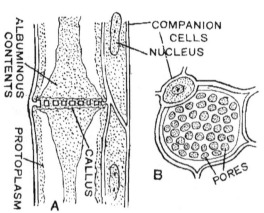

Fig. 28.—SIEVE-TUBE TISSUE.
A, Longitudinal section of a sieve-tube through a simple sieve-plate ; B, Transverse section showing the sieve-plate in surface view.

Usually the whole of the end-wall is perforated in this way to form a simple sieve-plate (fig. 28). Frequently, however, when the end-wall is not horizontal, but obliquely inclined, we can recognize on it a number of areas perforated in this way, the whole structure forming a compound plate (fig. 29). Less frequently in Angiosperms the sieve-plates are formed on the lateral walls. Inside the sieve-tube there is a lining layer of protoplasm, but no nuclei. The sieve-plates are covered on both surfaces by a peculiar substance (probably a modification of cellulose) called **callus**, which also lines the perforations of the sieve plates. Callus is abundantly developed in the autumn, sometimes to such an extent that the pores of the sieve-plate are completely stopped up (*e.g.* the vine). The contents of the sieve-tubes (in addition to the protoplasm) are albuminous. This albuminous substance is found more especially aggregated in the region of the sieve-plates; it contains small starch-grains, and, owing to the presence of these, is stained slightly violet or purple by iodine. Similar structures, consisting of elongated prismatic cells, are found in Gymnosperms and Vascular Cryptogams. Their sieve-plates are, however, most abundantly developed on the lateral walls and are apparently unperforated. No starch is found in these.

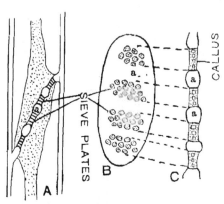

Fig. 29.—SIEVE-TUBE TISSUE—COMPOUND SIEVE-PLATES.
A, Longitudinal section; B, Oblique compound plate in surface view; C, In section. a = thickened portions of the wall between the sieve-plates.

Sieve-tube tissue discharges a vascular function. It serves for the *rapid* transport of elaborated food-material to the various parts of the plant from the leaves.

§ 28. (7) **Laticiferous Tissue (Milk-Tubes).** This is a tissue found only in certain groups of plants. It consists of long, branching tubes, containing a characteristic substance, often milky in appearance, and called **latex**. Two kinds of laticiferous tissue are recognized. The first kind consists of **vessels** formed by cell-fusion. Owing to the fact that the fusion takes place, not in definite longitudinal rows of cells, as is the case in wood-vessels, but in irregular series, the vessels not only are branched, but also anastomose (*i.e.* the branches run into each other) to form a network (fig. 19). The second kind consists of **cœnocytes** (p. 39). In the embryo of a plant which possesses these, certain peculiar cells can be recognized. In development these "cells" elongate and branch, but transverse septa are not formed in them. There is,

GENERAL HISTOLOGY.

however, repeated karyokinetic division of the nucleus, so that these structures are not elongated branched cells, but cœnocytes. Seeing that there is no cell-fusion, it is evident that the branches of these cœnocytes do not anastomose (fig. 30). In both vessels and cœnocytes the walls are somewhat thickened, but consist of cellulose; there is in both a lining layer of protoplasm with nuclei. Laticiferous vessels are found in many Compositæ (*e.g.* dandelion), Papaveraceæ (*e.g.* poppy), and Campanulaceæ (*e.g.* harebell). Laticiferous cœnocytes are found very typically in the spurges (Euphorbia).

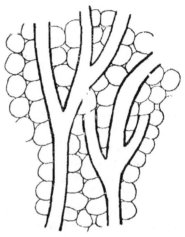

Fig. 30.—LATICIFEROUS "CELLS" (CŒNOCYTES) OF EUPHORBIA, RUNNING THROUGH THIN-WALLED PARENCHYMA.

The contained substance, the **latex**, presents different appearances in different plants. It is rarely quite watery (Banana); usually it is more or less milky (Euphorbia), occasionally thick and coloured (in *Chelidonium majus*, the greater celandine, a plant allied to the poppy, it has an orange colour). It consists of water containing various substances, either in solution or in suspension.

These substances are usually of the nature of excreted products, so that the tubes are often to be regarded simply as reservoirs of excreted matter. Of such substances opium, guttapercha, caoutchouc, are examples.

Frequently, however, the tubes contain nutritive substances as well. These may be nitrogenous or non-nitrogenous. The latex of Euphorbia, for example, contains elongated, rod-shaped or dumbbell-shaped starch-grains.

To some extent laticiferous tissue may discharge a vascular function, serving for the storage and transport of nutritive substances.

§ 29. (8) **Glandular Tissue.** This tissue consists of structures of various kinds in which secreted or excreted substances are produced. Many of these have been spoken of as "secretion-reservoirs." Although laticiferous tissue has been treated by itself, it is not clearly marked off from this glandular form of tissue. The substances produced are of very varied character, such as gum, mucilage, resin, tannin, ethereal oil (fig. 18), mineral crystals (figs. 23, 24), water, etc. Single cells ("sacs") containing such substances are frequently found scattered here and there through the tissue of plants—e.g. tannin or resin cells, cells containing raphides, etc. These isolated glandular cells are called **idioblasts.** Sometimes there are aggregations of glandular cells forming multicellular glands, e.g. the nectaries of flowers. The nectary consists of a group of sub-epidermal cells; the sweet substance secreted is poured out on the surface. Glandular hairs and other glandular epidermal structures must also be noticed here (epidermal glands). The hairs may be unicellular or multicellular. The secretion may be formed in any of the cells of a multicellular hair, but is usually found in the terminal cell, which is frequently more or less dilated or globular (*capitate glandular hairs*). Finally, the "secretion-reservoirs" may be of the nature of cavities or passages formed either lysigenously or schizogenously (pp. 32, 33). Lysigenous cavities, containing various kinds of ethereal oils, are frequently found in leaves (fig. 18) and also in many fruits (e.g. orange and lemon). Most **resin-passages** (fig. 31) are examples of such structures formed schizogenously. The resin-passage is usually surrounded by a layer of small thin-walled parenchymatous cells—the **epithelial layer** —by which the substance poured into the passage is formed.

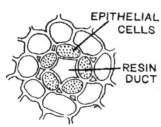

Fig. 31.—Resin-Passage. (Transverse section.)

C. Tissue-Systems.

§ 30. In plant-members the various forms of permanent tissue which we have just described are aggregated in various ways to form higher unities called **tissue-systems**. In all the higher plants there are three systems which, from their constant appearance in roots, stems, and leaves, stand out clearly and constitute systems of the first rank. These are (a) **the epidermal system**, (b) **the vascular system**, (c) the system of the **ground- or fundamental tissue**. It will be advisable to give a short general account of these before going on to consider the various arrangements met with in different types.

§ 31. (a) **The Epidermis—general characters** (figs. 32, 33). The epidermal system, or epidermis, is the outermost *protective* layer or membrane of stems, roots, and leaves. In many cases, as will afterwards be explained, a *true* epidermis is wanting altogether, its functions being often taken on by the outermost layer of ground-tissue. As aerial stems and leaves are the members most exposed to adverse external influences, it is on these that we naturally find the highest development of the epidermal system. Typically an epidermis consists of cuticularized tissue (p. 43), forming a single layer of cells. On roots and submerged parts it is not cuticularized. Sometimes it consists of several layers. This is the case at the apex of most roots where the many-layered epidermis forms a protective structure called the root-cap (fig. 65). An epidermis of several layers is also found in a few stems and leaves. For example, in the leaf of the indiarubber plant (fig. 13) it consists of three layers of small cells, with here and there a larger cell containing a cystolith. In the epidermis of aerial parts we have to recognize *ordinary epidermal cells, stomatal* or *guard-cells,* and various *epidermal outgrowths.*

§ 32. The **ordinary epidermal cells** of a typical one-layered epidermis are always more or less flattened or tabular. Their outline, when seen in surface view, is very various. As a general rule, however, in *long* members they are considerably elongated in the direction of the length of the member, *e.g.* stems and many monocotyledonous leaves (fig. 32, A); while, in members as broad or nearly as broad as long, they are not elongated, but have an extremely wavy outline, *e.g.* most dicotyledonous leaves (fig. 32, B). As already indicated (p. 26), the outer layers of the external walls are cuticularized to form the protective cuticle, which

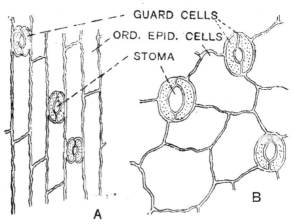

Fig. 32.—PORTIONS OF EPIDERMIS.
(Surface view.) A, Of elongated members; B, Of short members.

serves to prevent undue evaporation from the tissues and ward off the attacks of such pests as insects and Fungi. The cells, except sometimes in the old condition, have protoplasm, nucleus, and vacuole; but, in the great majority, of Flowering Plants, these ordinary epidermal cells have no chloroplasts. These, however, are present in the epidermal cells of Vascular Cryptogams.

§ 33. **Guard-Cells and Stomata.** The stomatal or guard-cells are so called because they surround or guard the openings, known as stomata (figs. 32, 33), numerously developed in the epidermis of aerial parts. These stomata communicate with the system of intercellular spaces in the underlying ground-tissue, and serve, as we shall see later, as a means of gaseous interchange between the plant and the atmosphere. Usually each stoma is surrounded by two guard-cells—one on each side. The guard-cells are crescentic in form. The aperture is formed by the splitting of the common wall between them. They always contain protoplasm, nucleus, and numerous chloroplasts. Their walls are thickened; the thinnest in each guard-cell is that which is farthest from the pore. The guard-cells can alter their form, and thus increase or diminish the size of the opening. In this way they have an important part to play in regulating the amount of water-vapour passing out of the plant in the process of transpiration. Sometimes additional modified epidermal cells (*subsidiary cells* of the stoma) lie outside the guard-cells.

§ 34. **Position of Stomata.** Stomata may be developed on all *aerial* leaf and stem structures — even on the ovary and anthers of the flower. Except in the case of a few Bryophyta, they are confined to the Vascular Cryptogams and Flowering Plants. They are not developed on roots or aquatic members. On green foliage leaves, where they are most numerously developed, their number and position depend largely on the position and direction of the leaf. If the leaf is expanded horizontally (bifacial leaves, p. 9), they are usually confined to the lower surface.

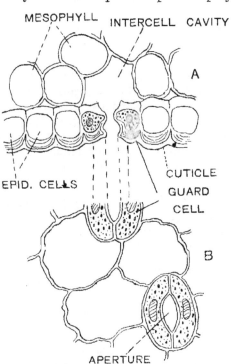

Fig. 33.—Epidermis and Stomata.
A, Section; B, Surface view.

On floating leaves, *e.g.* leaves of the water-lily, they are found on the upper surface. In some bifacial leaves, but more especially on vertical leaves (isobilateral leaves, *e.g.* the Iris), they are about equally distributed on both surfaces.

§ 35. **Epidermal Outgrowths.** Frequently there are outgrowths of the ordinary epidermal cells. These are of the nature of hairs or **trichomes** (*e.g.* fig. 51). They are of varied character—sometimes unicellular, sometimes multicellular. They may end in a sharp point (*spike-hairs*), or in a rounded knob (*capitate hairs*). They may be membranous and attached by their surface to the epidermis (*scale-hairs*). Elongated membranous hairs attached by one end are called *ramenta*. The hairs may be *branched*. In many cases they are *glandular*. Their function is chiefly protective, sometimes absorbent (*e.g.* root-hairs). The glandular hairs especially serve as a protection, their secretion either preventing the injurious action of moisture, or tending to ward off enemies, such as insects and Fungi. **Root-hairs** *are always unicellular*.

Emergences. Stronger outgrowths are often found on the surface of the plant. They differ from trichomes in containing a core of ground-tissue (occasionally also vascular tissue) and not being outgrowths of the epidermis alone. Such outgrowths are called emergences. Frequently they are of the nature of **prickles**, as *e.g.* in the rose (fig. 76). Other examples are found in the *membranous* outgrowths, called **ligules**, of many leaves, *e.g.* the leaves of grasses (fig. 74, E), the petals of the pink (fig. 97, B).

§ 36. **Water-Pores or Water-Stomata.** Other openings, closely resembling stomata in appearance, but differing from them in important respects, are frequently found on leaves. They are called *water-pores* or *water-stomata*, because, instead of giving out water-vapour, they excrete drops of water. They are on the average larger than stomata, and their guard-cells have no power of altering their shape. They are usually developed in groups on the upper surfaces of leaves, frequently on leaf-teeth or leaf-apices. These groups are often found over the fine terminations of the veins of the leaf, *i.e.* the terminations of the vascular system, and are associated with a peculiar glandular tissue (*epithem tissue*) which is found at the ends of the veins and belongs to the vascular system. These glands are often called *chalk-glands*, because the water which is given out by the water-pores usually contains calcium carbonate.

§ 37. (*b*) **The Vascular System.** This is the system of tissue which serves for the rapid transference of nutritive fluids in the plant. Typical vascular tissue is found only in the Vascular Cryptogams and Flowering Plants (p. 18). It is *a continuous system in stem, leaf, and root*. Usually it consists of a varying number of strands or bundles— **vascular bundles**—running longitudinally through stem and root, and passing out into the leaves at all levels. These bundles consist of parts (made up of various tissues) called **phloem** or **bast** and **xylem** or **wood**. In stems and roots showing secondary increase in thickness owing to the activity of a cambium (p. 40), as in trees, this primary arrangement in bundles is modified, and stout cylinders of phloem and xylem are produced. At present, however, we shall confine ourselves to the general characters of a vascular bundle.

§ 38. **The Vascular Bundle** (*e.g.* figs. 51, 52, 59, 60). Vascular bundles may consist of xylem only, or of phloem only, as in roots; or, as in stems and leaves, of both xylem and phloem. In the latter case they are called **conjoint bundles**. The xylem essentially contains tracheal tissue (p. 44). Associated with this, however, there are parenchymatous cells (either thin-walled or thickened and lignified—p. 42) called **wood-parenchyma**, and frequently also sclerenchymatous fibres, called **wood-fibres**. The phloem essentially contains sieve-tubes, but associated with it is a certain amount of thin-walled parenchyma —the **phloem-parenchyma**. Attached to the outer side of the phloem in many bundles there is a bundle of fibrous sclerenchyma. This really belongs to the ground-tissue, although it is called the **hard bast**. The phloem proper is distinguished from this as the **soft bast**. When a bundle contains a marked amount of sclerenchymatous tissue, **or** has a stereid bundle (p. 43) associated with it, it is called a **fibro-vascular bundle**. At the apex of stem or root the bundles pass into the meristematic tissue, from which they are differentiated. In the leaves they spread out and form the veins; the veins end in various ways.

§ 39. (c) **The Ground-tissue System** includes all tissues not belonging to the epidermal or vascular systems. It is evident, therefore, that it consists of many different kinds of tissue. The most abundant tissue is thin-walled parenchyma ; but associated with this are the other tissues in varying amount—sclerenchyma, collenchyma, laticiferous tissue, and glandular tissue. Very frequently this system is marked out into distinct regions, such as the *pith*, the *cortex*, the *medullary rays*, the *hypodermis*, the *endodermis* or bundle-sheath, the *pericycle*. All these will be described in due course.

PART II.—THE ANGIOSPERM.

CHAPTER III.

SEED AND EMBRYO.

§ 1. Before proceeding to a detailed consideration of the form and structure of the various members of the adult plant, it will be advisable to examine their earliest forms as found in the embryo. These embryos are present in the seeds of Flowering Plants, and we have therefore to consider the structure of one or two typical seeds. The student should make a careful practical examination of these with the help of the following figures and descriptions:—

§ 2. **The Sunflower Seed** (fig. 34). The so-called sunflower seeds, obtainable of any seedsman, are really fruits, each containing one true seed. Before examination they should be soaked in water for some time. The wall of the fruit is called the **pericarp**. It is dry and thick, and can readily be removed by means of a penknife or scalpel. The seed which lies inside is invested by a thin yellowish or brownish membrane, constituting the seed-coat and called the **testa**. The removal of this discloses a rather fleshy embryo plant, pointed at one end. The pointed end is called the **radicle**. The greater part of the embryo above the radicle can readily be split into two lobes. These are called the **cotyledons**, and are thick and fleshy because of the large amount of food-material which they contain. If a thin section of a cotyledon be examined under the microscope, the embryonic cells can readily be made out. They are filled with large numbers of rounded aleurone grains (they are stained brown or yellow by iodine). A large amount of oil also is present. If the cotyledons be gently separated there will be found towards the base, and lying between them, a small pointed structure known as the **plumule**. The various parts of the embryo can be examined in a longitudinal section. If the fruit be placed in the soil, under proper conditions, the seed begins to *germinate*. The embryo grows and develops into a seedling plant. This growth takes place at the expense of the food-material (oil and proteid grains) stored up in the cotyledons. These insoluble food-substances are rendered soluble by means of ferments (p. 31); they are, in fact, converted into soluble compounds by a process of digestion. The soluble compounds diffuse to the growing apices of the plumule and radicle, and are made use of as food-material by the protoplasm. During these changes the pericarp and testa are burst asunder. The radicle elongates, grows downwards into the soil, and forms the root. The part of the radicle immediately beneath the cotyledons also elongates, and grows upwards, carrying with it the cotyledons, which increase in size, turn green in the sunlight, and are then readily recognizable as *leaves* of very simple form. The upper part of the radicle which by its elongation has carried up the cotyledons, and which (after germination) is found between the cotyledons and the surface of the soil, is called the hypocotyledonary portion of the axis, or simply the **hypocotyl**.* The plumule

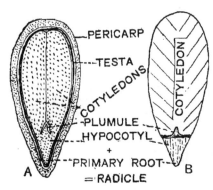

Fig. 34.—FRUIT AND SEED OF SUNFLOWER. A, Longitudinal section of fruit; B, Embryo with one cotyledon removed.

* As we shall see later, the hypocotyl is a portion of the axis which is intermediate in structure between stem and root.

also turns green in the presence of sunlight, and eventually develops into the stem and leaves of the sunflower plant. From the above it is evident that the plumule must be regarded as the embryonic *shoot*, the cotyledons as the embryonic *leaves*, and the radicle as the embryonic *root* (and hypocotyl). In the sunflower, as in most seeds, the plumule is an extremely small conical structure, showing no trace of young leaves. In some plants, however, it is large (*e.g.* the almond), and bears little outgrowths which are easily recognized as tiny undeveloped leaves. The axis of a plumule is called the epicotyledonary portion of the axis, or simply the **epicotyl**. Together with the radicle it constitutes the axis of the embryo, just as stem and root constitute the axis of a full-grown plant. In the sunflower there are two cotyledons. This is characteristic of the *Dicotyledons*, the group of Angiosperms to which the sunflower belongs. Where, as in the sunflower, the cotyledons come above ground and form the first foliage-leaves of the plant, they are said to be **epigeal**.

§ 3. **Seed of the Broad Bean** (fig. 35). This is a true seed, the pod of the bean being the fruit. The seed, as before, should be soaked first in water. The seed-coat here is double. The outer covering is the testa. There is a thinner membrane within this called the **endopleura**, or tegmen. This is absent from the seeds of most Flowering Plants. On the testa at one end of the seed there is an elongated scar of a dark colour, the **hilum**. This is the point where the seed has broken away from its stalk. On gently pressing the *soaked* seed, a drop of moisture will be seen to exude from a minute aperture—the **micropyle**—situated at one end of the hilum. Hilum and micropyle are present, but not recognizable in the sunflower. Inside the seed-coat there is a large embryo plant. This consists, as in the sunflower, of a radicle, a plumule, and two cotyledons. The **radicle** is short and blunt, and is situated to one side. Its tip lies close to the micropyle. The **cotyledons** here are much more massive than in the sunflower, because of the larger amount of food-material stored up in them. The **plumule** is larger, and is found, as before,

between the cotyledons. A striking difference between the
bean and sunflower is found in the behaviour of the cotyledons
at germination. In the bean the hypocotyl remains short,

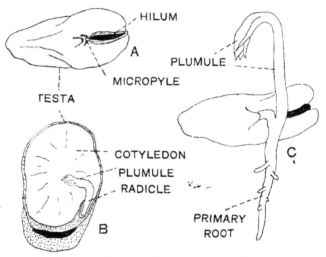

Fig. 35.—SEED AND EMBRYO OF BROAD BEAN.
A, Entire ; B, One cotyledon removed ; C, After germination.

and the cotyledons do not come above ground, but remain
inside the seed-coat, and simply supply food-material to the
young plant. Such cotyledons are said to be **hypogeal.***

§ 4. **Seed of the Ash** (fig. 36). Ash-trees towards the
autumn are found to bear dense tufts of brown-coloured
fruits. If one of these fruits be examined, it will be seen
that the *pericarp* towards the free end has developed into a
tough membrane, or wing. The other end is somewhat
swollen, and contains a single seed, which is readily exposed
by removal of the pericarp. The brown outer layer of
the seed is the *testa*; it is firmly adherent to the mass
contained within. If, however, the seed, after being
thoroughly soaked in water, be squeezed, a fairly large
embryo will be expressed at one end. It consists of two
thin membranous *cotyledons*, a very long *radicle*, and a
minute *plumule*, similar in position to that of the sun-

* The cotyledons of the French bean are epigeal.

flower, but recognizable only in section under the microscope. If now the seed, from which the embryo has been pressed, be cut across, a white substance will be found lying against the inner side of the testa. This is a tissue containing a

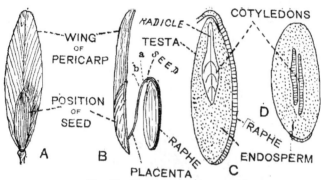

Fig. 36.—FRUIT AND SEED OF ASH.
A, Entire fruit ; B, Part of Pericarp removed ; C, Longitudinal, and D, Transverse, sections of seed. *a*, Funicle ; *b*, Aborted ovules. Terms not given in § 4 are explained later.

store of food-material, and it is known as the **albumen**,* or **endosperm**. In the middle of the endosperm there is a space, in which the embryo lies. Seeds like those of the ash, in which, in addition to an embryo plant, there is a store of endosperm, are said to be **albuminous** or **endospermic**. In the sunflower and bean food-substance is also stored up, but it is in the cotyledons, and not in a special layer of endosperm tissue. Such seeds are **exalbuminous** or non-endospermic. In the ash, the cells of the endosperm contain granular proteid substance, but, in addition, the walls are somewhat thickened, and represent a storage of carbohydrate in the form of *cellulose*. The cotyledons remain inside the seed-coat for some time during germination to absorb the food-material, but eventually come above ground (epigeal).

§ 5. **Maize** (fig. 37). The so-called maize seed is really a fruit containing a seed. *Pericarp* and *testa* are both thin, and fused together to form a single membrane. On one

* The term albumen here must not be regarded as indicating any definite chemical compound. It came to be applied in this connexion by *analogy* with the albuminous white substance in a hen's egg.

side there is an oblong area of a light colour. The embryo lies immediately underneath this. If the fruit be cut longitudinally through the middle of this area, the embryo will be seen in section lying to one side of a mass of *starchy endosperm*. If the cut surface be moistened with iodine solution, the endosperm will be stained blue, and the embryo more clearly shown. The outermost layer of the endosperm (just under the testa) contains aleurone grains, and is called the **aleurone layer**. The embryo consists of a large *plumule*, a *radicle* enclosed in a sheath, and a single massive *cotyledon* (called the **scutellum**), which lies against the endosperm. At germination the scutellum, or cotyledon, remains behind in the seed and absorbs the endosperm (hypogeal). The ferment which converts the starch into sugar is secreted by the outermost layer—**epithelial layer**—of the scutellum. This is the seed of a **monocotyledon**. In this division of Angiosperms, as the name indicates, the embryo has only one cotyledon.

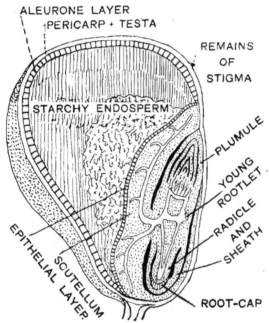

Fig. 87.—Fruit of Maize cut Longitudinally.

§ 6. From these examples the student will have obtained a general idea of the parts of the embryo plant and the names applied to them. We now pass to the consideration of the fully developed members—stem, leaf, root—of the Angiosperm.

CHAPTER IV.

THE STEM OF THE ANGIOSPERM.

§ 1. The shoot in the Angiosperm is differentiated into stem and leaf structures. These are very highly specialized and show an immense variety of forms. For this reason it will be convenient to study them separately. The present chapter is therefore devoted to the stem—its general external characters and internal structure.

A. EXTERNAL CHARACTERS.

§ 2. **Nodes and Internodes.** We have already stated that the plumule grows upwards into the sunlight, and develops into the leaf-bearing stem of the plant. As growth goes on the stem not only produces members unlike itself, the leaves, but may also give rise to similar members, *i.e.* it may branch. In the fully grown part of most stems, the leaves are separated by intervals from each other. The points at which one or more leaves are given off are called the **nodes** of the stem, and the regions between these the **internodes** (fig. 38).

§ 3. **General Descriptive Terms.** Usually stems in transverse section are quite circular, and are described as **cylindrical**. Others are marked by alternate ridges and furrows, and are said to be **angular**. Thus the stem of the white dead-nettle is quadrangular or square. Some stems are flattened. In rarer cases we meet with globular or altogether irregular stems. The stem may be either **herbaceous** or **woody**. In some plants, as in the wall-flower, it is herbaceous above and woody below. A typical erect herbaceous stem like that of the foxglove or lily is called a **caulis**. A thick, woody stem, as in trees, is called a **trunk**. Some herbaceous stems are more or less dilated or expanded at the nodes. This is due to the arrangement of vascular tissue at these points. The stems appear jointed, and hence the terms **jointed** or **articulated** applied to them. Such stems are called **culms.** Examples are met with in the pink and in grasses. The stem may be more or less **hairy.** It may be **prickly** or **spiny.** If there are no hairs, and the stem is quite smooth, it is described as **glabrous**; if, in addition, it is more or less shiny, with a bluish colour, it is said to be **glaucous.**

§ 4. **Buds** (fig. 38). The growth in length of the main stem, or a branch, takes place towards the apex. At the extreme apex the internodes have not yet elongated; the young leaves, which are just in the course of development, are crowded together and closely overlap the growing apex of the stem or branch. This compact structure, which we nearly always find at the apex of a stem, is called a **terminal** bud. As growth takes place the internodes gradually elongate, and the leaves become separated. It is important to notice that the branches of a stem make their first appearance as buds. These buds, which, with reference to the stem on which they are borne, are called **lateral** buds, are situated in the **axils** of the leaves, *i.e.* in the angle between the leaf and the upper part of the stem. The *axillary* position of the buds should be carefully noticed. In the Angiosperms it is the rule that each leaf has a bud in its axil. Very frequently, either naturally or owing to the influence of external conditions, only some of these buds develop into branches; the others remain quiescent. In circumstances of necessity, however, as, for example, when the main stem and chief branches have been destroyed, these **dormant** buds become active, and give new life to the plant. The shoots produced from such buds are said to be **deferred.** The shoots developed late on the stems of trees have sometimes this character.

The **normal** axillary buds are developed in acropetal order (p. 10). Buds which are developed out of their proper order, or without any relation to the leaves, are called **adventitious.** The shoots of Pollards and on the trunks of many trees (*e.g.* Elm) arise from such buds. They may also be developed on leaves or roots. If the leaf of Begonia, for example, be artificially wounded and laid on the surface of the soil, adventitious buds are

THE STEM OF THE ANGIOSPERM. 63

developed from the wounded surface, and produce new plants. In the dandelion they spring from the root.

Sometimes more than one bud is developed in the axil of a leaf. These are called **accessory** buds. Examples are found in the walnut, ash, and some willows.

A bud, then, as found in the Flowering Plant, may be

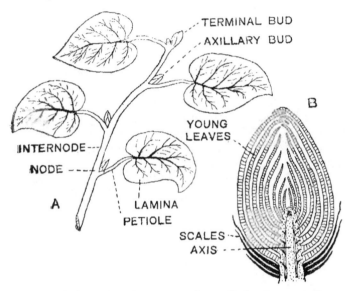

Fig. 38.—A, Twig with Leaves and Buds; B, Longitudinal Section of a Bud.

defined as a rudimentary or embryonic shoot, consisting of a short axis in which the internodes have not yet elongated and in which the young leaves are closely crowded together and overlap the apex. These buds can be recognized on plants at all seasons, but are most noticeable, and are seen in the greatest variety, in winter. In many buds the young leaves are all of the same kind and in course of time develop into green foliage-leaves, but frequently, especially in winter buds, only the central leaves of the bud are of this kind, while the outer ones are small and scaly—*scale-leaves*—and serve as a protection against cold and the loss of water (fig. 38, B). In many winter buds also the loss of moisture, which would be injurious to them, is prevented by the corky nature of the scales, by the secretion of resinous substance (*e.g.* the horse-chestnut), or by the development of a covering of hair (many willows).

§ 5. **Branching of the Stem.** The branching of the stem in the Angiosperm is probably always **lateral** (p. 10) ; in other words, the branches arise as lateral buds in the axils of the leaves. The young leaves and their axillary buds originate as little protuberant outgrowths just below the extreme tip of the parent-axis. The branching may be **racemose** or **cymose** (p. 11). In indefinite or racemose branching (fig. 3, B), at each node there may be either one or a series (called a whorl) of two or more branches, according to the number of buds developed (which will depend largely on the number of leaves). Definite or cymose branching, if only one daughter-axis is given off at each branching, is said to be **uniparous** (fig. 39, A—D); if two, **biparous** (fig. 39, E); if more than two, **multiparous**. The biparous cymose form of branching, owing to the abortion of the growing-point of the parent-axis, frequently resembles a dichotomy, hence the name *false dichotomy* often applied to it. The abortion of the mother-axis may be natural or due to accident. Examples are seen in the lilac and mistletoe. In uniparous cymose forms the successive daughter-axes may be developed right and left alternately—the **scorpioid** or **cicinnal** form (fig. 39, A); or always on the same side—the **helicoid** or **bostrycoid** form (fig. 39, C). In these two forms the branching would present a zigzag or spirally coiled appearance respectively, if the branches retained the position in which they are developed. But in nature the branching becomes straightened out (fig. 39, B, D), and the basal portions of the successive daughter-axes constitute what *to all appearance* is a simple parent-axis, but *really* a false axis or **sympodium**. The scorpioid form resembles a typical raceme ; the helicoid form a one-sided raceme. These *sympodial* forms are distinguished from true racemose ones by the position of the leaves, which, it should be noticed, are given off on the opposite side from what are *apparently* lateral branches.

The student is advised not to be content with a theoretical

THE STEM OF THE ANGIOSPERM. 65

knowledge, but to make a careful practical study of branching by examination of a large series of plants. The racemose (monopodial or indefinite) is by far the commoner in purely vegetative parts of stems; but cymose branching is frequently met with in trees—*e.g.* the uniparous form in the elm and lime. Types of branching are most easily recognized in herbaceous plants; in many woody plants (shrubs and trees), owing to the frequent loss or injury to which they are exposed, branching cannot be studied

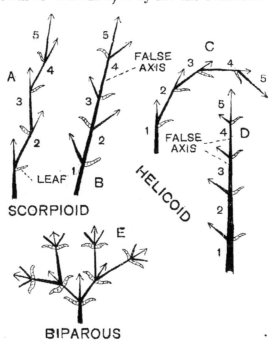

Fig. 39.—Forms of Cymose Branching.
A—D, Uniparous.

in its primitive condition, and the premature attempt to diagnose the more complicated forms of branching will inevitably lead to confusion.

§ 6. **Forms of Stems.** We have already indicated that the various parts of plants are *adapted* to the performance of certain functions. The ordinary functions of a stem are (*a*) to bear the leaves and keep them expanded or exposed in such a way that they may best carry on their functions; (*b*) to give passage to various nutritive solutions passing between roots and leaves. These functions may, however, be carried on in many different ways, according to the mode of life of a plant, or peculiarities in its environment.

Stems in different plants must have a structure and organization adapted to the conditions in which they live. In addition to this stems may take on special functions. Thus they may serve as organs of vegetative propagation, or as storeplaces of nourishment; they may be specialized to act as protective organs, or to perform the functions usually carried on by other members. The student will understand, then, that stem-structures assume an immense variety of forms, according to their special adaptations. Some (*e.g.* foxglove, sunflower, lily) grow straight up, and are self-supporting; they are said to be "**erect.**" This is the typical form of stem. Others are **weak stems**, incapable of themselves of growing erect. Most stems are aerial, but many are buried in the soil, and are called underground or subterranean stems. Some are herbaceous; others are woody. Most woody plants form shrubs or trees, and are perennial, *i.e.* persist for a number of years. Herbaceous plants may be **annuals**, living only during one season, and dying down in the autumn; others are **biennial** (*e.g.* the turnip), living during two seasons. In the first year they produce only vegetative shoots; in the second, flowers, fruits, and seeds. Many herbaceous plants, however, which die down in the autumn, perennate by means of their underground stems. Some small shoots or twigs have only a limited development, and are called "**dwarf shoots**" —for example, in the apple, where they are the twigs producing the flowers. Finally, in some plants, the stem is extremely short ("*reduced stems*") and the leaves *appear* to come off from the top of the root. Good examples are seen in the daisy and dandelion. We will now consider some of the more important of these forms.

§ 7. **Weak Stems.** Stems which do not rise but trail on the surface of the ground are said to be *prostrate* or *procumbent*; if they rise towards their apices, *decumbent*; if they gradually incline upwards from the base, *ascending*. Many weak stems, however, although unable to grow erect themselves, make their way upwards by attaching themselves to surrounding objects. These are known as **climbing** and **twining** plants. Climbing is effected in various ways.

The ivy, for example, climbs by means of adventitious roots; these roots, developed on the stem, fix themselves to the trunk or wall on which the plant climbs. The pea, the passion-flower, the vine, etc., climb by means of special organs called **tendrils**. These tendrils, as we shall see later, may be specialized stems, leaves, or parts of leaves. The Virginian creeper climbs by means of adhesive, sucker-like discs; clematis, by means of its leaf-stalks or petioles, which act the part of tendrils, and, in fact, are often spoken of as *petiole-tendrils*. The blackberry supports itself by means of prickles. As distinguished from these, **twining plants** achieve the same result by themselves twining round some support, as, for example, the hop, convolvulus, and others. The object in all cases of climbing or twining is, of course, to enable the plant to reach the light, which is necessary for the formation of chlorophyll, and for the assimilation of carbon dioxide (p. 12).

§ 8. **Runner, Offset, Stolon, and Sucker.** Many plants give off highly specialized shoots, serving chiefly for purposes of vegetative reproduction. Of these, the runner, offset, stolon, and sucker are the commonest. The **runner** (fig. 40) is a very slender shoot running along the surface of the ground, and attaining a considerable length. It arises in

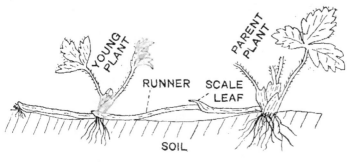

Fig. 40.—RUNNER OF STRAWBERRY.

the axil of a leaf, at the level of the soil. At intervals it produces small leaves, with a bud in the axil of each. From the bases of these buds adventitious roots pass down into the soil, and in this way new plants are formed. The

68   BOTANY.

strawberry gives a good example. The **offset** (*e.g.* house-leek, fig. 41) resembles the runner in origin, but is shorter and stouter. The **stolon** is a specialized branch given off above the level of the soil.

Fig. 41.—Offset of House-Leek.

It grows downwards, and the surface in contact with the soil gives off rootlets (*e.g.* currant, gooseberry). The **sucker** (fig. 42) differs from this in that it is given off *below* the level of the soil; it grows upwards, develops roots and aerial shoots. These suckers are white or pink in colour, and resemble roots. They are distinguished as stems, however, by their axillary development and the possession of scale-leaves. Good examples are seen in the mint, dead-nettle, and rose.

§ 9. **Bulbils.** These may be described as axillary buds, which become large and fleshy owing to the storage of food-material in their leaves. They differ also from ordinary buds in the fact that they separate from the parent plant, fall to

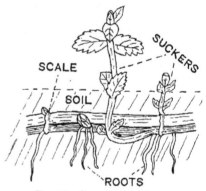

Fig. 42.—Suckers of Mint.
(From Green, modified.)

the ground, and produce new plants, thus serving for reproduction (*e.g.* Lesser Celandine, some Lilies). Similar structures may take the place of flowers (Onion, some grasses, etc.). In plants producing them seed formation is usually uncertain.

§ 10. **Underground Stems.** There are several underground forms of stems which deserve special mention. They serve either for vegetative reproduction or as a means of perennation.

The **Rhizome** (incorrectly called the *root-stock*) is a stout

THE STEM OF THE ANGIOSPERM. 69

elongated underground stem more or less filled with food-material (fig. 43). Students are very apt to mistake the rhizome for a root. It is distinguished by the presence of leaves and buds (and also by internal structure). The leaves may be large foliage-leaves, but more frequently the rhizome bears only small brown scale-leaves, the foliage-leaves being borne on the aerial shoots developed on the rhizome. The rhizome may be horizontal and dorsiventral (p. 9), or it may run more or less obliquely through the

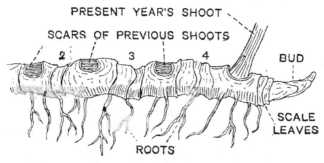

Fig. 43.—Rhizome (Sympodial) of Solomon's Seal.
The numbers indicate the segments of the sympodium.

soil. Adventitious roots are given off from its surface. It usually branches freely, and if these branches become separated they form distinct plants. The aerial branches may be given off after the racemose type of branching, in which case the rhizome on which they are borne is a simple parent-axis with a persistent apex (*e.g.* the wood-sorrel). In other cases the apex of the rhizome grows up into an aerial shoot, and the growth of the rhizome is continued by a lateral bud developed in the axil of a scale-leaf. Here the rhizome is sympodial in development, being made up of the persistent basal portions of the successive growths, as indicated in fig. 43. Examples of sympodial rhizomes are found in the wood-anemone and Solomon's seal. As a rule, the scars of leaves and branches can be readily detected on a rhizome.

The elongated sympodial underground shoots developed in many grasses and sedges (*e.g.* the sand-sedge) are best regarded as very slender forms of the sympodial rhizome. By some they are distinguished as **creeping stems**, but this term, if used at all, is more appropriately applied to *prostrate* stems (p. 66), like those of the ground-ivy, which run on the surface of the ground, and give off roots at the nodes. Used in this sense, the creeping stem would differ from the runner in being the whole prostrate stem of the plant, and not a special lateral reproductive branch.

The **Corm** (figs. 44, 45) is an underground stem which may be regarded as a very much abbreviated one-jointed form of rhizome. It consists of a massive swollen stem called the **disc**, on which are a number of loose more or less sheathing scale-leaves. The size of the disc is due to the large amount of food-material stored in it. One or more buds are present in the axils of the leaves, sometimes towards the apex of the disc (*e.g.* crocus, fig. 44), sometimes towards the base (*e.g.* Colchicum, the meadow-saffron, fig. 45). In the spring these buds develop at the expense of the stored food-material, and grow up into aerial flowering shoots. Adventitious roots are developed from the base of the bud, and pass down into the soil. During

Fig. 44.—CORM OF CROCUS. (Longitudinal section.)

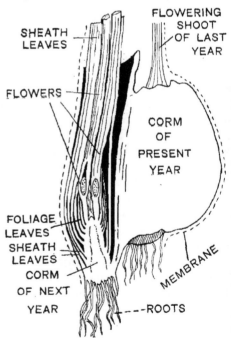

Fig. 45.—CORM OF COLCHICUM. (Longitudinal section.)

THE STEM OF THE ANGIOSPERM. 71

the summer the surplus amount of plastic substance is stored up in the basal portion of the new stem, and thus a new corm is gradually formed, which will in the same way produce new plants the following year. Thus the corm represents the basal underground part of a stem, laden with food-material and bearing buds and scale-leaves. Corms are *popularly* spoken of as bulbs.

The **Bulb** (fig. 46) may be regarded as a specialized underground bud. It has a structure somewhat similar to the corm, but the stem or disc is comparatively small, and the food-material is stored up in the leaves of the bud, which are therefore large and fleshy, and overlap the disc. The bud develops at the expense of the stored material into an aerial flowering shoot. Adventitious roots are given off from the base. One or more small buds are present in the axils of the innermost leaves. These, later on, gradually enlarge, owing to storage of food-material in their leaves, and form new bulbs which will repeat the process next year. In some bulbs the leaves are large, and completely ensheath each other. The withered and shrivelled-up remains of these invest the new bulb when formed like a tunic. Hence the name **tunicated** bulb (*e.g.* the onion, hyacinth). In others the fleshy scales simply overlap at the margins. These are called **scaly** or **imbricated** bulbs (*e.g.* lilies). True bulbs and corms are found only in monocotyledonous plants. In monocotyledonous stems there is (with a few exceptions) no secondary increase in thickness by the activity of a cambial layer. These bulbs and corms supply another means of reproduction and perennation.

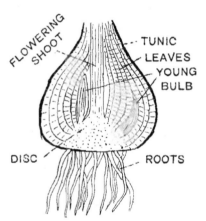

Fig. 46.—Tunicated Bulb. (Longitudinal section.)

The **Stem-tuber** (fig. 47) is a swollen underground stem,

or part of a stem, laden with food-material, and serving for purposes of vegetative reproduction, e.g. in the potato and Jerusalem artichoke. In the potato the tubers are borne on slender underground shoots, which are recognized as such, not only by their internal structure, but also by the fact that they bear scale-leaves. The tubers make their appearance either at the apex of a shoot, or in the axils of the scale-leaves, and, instead of developing into normal branches, become enormously dilated by the deposition of starchy food-material. The tuber, however, is readily distinguished as a modified stem-structure, not only by its position of development, but also by the possession of buds, known as the "*eyes*." When a tuber, or part of a tuber, is placed in the soil under proper conditions, the buds or "eyes" develop at the expense of the stored food-material, and produce new plants.

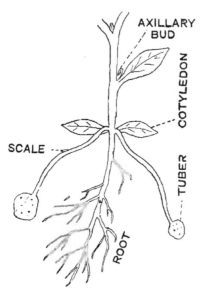

Fig. 47.—Seedling of Potato with Developing Tubers.

§ 11. **Stem-Tendrils, Spines, and Cladodes.** In many cases, stem-structures undergo more complete specialization than those indicated above. The most striking of these are the "modifications" into tendrils, spines, and cladodes. These are quite unlike ordinary stems, and assume forms which are also met with in morphologically different (dissimilar) members.

Stem-tendrils are highly specialized climbing organs. They are very slender, usually branch, may bear small scaly leaves, and are sensitive to contact. When in their growth they touch a suitable support, they twine round it and raise up the part of the plant on which they are developed. Their morphological value is recognized by their position. Sometimes they represent lateral shoots. In this case they arise distinctly in the axils of leaves, e.g. the passion-flower and white bryony. Others, however, represent the modified apices of the successive shoots of a cymose branching. In this case they do not appear in the axils of leaves, but are placed on the opposite side of the sympodial axis from the leaves (fig. 48, and cf. fig. 39, B, D).

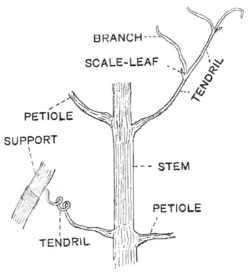

Fig. 48.—TENDRILS OF THE VINE.

Stem-spines, or **thorns** (fig. 49), are aborted branches which have become hard and sharp-pointed. They have lost the apical growing-point, and are adapted to function as protective structures. Examples are seen in the sloe and hawthorn. Their stem-nature is recognized as before by their structure, by their position in the axils of leaves, and also by the fact that though the terminal bud is aborted, they may bear lateral buds. It is interesting also to notice that in the plum, which may be regarded as the cultivated form of the sloe, these structures are represented by leafy, flowering shoots. The student must carefully distinguish between spines and prickles. The latter are irregularly developed, i.e. have no definite relation to the leaves, have no vascular tissue, and are very readily broken off.

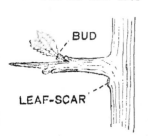

Fig. 49.—SPINE OF THE SLOE.

A cladode (phylloclade or cladophyll) is a branch which has taken on the appearance and general functions of a leaf. Sometimes the whole stem may be so modified, as in the duckweed, where it consists of a series of flattened leaf-like segments giving off branches of similar character. Usually, however, only the small lateral branches are so modified, as in *Ruscus aculeatus*, the Butcher's Broom (fig. 50). These branches are very like leaves externally, but they bear flower-buds, and arise in the axils of small, scaly structures which are the true leaves.

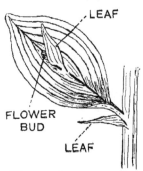

Fig. 50.—Phylloclade of Ruscus.

§ 12. **The Torus or Thalamus.** We have already mentioned (p. 7) that the flower is a specialized shoot. That part of the floral axis which bears the floral leaves (sepals, petals, etc.) is called the *thalamus* or *torus*. It presents certain peculiarities, and must therefore be regarded as a special form of stem. It will be considered fully in the chapter on the flower.

§ 13. **Summary.** We may summarize the general distinctive characters of the stem in the following statement: stem-structures *tend* to grow *upwards, towards the light*; they usually end in a bud and bear leaves, lateral buds, and often reproductive organs; lateral branches (in Flowering Plants) arise in the axils of leaves (for development and internal structure see p. 93). We cannot, however, regard this as of the nature of a definition, distinctly marking off stems from leaves and roots, for these general characters are not absolute. Thus we have seen that some stems (*e.g.* rhizomes) remain under ground and partake of the functions of roots; others have lost their terminal bud; in a few cases, again, buds are developed on roots and leaves. At the same time the student must notice these characters carefully; for it is by attention to these that he can as a rule recognize members which, however modified they may

THE STEM OF THE ANGIOSPERM. 75

be, have the morphological value of stems. In this way, as already indicated, the rhizome, the sucker, the tubers of the potato, and the spines of the sloe, etc., can all be recognized as stem-structures.

B. INTERNAL STRUCTURE.

I. THE DICOTYLEDON.

§ 14. Primary Structure. The primary arrangement of tissue characteristic of the dicotyledonous stem may be studied in the sunflower (Helianthus) or other herbaceous types. Fig. 51 represents, diagrammatically, a portion of a transverse section of a well-developed internode of the sunflower. On the outside is the **epidermis** (p. 49). The **fibro-vascular bundles** (p. 53) are seen to be arranged in a ring. Owing to this characteristic arrangement, the **ground-tissue** (p. 54) is marked off into (*a*) a central region,

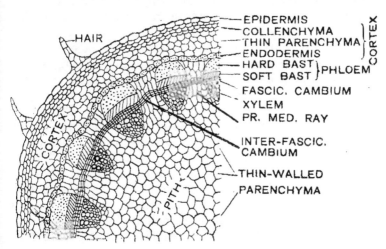

Fig. 51.—PART OF A TRANSVERSE SECTION OF STEM OF SUNFLOWER.
The Inter-fascicular Cambium is not found in Young Internodes.

the **medulla** or **pith**, (*b*) an outer region between the epidermis and the vascular ring, the **cortex**, and (*c*) a number of strands running between the bundles from pith to cortex, the **primary medullary rays.** The pith, medullary rays, and inner region of the cortex, consist chiefly of thin-walled *parenchyma* (p. 41). The hypodermal region of the cortex (the **hypodermis**), the region immediately under the epidermis, consists of *collenchyma* (p. 42). In the cortical parenchyma, sometimes also in the pith and medullary rays, small *resin-passages* can be detected, each with its *epithelial* layer (p. 48). If the section be stained with iodine it is found that the cells of *the innermost cortical layer* contain starch-grains. This layer is then distinctly marked off from the rest of the cortex. It is the **endodermis** or **bundle-sheath** (here a *starch-layer*).

Fig. 52 represents the structure of one of the fibro-vascular bundles in transverse and also in longitudinal section. The bundles are **conjoint** (p. 53). The **xylem** or *wood* is internal; the **phloem** or *bast* external; and there is between them a strip of meristematic tissue, the **fascicular** or **intra-fascicular cambium.** Bundles having the xylem and phloem placed side by side in this way are called **collateral.** When a cambium is present, so that further growth (secondary growth) may take place, they are said to be **open.** The bundles of the sunflower, then, are collateral and open. The *primary* xylem contains annular, spiral, reticulate, and pitted vessels. They are more or less arranged in radial rows, and amongst them is a tissue consisting of *wood-fibres* (p. 53) and *wood-parenchyma* (p. 42). The smallest (annular and spiral) vessels are found in the region of the primary xylem which lies next the pith. This is the region of the **protoxylem.** As it consists of lignified tissue the wood turns brown under the action of iodine. So also the **hard bast**, which is sclerenchymatous. The **soft bast** consists of sieve-tubes and thin-walled parenchymatous cells. Some of these cells are small and closely associated with the sieve-tubes; they are known as **companion cells** (fig. 28). Others are larger and called **cambiform cells.** The cambium is a single layer of thin-walled elongated cells which in transverse section are seen to be more or less four-sided and flattened. Towards the beginning of secondary growth, owing to the division of its cells, it seems to consist of several **layers.**

§ 15. Longitudinal Course of the Bundles.

Fig. 53 represents diagrammatically the longitudinal course of the bundles in a Dicotyledon. Tracing one of these upwards, we find that it runs through one or two internodes, and then bends out into a leaf. At the point where this bending

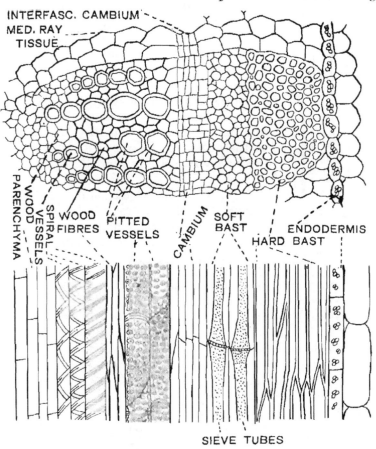

Fig. 52.—Fibro-vascular Bundle—Stem of Sunflower.
(Transverse and radial longitudinal sections.)

takes place, another bundle arises, runs upwards through one or two internodes, and passes out to a higher leaf. So with all the bundles of the ring. Or we might express it in another way, and say that bundles pass into the stem

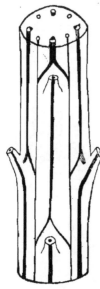

Fig. 53. — Diagram showing generally the longitudinal course of the bundles in a dicotyledonous stem.

from the leaves, run downwards in the stem, and finally join on to bundles entering the stem from older leaves. The bundles are **common bundles**, *i.e.* they are not confined to the stem, but are common to stem and leaf. The upper part of the bundle, running obliquely through the cortex towards the leaf, is called the **leaf-trace**. In the stem all the bundles run parallel to the epidermis, and at an equal distance from it. That is why, in transverse section, they form a ring. There is frequently considerable branching and intercommunication of the bundles at the nodes. It follows that the primary medullary rays are of limited height.

§ 16. **The Apical Meristem and Development of Tissues.** Having now described the arrangement and different regions of permanent tissue found in the fully grown herbaceous stem, let us see if we can trace any relation between these and the apical meristem from which they are derived. A longitudinal section (figs. 54, 55) through the apical bud of a dicotyledonous stem shows the apical meristem, and also the mode of origin of young leaves and branches. By examining such a section, and also a *regular* succession of transverse sections, we can trace the gradual differentiation of tissues.

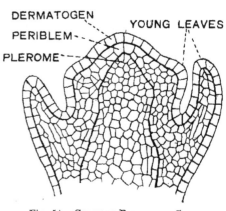

Fig. 54.—Growing-Point of a Stem.
(Longitudinal section.)

THE STEM OF THE ANGIOSPERM.

Three regions can be distinguished in the apical meristem. There is a single outermost layer passing right over the apex. If we trace this layer away from the apex into the region of permanent tissue, we find that its cells divide only by walls at right angles to the surface; there are no divisions parallel to the surface. Thus the layer remains single. It passes finally into the epidermis. It is the embryonic epidermis, and is called the **dermatogen**. Internal to this is the second region, the **periblem**. At the extreme apex this is usually a single layer; but, behind the apex, owing to the irregular division of its cells, it becomes many-layered. From it the cortical region of ground-tissue is developed. The

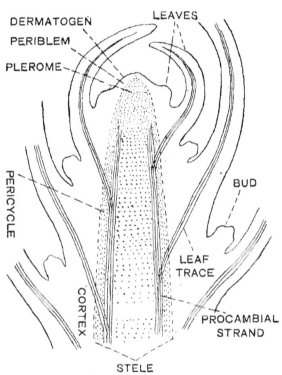

Fig. 55.—GROWING-POINT OF A STEM.
(Longitudinal section—diagrammatic.)

periblem, therefore, is the embryonic cortex. Its innermost layer becomes the endodermis, or bundle-sheath. The third region forms the core, or central part of the meristem. It is known as the **plerome**. From this region are differentiated all tissues lying within the endodermis, including the vascular bundles, pith, and medullary rays. Such a column or cylinder of permanent tissue differentiated from a plerome is called the **central cylinder**, or **stele** (Gr. στηλη, a column). The plerome, therefore, is the embryonic stele. In the stems of nearly all Flowering Plants there is a single plerome giving rise to a single stele—hence these stems are said to be **monostelic**. A stele, although it may present different appearances in different plants, always contains vascular tissue (*i.e.* the vascular system of tissue is developed in the stele) associated with a larger or smaller amount of ground-tissue (pith, medullary rays, etc.). The ground-tissue of the stele is called **conjunctive**, or **intra-stelar**, to distinguish it from the cortical ground-tissue, derived from the periblem, which is outside the stele, and therefore called **extra-stelar**. We must examine the differentiation of these tissues of the stele in greater detail.

§ 17. **Development of the Vascular Bundle.** Some little distance behind the apex of the stem, several longitudinal strands of elongated cells make their appearance *near* the periphery of the plerome. These are known as the **procambial strands** (fig. 55). The cells of these strands, although they have grown in length, are still meristematic. They become elongated, the divisions that take place in them being chiefly longitudinal. A transverse section (fig. 56) shows that they are developed in a ring in the dicotyledonous stem. These procambial strands develop into the vascular bundles; they are in fact the embryonic bundles. The differentiation of the procambial strand can be followed by examining sections farther and farther from the apex. The first xylem elements—

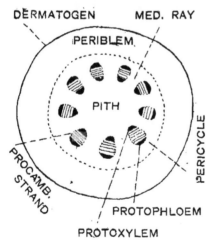

Fig. 56.—TRANSVERSE SECTION NEAR THE APEX OF A STEM, ILLUSTRATING THE DIFFERENTIATION OF TISSUE. (Diagrammatic.)

THE STEM OF THE ANGIOSPERM. 81

protoxylem, consisting of annular and spiral vessels—make their appearance on the inner side of the procambial strand; the first phloem elements—*protophloem*—on the outer side. Differentiation proceeds from these two points towards the centre of the strand. The xylem vessels formed last are pitted. The differentiation of the procambial strand, however, is incomplete. A layer of meristematic cells persists in the middle, between the xylem and the phloem, as the *fascicular cambium*. Being derived directly from the apical meristem, this is a primary meristem (p. 40). The groups of xylem and phloem thus developed from the procambial strands are described as primary to distinguish them from the secondary xylem and phloem formed later by the activity of the cambium. The protoxylem, as already indicated, is only the first-formed portion of the primary xylem. It should be particularly noticed that the primary phloem thus developed is the soft bast. The hard bast is not developed from the procambial strand.

§ 18. **Intra-stelar Ground-Tissue.** The *medulla* or *pith* is the ground-tissue differentiated from the central region of the plerome, *i.e.* inside the ring of procambial strands. It is the central ground-tissue of the stele. The *medullary rays* consist of ground-tissue developed between the procambial strands. The student must now very carefully notice that, in addition to these, there is a band of ground-tissue developed from the peripheral region of the plerome, *i.e.* outside the ring of procambial strands, and just internal to the endodermis (figs. 55, 56). This is the **pericycle**. In some Dicotyledons, this band consists entirely either of thin-walled tissue or of lignified tissue (sclerenchyma). Most frequently, however, the portions of the pericycle which lie just outside the procambial strands, and these only, are lignified, and consist of fibrous sclerenchyma. These lignified strands constitute the *hard bast* of the bundles. It will now be evident that, strictly speaking, the hard bast does not belong to the bundle at all, but is a lignified portion of the intra-stelar ground-tissue (although in practice it is convenient to regard it as part of the phloem). The portions of the pericycle which lie between the hard bast strands are parenchymatous, and are usually, in description, included in the medullary rays Thus a **pericycle** may be defined as the peripheral region of intra-stelar ground-tissue. It may be homogeneous, consisting of one kind of tissue only, or heterogeneous, consisting of different tissues. Occasionally, the pericycle of the stem is a *single* layer of parenchymatous cells (*e.g.* wall-flower).

§ 19. The relations between the regions of permanent tissue and the regions of apical meristem may be shown thus:—

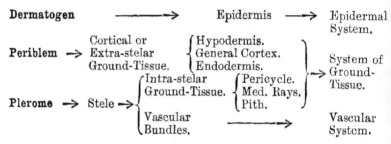

§ 20. **Summary.** The stems of most herbaceous Dicotyledons, and the young tender shoots of dicotyledonous shrubs and trees, have a structure agreeing in its general characters with that just described. These general characters may be summarized:—

(*a*) The apical meristem shows dermatogen, periblem, and plerome.
(*b*) The stem is monostelic.
(*c*) The bundles, in transverse section, are arranged in a ring, and thus the ground-tissue is divided into cortex, pith, and medullary rays.
(*d*) The bundles are collateral, and most of them common. The xylem contains typical vessels, wood-fibres, and wood-parenchyma. The phloem contains sieve-tubes, cambiform and companion cells. Hard bast is often present as a lignified portion of a pericycle.
(*e*) The bundles are open, so that secondary growth may take place.

§ 21. **Exceptional or Peculiar Arrangements.** In some stems, *e.g.* white bryony (*Bryonia dioica*), instead of a single ring of bundles in transverse section, there are two. The bundles of the two rings, however, alternate, *i.e.* lie on different radii of the stem. Sometimes (Cucurbitaceæ, *e.g.* white bryony) we find a modified form of the collateral bundle called the **bicollateral**. In this there are two phloem masses, one external as in the collateral bundle, the other internal, next the pith. It differs from the ordinary open collateral bundle simply in having an internal mass of phloem.

In some aquatic plants, *e.g.* Myriophyllum, the water milfoil, the vascular tissue of the stele has not separated into distinct collateral bundles. The bundles are all aggregated towards the centre. Hence there is no pith. The xylem consists of a single central mass, and is completely surrounded by a band of phloem. This vascular mass is surrounded by pericycle and endodermis. Till recently such a vascular mass was spoken of as a **concentric bundle**. In reality it is a simple *stele* consisting of a number of vascular bundles completely fused or not distinct from each other. We may still speak of it, however, as the **concentric arrangement**. The essential resemblance between this simple stele and the stele of the sunflower will be recognised if the student imagines the bundles of the sunflower to become aggregated and fused together in the middle of the stem so that the pith is obliterated.

Finally, there are a few dicotyledonous plants, *e.g.* Auricula, in which *several* plerome strands are found in the apical meristem. Each of these gives rise to a distinct stele, so that the stem contains a number of steles, and is said to be **polystelic**. These steles are not so strongly developed as is the single stele of the typical monostelic forms. They are simply concentric masses of vascular tissue, surrounded by pericycle and endodermis (cf. the single stele of Myriophyllum). It is by the presence of a pericycle and endodermis that we can, in the full-grown stem, recognize each as a distinct stele.

Simple steles like those of milfoil or Auricula are confined to the stem and do not themselves pass out to the leaves, although the leaves receive vascular strands from them. They are usually spoken of as *cauline* (or stem) bundles to distinguish them from common bundles.

§ 22. **Secondary Growth.** We have now described the *primary* structure characteristic of dicotyledonous stems. In herbaceous Dicotyledons it is *practically* the only structure to be recognized. On the other hand, in woody stems, where a larger amount of vascular tissue is required, this primary structure is completely modified by **secondary growth**. By secondary growth is meant the formation of new tissue owing to the activity of a cambium layer, so that the member in which it occurs increases in thickness.

The student must bear in mind that the cambium is a meristem. Its cells are capable of dividing and forming new cells, which are modified or differentiated into permanent tissue-elements. The new tissues thus formed are called *secondary* to distinguish them from the tissues differentiated from the apical meristem. In considering this process we have to study the formation, not only of secondary vascular tissue, but also of secondary ground-tissue (phelloderm) and secondary tegumentary tissue (cork and bark).

§ 23. **Initiation of the Process.** We have already seen that there is a layer of fascicular cambium between the primary xylem and phloem. When the process of secondary growth is about to begin, certain of the parenchymatous cells in each medullary ray also become meristematic. These strips of *secondary* meristem (p. 40) are called **interfascicular cambium** (fig. 51). They cross the medullary rays from one bundle to another, and join on to the fascicular cambium. In this way a complete ring of cambium—the **cambium ring**—is formed in the stem. Its formation can readily be studied in the older internodes of Helianthus, where there are the beginnings of secondary growth, or in the young green twigs of trees.

§ 24. **Division of the Cambial Cells.** The elongated cambium cells are flattened in the radial direction, and their end-walls are obliquely inclined (fig. 4, B). The method of division is as follows :— Each cell divides *tangentially* (*i.e.* by a wall at right angles to the radial direction) into an outer cell and an inner cell. Of these one continues as a cell of the cambium. The other may divide once or twice, but all the cells to which it gives rise are ultimately differentiated into permanent tissue. The cell which continues as a cambial cell increases in size and again divides. As before only one of the two cells is differentiated. And so on.

§ 25. **The Secondary Tissue** (fig. 57). The new cells formed by the cambium are given off on both sides— internal and external. Those given off on the inner side are modified into wood elements—**secondary xylem**; those on the outer side, into phloem elements—**secondary phloem**. It is evident, if the original position of the cambium ring be kept in mind, that the secondary xylem is laid down just outside the pith and primary xylem groups, and that, as a consequence of this, the cambium ring passes farther and farther from the centre of the stem, pushing in front of it the phloem-tissue both primary and secondary. In other words, the primary xylem and primary phloem become widely separated from each other, owing to the intercalation between them of the tissue formed by, and on either side of, the cambium. Thus the primary phloem

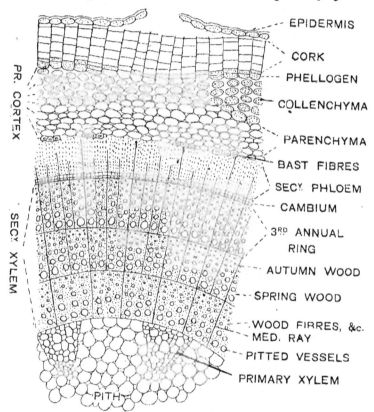

Fig. 57.—Part of a Transverse Section of a three-year-old Internode of a Dicotyledonous Stem.
(*e.g.* Elder.)

lies on the outer side of the secondary. Being thrust outwards as it is, and lying therefore on the circumference of a widening circle, it is subjected to lateral tension. For this reason it is very frequently spread out over the surface of the secondary phloem, and the original distinct groups of primary phloem are no longer recognized; but scattered groups of bast-fibres (representing the original hard bast, *i.e.* pericycle fibres) are frequently seen on the periphery of the secondary phloem.

Seeing that the interfascicular cambium as well as the fascicular produces this secondary tissue, there are no longer *wide* medullary rays running between pith and cortex. Certain cells of the cambium ring, however, instead of giving rise to wood and phloem elements, produce parenchymatous cells which form narrow medullary rays traversing radially the secondary wood and phloem. Strictly speaking, seeing that they are formed from the cambium, they consist of secondary tissue. Usually, however, those whose formation commenced at the beginning of secondary growth, and which therefore run (though they are very narrow) from pith to cortex, are still called primary, the term secondary being reserved for those whose formation began later, and which therefore start somewhere in the secondary wood and end somewhere in the secondary phloem.

The cambium ceases division during the winter. It renews its activity in the spring. Each year it forms a band of secondary wood and secondary phloem. The circular bands of secondary wood are distinctly marked off from each other, and are known as the **annual rings**. The wood formed in the spring differs somewhat from that formed in the autumn. The former, the **spring wood**, consists of large, well-formed elements; the latter, the **autumn wood**, of smaller, more strongly thickened and lignified elements. Larger elements are formed in the spring, because there is less pressure inside the stem then than there is in the autumn. It is because the spring wood of each annual ring lies just outside the autumn wood of the preceding ring, that the appearance of ringing is so distinctly marked. By counting the number of annual rings the age of the stem can be ascertained. Thus four annual rings would be found in a four-year-old internode. This appearance is not seen in the phloem. Usually the phloem forms a continuous circular band, traversed by narrow medullary rays. In some cases, however (*e.g.* the lime), owing to the medullary rays expanding tangentially, by growth and division of their cells, the phloem seems to be made up of a number of conical masses with the apices directed outwards. The primary phloem groups are found at the apices of these.

The secondary xylem consists of wood-vessels, wood-fibres (sclerenchymatous fibres and tracheides), and wood-parenchyma. *It contains only pitted vessels.* The secondary phloem frequently consists entirely of soft bast, the only bast-fibres present being those of the primary phloem spread out over the surface of the secondary phloem. Sometimes, however (*e.g.* lime), the second phloem consists of alternating layers of soft and hard bast. The **medullary rays** are vertical plates of parenchyma traversing the wood and phloem in a radial direction. Their cells are elongated radially. They are not usually more than one to a few cells wide. In height they vary from two to about fifteen cells. They must not be thought of as sheets of tissue running continuously from the base of the stem to the apex.

§ 26. **Duramen and Alburnum.** In old trees showing *many* annual rings, the central region of secondary wood becomes distinctly marked off from the peripheral region. This is due to the wood-cells (wood-parenchyma) losing their contents. The result is that the central region contains much less water, and with this there is associated a change in colour. It is darker and harder than the peripheral region. The central region is called the *duramen* or heartwood; the peripheral region the *alburnum* or sap-wood. Only the latter retains its full functional activity in the transference of watery solutions.

§ 27. **Phellogen—Formation of Cork** (fig. 57). It is evident that this internal formation of secondary wood and phloem must exert considerable pressure on the peripheral tissues (cortex and epidermis). The epidermis is stretched and eventually ruptured. To provide for this there is the formation of new tissue from another cambial layer developed in this region. This cambial layer, which arises as a secondary meristem, is the **phellogen** or **cork-cambium**. In the majority of cases (*e.g.* elder) it has a superficial origin in the outermost layer of the cortex, immediately underneath the epidermis. At some stage in the process of secondary growth, the cells of this layer become meristematic. The young cells given off to the outer side of the phellogen are suberized, and form a tissue known as **cork** or **periderm**. This tissue being impermeable to water cuts off the epidermis from nourishment. The epidermis dies and gradually peels off as the first bark of the tree. The cork is the secondary tegumentary tissue developed to replace the epidermis and carry on its functions. New cells may also be produced on the inner side of the phellogen. The tissue thus formed is parenchymatous and is added on to the primary cortex. It is the **phelloderm** or *secondary cortex*. This tissue, however, is frequently absent (fig. 57), or only sparingly developed, during the first few years of secondary growth.

Although the phellogen of the stem *usually* originates in the outermost cortical layer, it may arise in other layers. Thus in willows it originates in the epidermis itself. Sometimes it is the second or third layer of the cortex which becomes meristematic, *e.g.* laburnum. In clematis, the vine, and others, the first phellogen arises in the pericycle. In these last cases the first bark consists not only of the dead epidermis, but also of all cortical tissue external to the phellogen. It may be taken as a general rule that the deeper the origin of the phellogen, the earlier and more abundant is the formation of phelloderm.

§ 28. **The Bark** may be defined as all dead tissue lying outside an active cork-cambium. We have already indicated what the first bark consists of. The first phellogen formed may persist for a large number of years, *e.g.* the birch; in the beech it persists throughout the life of the tree. This is the case only where the first phellogen has a superficial origin. In such cases there may be a considerable formation of bark owing to the dying off of the older cork-layers. But in most cases this first phellogen dies, sooner or later in those plants where it has a superficial origin, early in those where it is deep-seated. It is replaced by a new or secondary phellogen developed in the deeper tissue. This produces a new cork-layer, and as a result all the outlying tissues (the original cork, etc.) die and are added to the bark. If the succession of secondary phellogens is rapid it often happens that the phellogen comes to lie close to the phloem. In some cases, even, the new phellogens may arise in the phloem, *e.g.* vine and Clematis.

In some trees the bark comes away in sheets, and is spoken of as **ring-bark**. This may be due either to the fact that the first phellogen is persistent, *e.g.* the birch, or to the fact that the successive phellogens appear in the form of regular rings. But in many trees the bark is given off in scales—**scale-bark**—*e.g.* the plane. This is due to the fact that the secondary phellogens do not arise as regular rings or layers, but in the form of little tangential strips abutting on the previous phellogen.

§ 29. **Lenticels** (fig. 58). In the young green shoot, the epidermis, as we have seen, has stomata allowing for the interchange of gases and water-vapour necessary to the life of the plant. When the cork-tissue is developed we usually (not always) find certain structures known as **lenticels**. These form small oval scars on the brown surface of the shoot (*e.g.* in the elder). Sections show that at these points the cork-cells are not in close con-

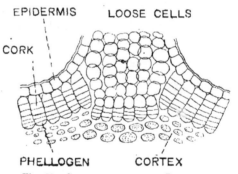

Fig. 58.—Section through a Lenticel.

tact, but have separated from each other and form a loose, granular, or powdery mass through which gases and vapours can readily pass. These lenticels are as a rule developed immediately underneath the stomata. Where a thick mass of cork is developed they form long passages or canals, filled with the powdery cork-cells, as, for example, in the ordinary cork of commerce. The lenticels are closed during winter by the formation of ordinary cork-tissue.

II. THE MONOCOTYLEDON.

§ 30. The Typical Arrangement. Fig. 59 shows the arrangement of tissues in the typical monocotyledonous stem as seen in transverse section. There is an epidermis and a large number of vascular bundles *scattered irregularly* through the ground-tissue. Owing to this **scattered arrangement** of the bundles, the ground-tissue is not marked off into pith and medullary rays. The ground-tissue consists chiefly of thin-walled parenchyma, but just under the

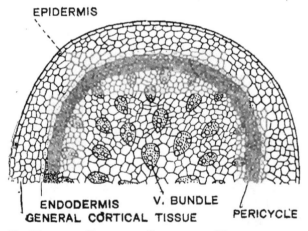

Fig. 59.—Half of a Transverse Section of a Monocotyledonous Stem. (Diagrammatic.) The tissue between the bundles is thin-walled intra-stelar ground-tissue.

epidermis there *may* be patches of collenchyma or sclerenchyma; and as a rule, also, there is a sheath of lignified (sclerenchymatous) ground-tissue round each bundle (fig. 60). In addition to this there is, in *many* monocotyledonous stems, a stout band of sclerenchyma, called the **strengthening zone**, developed just outside the region containing the bundles. This strengthening zone is the lignified sclerenchymatous **pericycle**. The layer of cells immediately outside it is the **endodermis**, which, however, is usually very faintly marked in Monocotyledons. The endodermis, as in

THE STEM OF THE ANGIOSPERM.

Dicotyledons, is the innermost layer of the cortical or **extra-stelar** ground-tissue. All the tissues within the endodermis belong to the stele, consisting of vascular bundles and **intra-stelar** ground-tissue. The monocotyledonous stem is **monostelic**.

While this arrangement is found in many monocotyledonous stems, *e.g.* Ruscus (the butcher's broom) and Asparagus, it should be carefully noticed that in others the pericycle and endodermis are not distinguishable from the rest of the parenchymatous ground-tissue, intra-stelar or extra-stelar—*e.g.* the Maize.

§ 31. **The Vascular Bundle** (fig. 60). The bundles are **collateral**. The xylem is directed towards the centre of the stem, and is usually more or less distinctly V-shaped. Large pitted vessels, one or more, are situated on each arm of the V. The protoxylem vessels occupy the apex of the V.

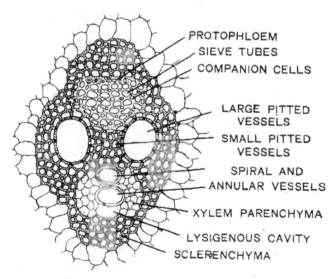

Fig. 60.—Transverse Section of Vascular Bundle of Maize.

In some plants, *e.g.* the maize, one or more of the annular vessels break down and give rise to an air-passage. The phloem lies between the arms of, but, as a rule, slightly outside, the V. It consists of sieve-tubes, cambiform and very regular companion cells. On its outer side small protophloem elements can often be recognized, but there is no hard bast. The reason of this will be evident if the student remembers that the hard bast in the Dicotyledon is a lignified portion of the pericycle. The bundles are **closed**, *i.e.* there is no cambium, and therefore no secondary growth.

§ 32. **Longitudinal Course of the Bundles** (fig. 61). The bundles are *common*. The leaves have usually a broad insertion, and from them a number of bundles can be traced into the stem. Their downward course in the stem is not parallel to the surface, but curved. They first run obliquely downwards towards the centre, and then bend outwards again towards the surface. After running through one or two internodes they join on to bundles passing in from older leaves. At all levels then we obviously have bundles situated at varying depths in the ground-tissue, and it is for this reason that the transverse section shows a scattered arrangement.

Fig. 61.—Longitudinal Course of the Bundles in the Monocotyledon. (Diagrammatic.)

§ 33. **Apical Meristem and Differentiation of Tissues.** In the apical meristem *dermatogen, periblem*, and *plerome* can be distinguished as in the Dicotyledon. The dermatogen gives rise to the epidermis, the periblem to the cortical or extra-stelar ground-tissue, and the plerome to the stele. As already indicated, the endodermis (innermost layer developed from the periblem) and hypodermal tissue (collenchymatous or sclerenchymatous) may or may not be distinctly marked off. The pericycle may or may not be sclerenchymatous. Scattered procambial strands appear in the plerome. Differentiation of vascular tissue takes place as in Dicotyledons, but is complete, so that no cambium is left.

§ 34. The student should very carefully compare the monocotyledonous characters above described with those of Dicotyledons (p. 82).

§ 35. **Modifications.** Sometimes the bundles are not *irregularly* scattered, but confined to particular regions of the ground-tissue. In the black bryony (*Tamus communis*), for example, they run in the region of ground-tissue immediately inside the strengthening zone. In this case there is a superficial resemblance to the dicotyledonous arrangement. In grasses the central region of the ground-tissue in the internodes has been absorbed, so that the internodes are hollow, and the bundles therefore run in the (intra-stelar) ground-tissue near the epidermis. Finally, in a few Monocotyledons—Yucca, Dracæna, etc.—there is a form of secondary growth. It is only in these few forms that we meet with secondary growth among Monocotyledons. Certain examples of monocotyledonous *trees* may occur to the student, such as the palms. In these, however, there is no secondary growth. The whole of the tissues of the stout palm-stem are derived from a *huge* apical meristem. In these palms there is the typical scattered arrangement, though the tissues undergo much thickening and lignification.

§ 36. **Exceptional Secondary Growth.** In Yucca, Dracæna, and a few others, there is a form of secondary growth. In the primary condition of the stem there is the typical scattered arrangement of common bundles which are closed. A cambium originates *in the pericycle* entirely as a secondary meristem. It gives rise to new tissue on the inner side only, and this tissue is differentiated into new secondary bundles with intervening ground-tissue. The new bundles are *cauline* (true cauline *bundles*—not steles). A phellogen (secondary meristem) also develops beneath the epidermis, and produces cork.

III. GENERAL.

§ 37. **Origin of Lateral Branches.** In both Dicotyledons and Monocotyledons axillary buds have a superficial origin from the apical meristem of the parent-stem. They arise as little protuberances of dermatogen and periblem only (fig. 55). The plerome of the parent-axis takes no part in their formation. For this reason their development is said to be **exogenous**. As the axillary protuberance increases in size a plerome (derived from the periblem of the parent-axis) differentiates, and becomes connected with the plerome of the parent. Young leaves begin to grow out and overlap the apex. Thus we have an axillary bud which in all respects reproduces the structure of the apical bud of the parent-axis.

§ 38. **Cork-Formation in Wounds.** When a stem (or other member of a plant) is injured, the outermost uninjured layer of living ground-tissue forms a meristem (phellogen), producing a cork-layer which protects the wounded surface. This power of healing wounds is possessed by many Monocotyledons as well as by Dicotyledons.

CHAPTER V.

THE ROOT OF THE ANGIOSPERM.

§ 1. **General Characters.** The root may be defined as that member of a plant which tends to turn downwards, away from light and towards water; which, as an almost invariable rule, bears neither leaves nor buds; and which usually has at the apex a protective cap of tissue called the **root-cap**. The internal structure and development, also are characteristic. As already mentioned, it is by consideration of these characters that true roots can be distinguished from root-like stems.

A. EXTERNAL CHARACTERS.

§ 2. **Tap and Adventitious Roots.** As already explained, the terminal portion of the radicle is the embryonic or primary root. In the majority of *Dicotyledons* the primary root, at germination, elongates, grows down into the soil, branches, and forms the root-system of the plant. This is known as a *tap-root system*. The elongated parent-root is described as a **tap-root**, and the branches, if developed in regular acropetal succession, as *normal secondary roots*. The branching is invariably lateral. Where an *elongated* tap-root bears normal secondary roots the branching is *racemose* (fig. 47). Where the parent-root remains short, and the normal branches form the extensive root-system, the branching may be compared with the *cymose* type (fig. 62). Instead of normal roots, however, we may find **adventitious roots**. These are roots developed (*a*) on other roots, but not in the normal order of acropetal succession; (*b*) on stems;

(c) in a few cases on leaves. Adventitious roots also are common in dicotyledonous plants, more especially in those with rhizomes, runners, trailing stems, etc. (*e.g.* figs. 42, 43). In *Monocotyledons*, in nearly all cases the roots are adventitious. In this group the primary root elongates only very slightly at germination, or not at all, and adventitious roots are developed from the base of the stem (fig. 37).

§ 3. **Functions and Adaptations of Roots.** Like stems, roots have a form and organization adapted to their mode of life and the conditions in which they live. Only where they are exposed to light do they contain chlorophyll, and help to a slight extent in carbon-assimilation. Being *usually* buried in the soil, they are not exposed to such a diversity of influences as stems; their environment being less complex and more uniform, they naturally show less variety in form and adaptation. At the same time, the functions of an ordinary root—(*a*) the fixation of the plant, (*b*) the absorption of nutritive solutions from the soil—may be carried on in many different ways, according to the nature of the soil or the needs of the plant. We find also that roots may take on special functions. They may function as storeplaces of nourishment, or as climbing organs. Sometimes they are aerial, sometimes aquatic. In a few cases roots are more highly specialized as floats, tendrils, or spines. For these reasons the forms and adaptations of roots are by no means few.

§ 4. **Forms of Tap- and Normal Branch-Roots.** The most typical form, found more especially in herbaceous Dicotyledons, is *the fibrous branching tap-root.* Here both the main or tap root and the normal branches are elongated and slender, more or less resembling fibres (fig. 47). Such roots are found only in "deep-feeding" plants. A modification

Fig. 62.—SHORT PRIMARY ROOT WITH FIBROUS NORMAL BRANCHES.

of this is the short stout primary root, with an extensive system of fibrous normal branches (fig. 62). Such roots are found in "surface feeders." Sometimes a tap-root is very strongly developed, owing to the deposition of food-material in it. Thus (fig. 63), in the carrot and others it is stout at the base and tapers towards the tip—

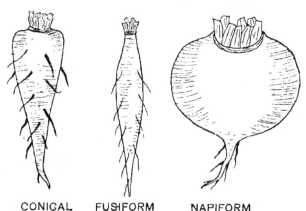

CONICAL   FUSIFORM   NAPIFORM
Fig. 63.—Forms of Thickened Tap-Root.

the **conical** tap-root; in the radish or cabbage it is stout towards the middle and tapers towards the base and apex—the **fusiform** tap-root; in the turnip the base is massive and rounded, passing abruptly into the tapering apical portion—the **napiform** tap-root. Sometimes normal secondary roots become swollen and **tuberous**, *e.g.* the dahlia. These *normal root-tubers* are developed in many other dicotyledonous plants. Roots whose branches are covered with little rounded outgrowths or tubercles are said to be **tubercled**.

§ 5. **Forms of Adventitious Roots.** Adventitious roots are usually slender and **fibrous**, as in grasses. But, frequently, owing to storage of food-material, they become **tuberous**, as in many orchids. These root-tubers may be simple and undivided; or double, *i.e.* branched into two (**double tuber**); or branched in a finger-like manner (**palmate tuber**, fig. 64). They are developed adventitiously from buds produced at the base of the season's shoot.

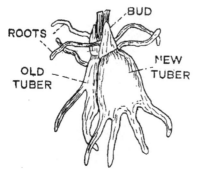

Fig. 64.—PALMATE TUBER OF AN ORCHID.

In the following year the buds develop into new aerial shoots at the expense of the material stored up in the tubers. Sometimes adventitious roots are *aerial*, as in many orchids. Occasionally these aerial roots function as climbing organs, *e.g.* the ivy (*root-tendrils*). Many plants have what are called **parasitic roots.*** These plants, instead of deriving their food-material in the usual way, send "suckers" into other plants and absorb their nutritive juices.

§ 6. **The Root-Hairs** (p. 52) are developed on roots a short distance behind the root-cap. They do not persist behind this region. Besides acting as absorbing organs they play an important part in the fixation of the plant, as the particles of soil firmly adhere to them.

B. INTERNAL STRUCTURE.

§ 7. **The Apical Region.** Fig. 65 represents diagram-

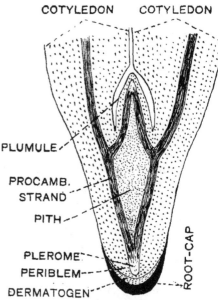

Fig. 65.—LONGITUDINAL SECTION OF EMBRYO OF SUNFLOWER (LOWER PART ONLY).
(Diagrammatic.)

* According to some these are not true roots, but structures of the nature of emergences (p. 52). They differ from simpler emergences in having a core of vascular tissue, and from most roots in being usually exogenous in origin.

matically a *median* longitudinal section of the radicle of the embryo of the almond or sunflower. See also fig. 66. Covering the apex is the **root-cap**, which, as has already been indicated (p. 49), is a many-layered epidermis. Beneath this is the meristem, which gradually passes farther back into the older tissue of the root. The meristem shows the same regions as in the stem—**dermatogen, periblem,** and a single **plerome** or embryonic stele. In roots, the dermatogen cells *usually* divide by both perpendicular and tangential walls, so that at the apex of the root it forms the many-layered root-cap. The periblem, as in the stem, gives rise to the cortical or extra-stelar ground-tissue. The plerome gives rise to the single stele; the root, like the stem, is **monostelic. Procambial strands** make their appearance in it, and, on further development, these are *completely* differentiated into vascular bundles— some into xylem bundles, others into phloem bundles. The differentiation of both xylem and phloem bundles begins on the outer side of the procambial strands, so that the protoxylem and protophloem both lie to the outside. In most Dicotyledons the root-cap tissue shades off farther back into a single layer, which produces the root-hairs.

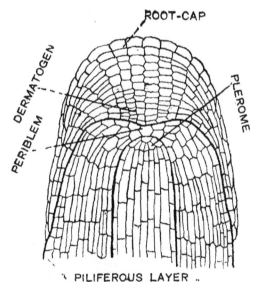

Fig. 66.—LONGITUDINAL SECTION THROUGH THE TIP OF THE RADICLE OF A MONOCOTYLEDONOUS EMBRYO.

In most monocotyledonous roots, as can readily be recognized in a similar section of the radicle of the maize, the same structures are seen; but here the tissue of the root-cap peels off completely, so that behind the apex the superficial layer is the outermost layer derived from the periblem (fig. 66).

Practically all roots are monostelic. Only one or two exceptional cases of **polystely** (p. 83) have been recently described in a few palms. It is only recently that the root-cap has come to be generally recognized as epidermis. Previously it was regarded as a distinct tissue, and the layer of meristem from which it is derived was called the **calyptrogen**—a term no longer necessary.

§ 8. **Primary Structure of Roots.** A transverse section of a monocotyledonous, or of a *young* dicotyledonous, root (figs. 67, 68) shows a varying number of vascular strands, or bundles, more or less aggregated towards the centre. These bundles, developed from procambial strands, are not conjoint, but consist of phloem only, or xylem only. The **xylem** and **phloem bundles** are equal in number, and alternate with each other, so that they are situated on different radii of the transverse section. They are separated by *conjunctive* or *intra-stelar ground-tissue*. It is important to notice that in the xylem of roots the protoxylem elements (annular and spiral) lie towards the periphery, not towards the centre, as in stem-bundles. This is one of the characteristic structural distinctions between stems and roots. In many roots all the xylem bundles fuse or meet in the centre of the root in a number of large pitted vessels; in this case there is no medulla or pith. In others, the xylem bundles are distinct, and the centre of the root is occupied by a parenchymatous (or, in some roots, sclerenchymatous) **pith**, representing the central region of the stele. This vascular mass is surrounded by two special layers of cells. The inner layer consists of parenchymatous cells with protoplasmic contents, and is the **pericycle** or *pericambium*.* It is the outermost layer of the stele (cf. the stem). In the roots of Angiosperms it is usually a single layer. The outer of the two layers is the **endodermis** or *bundle-sheath*, and is the innermost layer of cortical or extra-stelar tissue developed from the periblem (cf. stem). Its cells in transverse section are four-sided,

* The term originally applied to the pericycle of the root.

and slightly elongated tangentially. *Typically* the radial walls of the cells have a wavy, uneven character, and therefore, as seen under the microscope, appear less definite

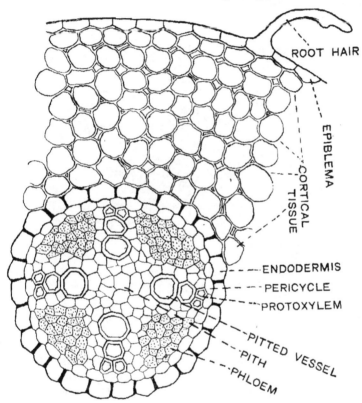

Fig. 67.—Transverse Section of a Young Dicotyledonous Root with Tetrarch Stele.

and somewhat darker than the others (fig. 67). Frequently the radial and inner walls are strongly thickened and cuticularized (fig. 68). Outside the endodermis lies the parenchymatous cortical tissue. The outermost layer of the root is called the **piliferous layer,** or **epiblema.** These terms are used instead of epidermis, because, as we have seen, this outermost layer has not a constant morphological value,—sometimes being a true epidermis (Dicotyledons generally), sometimes the specialized outermost layer of the cortical tissue (Monocotyledons generally). Sometimes the layer internal to the epiblema consists of large cells, or is otherwise distinguished; it is called the *exodermis*.

§ 9. The vascular cylinder in roots was formerly regarded as a compound vascular bundle, and described as a *radial vascular bundle*, because the xylem forms a number of rays with alternating groups, or patches of phloem. It was placed on a level with, and compared with, the collateral bundle of the stem. We have now to recognize that the vascular cylinder of the root is a stele, containing a number of bundles (in addition to intra-stelar ground-tissue), and is, therefore, comparable, not to a single collateral bundle, but to the whole stele of the stem. The steles of root and stem must be regarded as homologous morphological units.

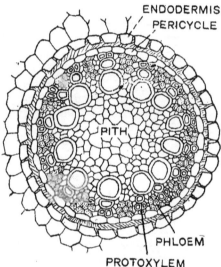

Fig. 68.—Transverse Section of the Central Part of the Root of Iris, showing the Polyarch Stele.

§ 10. **Monocotyledonous and Dicotyledonous Roots.** While in monocotyledonous and *young* dicotyledonous roots the *general arrangement*, as described, is the same, there are several very characteristic points of difference:—

(a) In *Dicotyledons* (figs. 67, 69, 70) the number of xylem bundles *usually* varies from two to five, although there may be more than five. In *Monocotyledons*, while a limited number—about five to eight—is sometimes found (*e.g.* root of leek), there are usually many more than this—as many as twelve to twenty (*e.g.* roots of iris, or maize, fig. 68). Where there are only two xylem (and two phloem) bundles, the stele is described as *diarch*; where three, *triarch*; four, *tetrarch*; five, *pentarch*; many, *polyarch*.

(b) In both Dicotyledons and Monocotyledons the differ-

THE ROOT OF THE ANGIOSPERM. 103

entiation of procambial tissue is *complete*; but, in most Dicotyledons, a cambium, and later a phellogen, arise as secondary meristems, and secondary growth takes place; in these the structure above described is only the *primary* structure. In Monocotyledons there is no secondary growth, and the same structure can be recognized in all the fully developed regions of the root.

(c) In Monocotyledons the pitted vessels are large and nearly circular in transverse section; in Dicotyledons they are usually much smaller, and more or less polygonal.

§ 11. **Secondary Growth in the Dicotyledon** (figs. 69-71). When secondary growth is about to begin, certain conjunctive cells lying on the inner side of each phloem bundle become meristematic (figs. 69, 70). Thus strips of **cambium**, equal in number to the phloem bundles, make their appearance. These gradually extend outwards between the xylem and phloem bundles, *owing to more of the conjunctive parenchymatous cells becoming meristematic.* The curved strips of cambium thus produced come into contact with the pericycle on each side of the protoxylem. These pericycle-cells now become meristematic, and, in this way, the cambium strips are united and become continuous round the tips of the protoxylem groups. Thus a continuous wavy band of cambium is

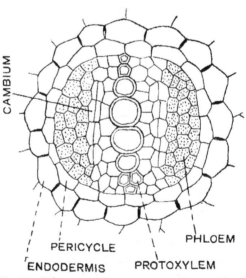

Fig. 69.—Transverse Section of the Diarch Stele of a Dicotyledonous Root.
The origin of the Cambium is shown.

formed, running internal to the phloem bundles external to the xylem. It should be recognized that this cambium is entirely a secondary meristem, arising partly from

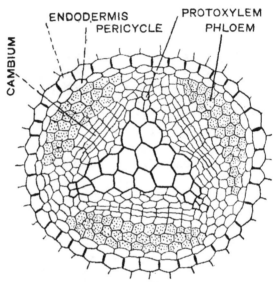

Fig. 70.—TRANSVERSE SECTION OF TRIARCH STELE ROOT OF ELDER.
Secondary growth is beginning.

parenchymatous cells between xylem and phloem, partly from the pericycle.

The cambium-cells divide exactly as in the stem. The **secondary xylem** (fig. 71) is laid down around the pith (if present) and the primary xylem bundles. The **secondary phloem** is formed outside the cambium, and, together with the primary phloem and other tissues, is gradually thrust outwards as the cambium adds to the secondary xylem. The cambium-cells on the inner side of each primary phloem bundle are the most active, and, owing to this, the cambium-layer as a whole, which was at first a wavy band (in transverse section), soon becomes circular. The cambium-cells lying just outside the primary xylem bundles, instead of giving rise to secondary wood and phloem, usually produce strands of parenchyma—the *main*

THE ROOT OF THE ANGIOSPERM.

(also called primary) *medullary rays*—radiating outwards through the secondary wood and phloem from the tips of the protoxylem groups. If a very compact secondary wood is formed, it may be difficult to detect the primary xylem bundles, or the main medullary rays. Small secondary medullary rays also are formed from the cambium-cells. If the primary structure and the subsequent development

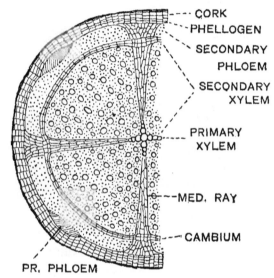

Fig. 71.—Transverse Section of Dicotyledonous Root after Secondary Growth.
(Diagrammatic.)

be borne in mind, it is evident that the primary phloem bundles should be found just outside the secondary phloem on radii alternating with the primary xylem bundles; but, frequently, as in the stem, they are more or less disorganized and mixed with the secondary phloem.

If there is any considerable secondary growth, the pericycle, sooner or later, becomes completely meristematic, and forms a **phellogen**, or *cork-cambium* (a secondary meristem). This phellogen produces cork externally, and usually also, internally, a considerable amount of phelloderm (as in most deep-seated phellogens). Lenticels may be developed. The endodermis and cortical tissue die, and are

106 BOTANY.

given off as *bark*. It is comparatively rarely in roots that the phellogen has a superficial origin.

§ 12. **Development of Lateral Rootlets** (fig. 72). Normal root-branches are usually developed, in Angiosperms, entirely from the pericycle. The cortical tissue of the parent-root takes no part in the formation of the tissues of the lateral branch. This development from a deep-seated layer is called **endogenous**. The development begins some little distance behind the apex of the parent-root,

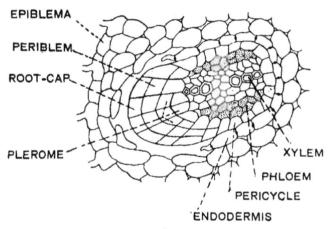

Fig. 72.—Transverse Section of a Dicotyledonous Root (with Diarch Stele), showing the Development of a Lateral Rootlet.

but before secondary growth has set in. The young lateral roots make their appearance in the pericycle just outside the protoxylem groups, so that the number of longitudinal rows of lateral roots usually corresponds to the number of xylem bundles in the stele. Thus, if there are four xylem bundles, there will usually be four longitudinal rows of normal lateral branches. When development begins, the cells of the pericycle divide, *i.e.* become meristematic, and produce a growing-point, which soon shows a distinction into dermatogen, periblem, and plerome. The young branch-root gradually elongates, boring its way through the overlying cortical tissue till it reaches the surface of the parent-root. At first the endodermis and, it may be, one or two layers of cortical cells form a sort of cap over the apex of the developing rootlet. This is known as the "*digestive sac*" because it secretes a ferment which disorganizes or digests the walls of the overlying cells and so enables the young root to reach the surface. The structure of the lateral root is identical with that of the parent-root. In Dicotyledons it is evident, considering the points at which the rootlets are developed, that after secondary growth has begun they will be seen (in a transverse section) radiating out from the tips of the primary xylem bundles, and therefore running as it were through the main medullary rays.

Adventitious Roots are developed similarly. If they are developed from the stem, they originate in the pericycle of the stem.

§ 13. **Exceptional Cases.** Not uncommonly lateral rootlets are developed from the pericycle opposite the phloem bundles, *e.g.* in many grasses where the pericycle is wanting opposite the protoxylem, and many Umbelliferæ where an oil-duct lies in the pericycle opposite each protoxylem group. Often when the stele is diarch there are four rows of lateral rootlets, two being developed opposite the phloem.

§ 14. **Secondary Growth in Dracæna.**—Exceptional secondary growth is found in the root of the monocotyledonous plant Dracæna. The meristem-ring originates as a secondary meristem either in the pericycle or in the cortical tissue, or partly in one, partly in the other. There is also cork-formation, the phellogen originating in the superficial cortical tissue beneath the epiblema. There are a few other monocotyledonous plants in which a similar formation of cork occurs in the root, although there is no secondary formation of vascular tissue (*e.g.* the Iris).

§ 15. **Functions of the Pericycle (Pericambium).**—The student must be careful not to confuse the pericambium with the cambium-layer. The cells of the pericycle have a great capacity for remaining or becoming meristematic. In both Monocotyledons and Dicotyledons, as we have seen, lateral roots originate in this layer, and, in most Dicotyledons, it helps in the formation of the cambium-layer, and later gives origin to the phellogen.

§ 16. Transition from Root to Stem—the Hypocotyl
(p. 56). We have already stated that the vascular system is continuous in root and stem. It is evident that the transition from the arrangement characteristic of the root to that characteristic of the stem is effected in that region of the axis which lies between typical stem and typical root. This region is the *hypocotyl*. The transition is effected in different ways; but, in many cases, if we trace the vascular tissue from the root to the hypocotyl, we find that each xylem and phloem bundle divides radially into two. These xylem and phloem bundles fuse in pairs to form the conjoint bundles passing up into the stem. In this process the phloem bundles practically retain their position with the protophloem towards the exterior, but the xylem bundles twist round so that they lie on the inner side of the phloem bundles with the protoxylem *internal*. In such cases there is the same number of *conjoint* bundles in the stem as there are xylem and phloem bundles in the root.

CHAPTER VI.

THE LEAF OF THE ANGIOSPERM.

A. EXTERNAL CHARACTERS.

§ 1. **Parts of the Leaf** (fig. 73). The leaf is a natural outgrowth on the stem, and arises as a morphologically dissimilar member. The axis of this outgrowth is spoken of as a **phyllopodium.** In typical cases the phyllopodium shows three regions :—(a) a basal region, the **hypopodium**; (b) an apical region, the **epipodium**; and (c) a middle region, or **mesopodium.** Usually the leaf is more or less flattened, owing to the development of a *membrane* or *wing* on one or more of these parts of the phyllopodium. The wing is nearly always present on the epipodium, and this region of the leaf (epipodium + wing) is called the **lamina** or *leaf-blade.* Sometimes, however, there is no wing, as in cylindrical leaves like those of the onion or stonecrop. Occasionally the epipodium is wanting altogether, as in many scale-leaves and in phyllodes (p. 124). The mesopodium *may* (fig. 87), but usually does not, have a wing. This region of the leaf is called the leaf-stalk or **petiole.** As a rule it is more or less cylindrical, but, frequently, flattened or grooved on its upper surface. The petiole is absent

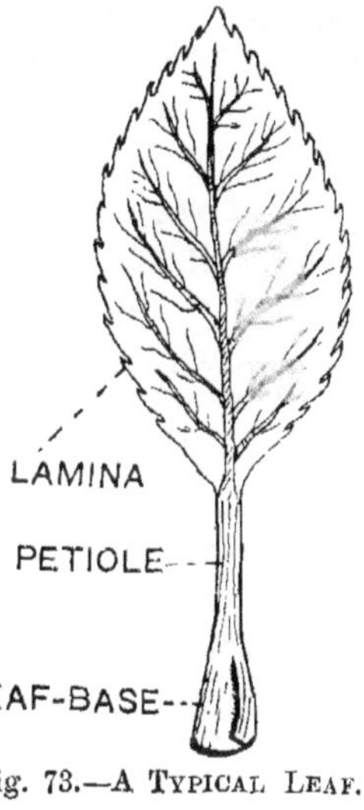

Fig. 73.—A TYPICAL LEAF.

in most Monocotyledons and many Dicotyledons. The hypopodium, the part joining on to the stem, may or may not be winged. This region is called the *leaf-base* or **vagina**. In many scale-leaves it is the only part of the leaf represented. It may be swollen and form a sort of cushion of tissue called the *pulvinus*. In many Dicotyledons, rarely in Monocotyledons, the leaf-base bears a pair of outgrowths called stipules, representing a development of its membrane or wing (fig. 74, B).

§ 2. **Various Types of Leaf-Structure.** Correlated with the diverse functions which have been taken on by leaves in adaptation to their surroundings, the forms of leaves are innumerable. Several well-marked types, however, are of general occurrence amongst the Angiosperms. They are as follows:—

(*a*) **Cotyledons.** These are the leaves of the embryo (p. 57). They function as nursing or feeding organs, serving either for the storing of food-material or its absorption. If they come above ground as the first assimilating leaves of the plant, they are much simpler in form than the foliage-leaves developed later.

(*b*) **Cataphyllary Leaves,** or *Scale-leaves*. Typically these are small brown membranous leaves, devoid of chlorophyll. They are developed on many underground stems (*e.g.* rhizomes), and form the protective scales of many buds (fig. 38, B). Their function is protective. They may serve to protect buds that are developed in their axils, or, in the case of bud-scales, they protect the inner undeveloped foliage-leaves of the bud. In most cases they represent leaf-bases, petiole and lamina being absent; but bud-scales may be the stipules of foliage-leaves (alder), or the stipules of scales (beech) or rudimentary laminæ (lilac). Sometimes scale-leaves function as reservoirs of storage material, as in many bulbs.

(*c*) **Foliage-Leaves.** These are the ordinary green leaves. They are the chief assimilating, respiring, and transpiring organs of the plant (p. 14). Chlorophyll is present because it is an essential factor in the assimilation of carbon.

(*d*) **Bracts** and **Floral Leaves.** These are the highly

THE LEAF OF THE ANGIOSPERM.   111

specialized leaves borne on the reproductive shoots (floral region). They have been specially adapted to carry on reproductive functions. These leaves will be fully considered in connexion with the Flower (Chapter IX.).

In §§ 3 to 15 we shall consider more especially the characters of ordinary foliage-leaves.

§ 3. General Descriptive Terms. If the leaf-base is winged and forms a sort of sheath clasping the stem half round at the insertion of the leaf, the leaf is said to be **semi-amplexicaul**; if it clasps completely round, **amplexicaul** (fig. 74, D). If the petiole is present,

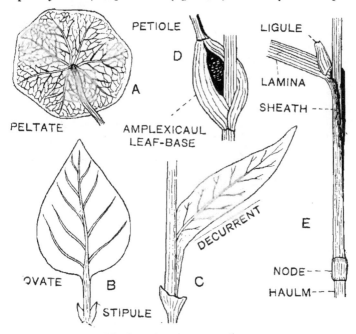

Fig. 74.—FORMS, ETC., OF LEAVES.
A, Peltate leaf of Garden Nasturtium; E, Ligulate leaf of a Grass.
(Haulm = Culm, p. 62.)

the leaf is **petiolate**; if absent, **sessile**. The leaf is **peltate** if the petiole is attached to its lower surface and not to its basal margin (*e.g.* garden Nasturtium—fig. 74, A). In *sessile* leaves, if the winged leaf-base (continuous with the lamina) clasps round the stem, the leaf is **auriculate** (fig. 75, B); if it fuses on the other side of the stem so that the stem seems to have grown up through the leaf, **perfoliate** (fig. 75, A). If there are two opposite leaves at the node, and their membranous bases fuse round the stem, they are said to be **connate** (fig. 75, C). If in a leaf the membrane runs vertically down the stem for some distance, the leaf is **decurrent** (fig. 74, C).

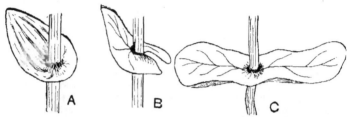

Fig. 75.—A, Perfoliate Leaf; B, Auriculate Leaf; C, Connate Leaves.

Sometimes, as in grasses, a membranous outgrowth (the ligule = an emergence, p. 52) is developed on the base of the lamina, and the leaf is said to be **ligulate** (fig. 74, E).

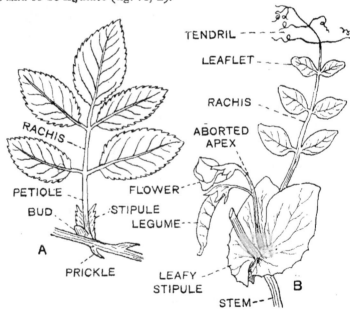

Fig. 76.—A, Compound Leaf of Rose with Petiolar Stipules; B, Part of Flowering Shoot of Pea, showing Compound Leaf in which the Upper Leaflets are modified into a Tendril. (From Green, modified.)

A leaf is **stipulate** or **exstipulate** according as stipules are present or absent. These stipules vary much in position, size, and form. Sometimes they are large, green, and leafy (fig. 76, B), showing the same development as an ordinary lamina (*e.g.* the pansy);

THE LEAF OF THE ANGIOSPERM.

frequently they are small and scaly; in some winter buds they form the outer protective scales (*e.g.* the alder). Occasionally the stipules are modified into spines (*Robinia pseudacacia*) or tendrils (*Smilax*).

Various kinds of stipules are recognized. If they run up the base of the petiole for some distance, they are called *petiolar* (rose—fig. 76, A). Where there is only one leaf at the node, if they run round to the other side of the stem and fuse there, an *opposite* stipule is formed; if their inner margins cohere between the leaf and stem, an *axillary* stipule is formed; if they cohere in both ways, a tubular sheath called an **ochrea** (fig. 77) is formed round the base of the internode (Polygonaceæ). Sometimes the stipules of opposite leaves (two at a node) fuse on each side to form *interpetiolar* stipules.

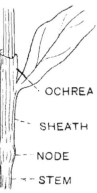

Fig. 77.—LEAF AND PORTION OF STEM OF *Polygonum*.

§ 4. **Insertion of the Leaf.** The point at which the leaf-base joins the stem is called the *insertion* of the leaf. Leaves are described as *cauline* or *ramal* according as they are developed on the main stem or on the branches. Leaves developed on very short "reduced" stems (p. 66) so that they *appear* to come off from a root are called *radical* leaves (*e.g.* dandelion, daisy, primrose).

§ 5. **Phyllotaxis** (the arrangement of leaves on a stem). Two kinds of phyllotaxis are recognized:—(*a*) **Spiral**, (*b*) **cyclic** or *whorled*. In spiral phyllotaxis the leaves are developed one at each node, and are said to be **alternate** (fig. 38, A). It is spoken of as the spiral arrangement, because, if an imaginary line were supposed to pass through the bases of the leaves in the order of their development, it would describe a spiral round the stem. In cyclic phyllotaxis two or more leaves forming a *whorl* (p. 64) are developed at each node; if two, the leaves are **opposite**; if more, **verticillate**. If in any one whorl the opposite leaves are placed immediately above those in the whorl below, so that there are only two *rows* of leaves on the stem, they are said to be opposite and **superposed**. Usually, however, they are placed at right angles, so that there are four rows of leaves; this is the opposite **decussate** arrangement.

Formerly the elementary student was distracted by the enormous importance attached to phyllotaxis. This incubus has now to a large extent been lifted from him. The following facts, however, may be interesting and worth attention. In spiral phyllotaxis the imaginary spiral line, following the order of development of the leaves, is called the **genetic spiral**. The angle of circumference between any one leaf and the next in order above it, in other words, the angle between the two *vertical* planes passing through these two leaves, is the **angle of divergence**. Thus, suppose the alternate leaves are arranged in two opposite vertical rows (as in grasses). Evidently the divergence, or circumferential distance between any two leaves taken in order, is $\frac{1}{2}$, *i.e.* the angle of divergence is 180°. The two vertical rows of leaves are called **orthostichies**. Again, suppose, calling a particular leaf No. 1, you pass through five leaves before coming to one, No. 6, lying immediately above No. 1, and that to reach No. 6 you have passed twice round the stem. Evidently the divergence is represented by $\frac{2}{5}$ (the whole circumferential distance divided by the number of leaves), and the angle of divergence is 144°. The whole course gone through from leaf 1 to leaf 6 constitutes a *cycle*. There are five rows of leaves or orthostichies. Thus to find the divergence simply divide the number of turns in a cycle by the number of leaves passed on the way or by the number of orthostichies. For example, in a divergence of $\frac{1}{3}$, it is leaf No. 4 which lies above No. 1, and only one turn of the circumference is gone through; there are three orthostichies.

The divergences common in plants may be arranged in two series:—(a) $\frac{1}{2}, \frac{1}{3}, \frac{2}{5}, \frac{3}{8}, \frac{5}{13}$. . . . (b) $\frac{1}{4}, \frac{1}{5}, \frac{2}{9}, \frac{3}{14}, \frac{5}{23}$. . . . The student should notice the peculiar relation existing between the members in each series. Each may be got by adding the numerators and denominators of the two preceding ones. Thus the series are easily remembered.

In cyclic phyllotaxis there is probably a number of genetic spirals running round the stem; thus, in the opposite decussate arrangement, two with a divergence of $\frac{1}{4}$.

§ 6. **Venation.** The vascular bundles which pass into each leaf from the stem branch out in the lamina and form the veins of the leaf. If the lamina is comparatively thin and membranous, we can recognize one or a number of chief veins or ribs as they give rise to projecting ridges on the under surface. But between these there are innumerable small veinlets running through the ground-tissue of the leaf and forming no projecting ridges. The character of the venation, *i.e.* the arrangement or appearance presented by the veins, depends chiefly on the number of prominent veins or ribs and the arrangement of the smaller veins or veinlets. Two chief types are recognized in Angiosperms :—(I) **Reticulate venation,** characteristic

THE LEAF OF THE ANGIOSPERM. 115

of dicotyledonous leaves, though occurring also in a few Monocotyledons; (II) **parallel venation**, found in Monocotyledons only. In reticulate venation the veinlets between the larger veins run together irregularly to form a network (fig. 74, A). In parallel venation the larger veins or the veinlets all run more or less parallel; no irregular network is formed (fig. 78). In both types the venation may be **unicostate** or **multicostate**, according as there is one chief vein (the midrib) or a number of chief veins. The former is also spoken of as the feather or pinnate type of venation. In multicostate venation the large veins may be *divergent* or *convergent* as they run towards the apex. Multicostate reticulate venation is usually divergent.

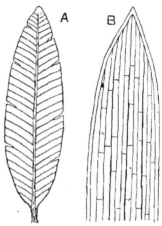

Fig. 78.—PARALLEL VENATION.
A, Unicostate; B, Multicostate.

I. **RETICULATE VENATION.**

{ 1. **Unicostate** (feathered or pinnately veined), fig. 73.
 2. **Multicostate**:

  { (*a*) **Divergent** (radiately or palmately veined), fig. 79.
   (*b*) **Convergent** (not common).

II. **PARALLEL VENATION.**

{ 1. **Unicostate** (feathered, pinnate or curved-veined), *e.g.*
   Banana and fig. 78, A.
 2. **Multicostate** (straight-veined):

  { (*a*) **Divergent**: Many palms.
   (*b*) **Convergent**: Grasses, lilies, fig. 78, B.

§ 7. **Branching.** The phyllopodium may branch. The branching is as a rule confined to the epipodium or region of the lamina. It is closely associated with and clearly indicated by the venation. In fact, the *chief* veins may be regarded as representing the branches of the epipodium.

Thus the unicostate type is clearly a racemose branch system; while the multicostate type is a cymose branch system, in which the median rib of the series (*a*, fig. 79) represents the parent-axis, and those to the side (*b*, *c*, *d*) a number of daughter-axes as strongly or almost as strongly developed. The appearance presented by the lamina depends chiefly on the extent to which the wing of the epipodium is developed *between the branches*. Sometimes it is completely developed, and the margin of the lamina is entire (fig. 74, B). Usually, however, it is not completely developed. The extent to which it is incomplete varies immensely. Sometimes there are only small irregularities or cuttings of the margin, as in figs. 73, 81 A; frequently larger indentations called *incisions* are produced between the veins or branches of the epipodium, *e.g.* figs. 82, 83. In many leaves the wing is not developed at all between the branches, and the lamina then consists of a number of distinct separate portions called **leaflets**, articulated at one point (fig. 83, A), or borne on a common stalk or *rachis* (fig. 76, A); such leaves are called **compound leaves**; all other leaves, in which the wing is present to some extent, however little, between the branches, are called **simple leaves**. The leaflets of compound leaves in many respects resemble simple leaves. They may even have structures resembling stipules and called *stipels*.

Fig. 79.—PALMATIFID LEAF, ILLUSTRATING MULTICOSTATE VENATION AND BRANCHING OF EPIPODIUM.

A **compound leaf** is one in which the lamina is broken up into a number of *separate* parts called leaflets, articulated at one point, or borne on a common rachis. A **simple leaf** is one in which the lamina is not split up into distinct leaflets. It is evident, remembering that the lamina is a branch system, that the nature of the incision in a simple leaf, and the number and arrangement of the leaflets in a compound one, will be correlated with the type of venation.

§ 8. Outline of the Lamina.

Many terms are in use to describe the simpler forms of outline presented by simple leaves or the leaflets of compound leaves. Only those most frequently used are given below.

A leaf is **subulate** (fig. 80, A) when it is narrow, firm, and hard, and gradually tapers from base to apex, ending in a sharp point, as in the gorse—here there is no distinction into petiole and lamina, and no wing is developed; **acicular**, if it is elongated and sharp-pointed, with distinct edges (fig. 80, B); **linear**, if elongated, flattened,

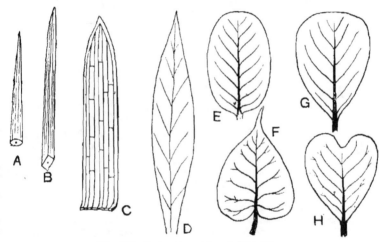

Fig. 80.—OUTLINES, ETC., OF LEAVES.
A, Subulate; B, Acicular; C, Linear; D, Lanceolate; E, Oblong; F, Cordate; G, Obovate; H, Obcordate. In E and G the apex is rounded or obtuse; in F, acuminate; in H, retuse.

and membranous, with parallel margins as in grasses (fig. 80, C); **lanceolate**, if elongated and gradually tapering towards base and apex (fig. 80, D); **oval**, or **elliptical**, if relatively shorter and broader, tapering towards base and apex (fig. 73); **oblong**, if of much the same length and breadth, but rounded at base and apex (fig. 80, E); **orbicular**, sub-rotund, or rounded, if it approximates towards the circular (fig. 74, A); **ovate**, if rounded off towards the base and pointed towards the apex (fig. 74, B); **obovate**, if the reverse (fig. 80, G); **cordate**, or heart-shaped if pointed at apex and notched at the base where the petiole is attached (fig. 80, F); **obcordate**, if notched at apex and tapering towards the petiole at the base (fig. 80, H); **reniform**, or kidney-shaped if notched at the base, more or less elongated transversely, and *rounded* at the apex (fig. 81, A); **spathulate**, if wide and rounded at the apex and *gradually* narrowing towards the base (fig. 81, B); **cuneate**, or wedge-shaped if similar to

118   BOTANY.

the spathulate form, but with a more or less acute or drawn-out apex (fig. 83, A); **sagittate**, if shaped like an arrow-head, with the two basal lobes directed backwards (fig. 81, C); **hastate**, or halbert-shaped if the two basal lobes are directed outwards (fig. 81, D). Sometimes

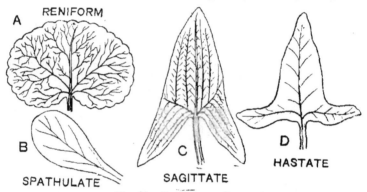

Fig. 81.—OUTLINES OF LEAVES.

the two halves of a leaf are not equally developed. Such leaves are *oblique*. They would be described as obliquely cordate, obliquely ovate, etc.

Sometimes the outline is best indicated by a combination of the above terms; thus, **ovate-lanceolate**, etc.

§ 9. **The margin** of a leaf or leaflet is **entire** if it is perfectly even and shows no irregularities (fig. 80); **serrate**, if it shows a number of sharp processes directed forward towards the apex (fig. 76, A); **dentate**, or toothed, if these processes project outwards and are not directed forwards (fig. 84, C); **crenate**, if the processes are rounded (fig. 81, A); **biserrate, bidentate, bicrenate**, if the processes themselves bear smaller secondary processes of the same kind; **spiny**, if the margin bears a number of hard, spiny processes resembling prickles (*e.g.* holly); **crisped**, or *curled*, if very wavy and irregular, as in the endive; **sinuate**, if the margin is more deeply indented, as in the oak (fig. 82, D). The sinuate margin forms a transition to the deeper cuttings of the margin, which are called incisions (§ 12).

§ 10. **The apex** of a leaf or leaflet may be **rounded** (or obtuse —fig. 80, E); if it comes to a point, it is **acute** (figs. 80 C, 76 A); if slender and very much drawn out, **acuminate** (fig. 83, A); if it seems to have been cut across, **truncate**; if it bears a *distinct* minute, pointed process, **mucronate** (fig. 82, F); if there is a rounded depression at the apex, **retuse** (fig. 82, A); if the depression is sharp, **emarginate**.

§ 11. **Hairs.** The leaf may be hairy. If the margin of the leaf bears a fringe of fine hairs, it is described as **ciliate**.

§ 12. **Incision of the Lamina.** In a *unicostate* leaf, if the incisions do not pass half-way down to the midrib, the leaf is **pinnatifid** (fig. 82, E); if rather more than half-way, **pinnatipartite** (fig. 82, C); if *almost* to the midrib, **pinnatisect** (fig. 82, B). Corresponding to these simple leaves, we have, where the incision is complete, the compound leaf of the

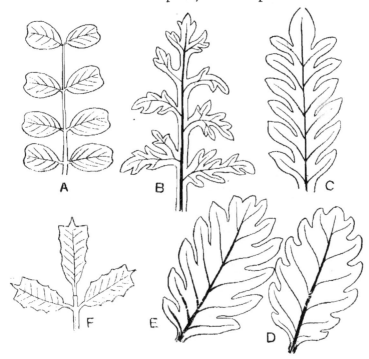

Fig. 82.—INCISION OF LAMINA.

A, Paripinnate compound leaf; B, Pinnatisect leaf; C, Pinnatipartite leaf; E, Pinnatifid leaf; D, Leaf with sinuate margin; F, Imparipinnate unijugate compound leaf.

pinnate type (fig. 82, A). Similarly, where the venation is multicostate, we may have **palmatifid** (fig. 79), **palmatipartite** (fig. 83, C), or **palmatisect** (fig. 83, B) simple leaves; and the corresponding compound leaf is of the **palmate** type (fig. 83, A).

The terms pinnati-fid, -partite, -sect, palmati-fid, -partite, -sect are also applied to leaflets of compound leaves. If the divisions of a simple leaf are again incised, the terms **bi-pinnatifid**, etc., are used ; or a pinnatipartite leaf may have divisions which are pinnatifid, etc.

When a unicostate leaf is incised in such a way that there is a large *rounded* terminal division with others which become gradually

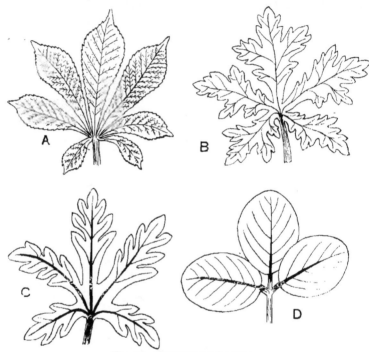

Fig. 83.—INCISION OF LAMINA.
A, Multifoliate palmate compound leaf ; B, Palmatisect leaf ; C, Palmatipartite leaf ; D, Ternate compound leaf.

smaller towards the base, the leaf is said to be **lyrate** (fig. 84, A). A **runcinate** leaf (*e.g.* dandelion, fig. 84, B) is a pinnatifid leaf in which there is a large pointed terminal lobe, and the apices of the smaller lobes behind are directed backwards. In a multicostate leaf, where, as already indicated, the branching is of the cymose type, only daughter-branches of the first order, as a rule, are given off in a cymose fashion, as in fig. 79 ; but occasionally these may again branch cymosely, as in fig. 84, C. This is known as a **pedate** leaf.

§ 13. **Compound Leaves.** Often a compound leaf is mistaken by beginners for a stem bearing leaves. The following points of difference should be carefully noticed :—
(*a*) a compound leaf has no apical bud or growing-point ;

THE LEAF OF THE ANGIOSPERM. 121

(b) it has a bud in its axil, and does not arise in the axil of a leaf; (c) it may have stipules, or an expanded sheath

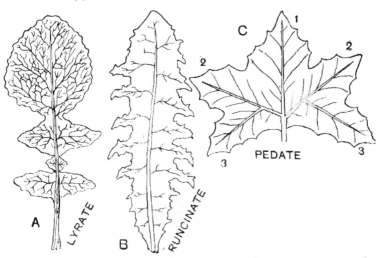

Fig. 84.—FORMS OF LEAVES.
In C the numbers indicate the branching.

at the base; (d) the *apparent* leaves (really leaflets) have no axillary buds.

There are certain special terms used in the description of compound leaves to which we must now refer. In the *pinnate leaf* the leaflets are borne on the common rachis or axis of the epipodium. Usually the leaflets are arranged in pairs, the leaflets of each pair being opposite each other. If an unpaired terminal leaflet is present, the leaf is described as **imparipinnate** (fig. 76, A); if no terminal leaflet is present, so that the number of leaflets is even, the leaf is **paripinnate** (fig. 82, A). If there is only one pair of leaflets, the leaf is *unijugate* (fig. 82, F); if two, *bijugate* (fig. 76, A), etc. Sometimes pairs of large leaflets alternate with pairs of small leaflets. Such a leaf is said to be *interruptedly pinnate*. The leaflets themselves may be completely incised. Here the secondary leaflets formed are called **pinnules**, and the leaf is said to be *bipinnate* (fig. 85). If these again are completely incised,

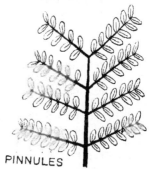

Fig. 85.—BIPINNATE LEAF.

the leaf is *tripinnate*. Usually, however, the upper leaflets in a bipinnate or tripinnate leaf are incompletely incised, and either pinnatifid or pinnatipartite.

In a *palmate* compound leaf the leaflets come off at the same point. If there are two leaflets, the leaf is *bifoliate*, or *binate*; if three, *trifoliate*, or *ternate* (fig. 83, D), and so on; if a large number, *multifoliate* (fig. 83, A). The ternate leaf resembles the imparipinnate unijugate leaf. Usually leaves with three leaflets are regarded as ternate, except where, as in fig. 82, F the secondary petioles come off at different points. Fig. 86 shows a biternate leaf.

Fig. 86.—BITERNATE LEAF.

In the orange there is a peculiar compound leaf, bearing only one leaflet. This is recognized as a compound leaf by the fact that the expanded lamina is distinctly articulated to the winged petiole (fig. 87).

§ 14. **Texture and Duration.** Leaves which are thin and membranous are described as *herbaceous*. Firm, thick leaves are called *coriaceous*. Some are succulent and *fleshy*. The leaves are *caducous* if they fall off very early; *deciduous*, if they fall at the end of each season; *persistent*, if they remain on the plant for more than one season. Plants with persistent foliage-leaves are *evergreens*.

§ 15. **Prefoliation.** This refers to the form and arrangement of young leaves in the bud condition. Here we have to consider (*a*) the form of the young leaves *in the bud, i.e.* the way in which they are folded or rolled on themselves—this is called the **ptyxis** of the leaf; (*b*) the relation between the different leaves in the bud, *i.e.* the manner in which they are arranged with regard to each other—this is called the **vernation** of the leaf. In flower-buds we speak of **prefloration**, including **ptyxis** and **æstivation**.*

Fig. 87.—COMPOUND LEAF OF ORANGE.

(*a*) **Ptyxis of the Leaf** (fig. 88). It is *plane* if there is no folding or rolling at all; *conduplicate*, if the right half is folded over on the left; *plaited*, or *plicate*, if there are numerous longitudinal

* The terms vernation and prefoliation, æstivation and prefloration, are variously used by different writers.

THE LEAF OF THE ANGIOSPERM.

folds; *crumpled*, if folded in all directions; *convolute*, if rolled from one margin to the other; *involute*, if rolled from both margins to the middle of the upper surface; *revolute*, if rolled similarly to

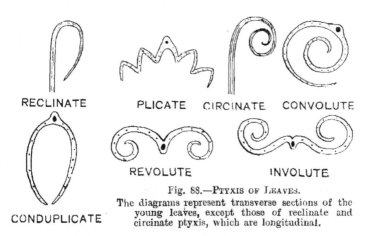

Fig. 88.—PTYXIS OF LEAVES.

The diagrams represent transverse sections of the young leaves, except those of reclinate and circinate ptyxis, which are longitudinal.

the middle of the lower surface; *circinate*, if rolled up from apex to base.

(*b*) **Vernation** (fig. 89). It is *valvate* if the young leaves touch each other laterally, but do not overlap; *imbricate*, if some overlap

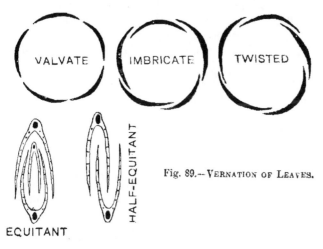

Fig. 89.—VERNATION OF LEAVES.

others, but not regularly; *twisted*, or *contorted*, if one margin of each leaf is directed inwards, and is overlapped, while the other margin is directed outwards, and overlaps the margin of the adjacent leaf.

§ 16. **Special Adaptations of Leaf-Structure.** There are many other striking modifications of leaf-structure. They are not to be distinguished, as regards the causes of their origin, from such highly specialized forms as the floral leaves already mentioned. Like the floral leaves, they have arisen simply as adaptations to special conditions. They are described separately simply because they are of much less general occurrence.

(a) **Leaf-Tendrils.** Leaves or parts of leaves frequently have the form of tendrils (p. 67). Thus in the pea the tendrils represent the leaflets of a compound leaf (fig. 76, B). In some species of pea all the leaflets are specialized in this way, and the functions of the foliage-leaf are taken on by the stipules, which are large and green. The stipules may be modified into tendrils, as in smilax. This is an adaptation for climbing purposes.

(b) **Leaf-Spines.** Similarly leaves or parts of leaves may take on the form of spines and serve as protective organs. The whole leaf may be thus specialized, as in the barberry, where the spine is triradiate. In *Robinia pseudacacia* the stipules are spiny. In the gooseberry the spines are developed on the pulvinus (p. 110).

(c) **Phyllodes.** In some foliage-leaves where the lamina is absent, the petiole develops a wing and takes on the appearance and functions of a lamina. Such flattened petioles are called *phyllodes*. They are distinguished from true laminæ by the fact that they are vertically expanded, with right and left surfaces; and also, in some plants, by the fact that in some of the leaves a true lamina is present above the flattened petiole. They are found in some acacias.

(d) **Pitchers, etc.** These are curious pitcher-shaped modifications of leaves found in certain orders of Flowering Plants, *e.g.* Nepenthes, the pitcher-plant. They may be regarded as long tubular peltate (p. 111) laminæ, and function as insect-traps. There is a little water in the pitcher, containing a small amount of a ferment secreted by the plant. Insects fall into the water, are drowned, and their bodies are digested by the ferment. The nitrogenous solutions thus produced are gradually absorbed. Thus there is an accessory method of nutrition in these plants.

There are several other interesting modifications of leaves for insect-catching purposes. Thus in Drosera, the sun-dew, the margin of the leaf is fringed by a series of knob-like glandular processes. When an insect touches these, they fold up towards the centre of the leaf, and the insect which adheres to the glandular secretion is made prisoner. Its body is then digested and the products absorbed. In Dionæa, Venus's fly-trap, the leaf is bilobed and the margin fringed with spiny teeth. There are three long hairs in the middle of each lobe. When an insect touches these, the two lobes fold up and the teeth interlock. The plants which feed on insects in this way are spoken of as "carnivorous plants."

[*Note :—For directions with regard to the description of leaves, see Appendix.*]

B. INTERNAL STRUCTURE OF THE LEAF.

§ 17. **The Petiole.** A stout petiole or leaf-stalk, when examined by itself, *i.e.* without its lamina, is apt to be mistaken by the student for a stem. As a rule, however, it is readily distinguished. In most cases the petiole is not perfectly cylindrical, but more or less flattened, often markedly grooved on its upper surface. In Angiosperms one or more collateral bundles pass from the stem into the leaf (p. 77). They are accompanied by a tissue continuous with the pericycle and endodermis. Usually, as they run through the petiole, they break up into a number of smaller collateral bundles, each of which becomes surrounded by pericycle and endodermis. These, as seen in transverse section, may be scattered more or less irregularly with their xylem portions towards the middle of the upper surface, or in a curved band. Occasionally they form a ring, as in the stem of the Dicotyledon—their xylem portions being towards the middle of the petiole. Even in this case it is usually found that the bundles towards the upper surface are smaller than those towards the lower surface.

The pericycle and endodermis, though present, are as a rule not very distinctly marked off from the surrounding ground-tissue. Sclerenchyma may be developed in the pericycle. The rest of the ground-tissue is mostly parenchymatous, but bands or ridges of collenchyma or sclerenchyma are frequently developed beneath the epidermis. The epidermis resembles that of the stem.

In dicotyledonous petioles a rudimentary cambium is present between xylem and phloem. It is only in a few exceptional cases that it becomes active and gives rise to secondary growth. Only in rare cases is there cork-formation (except in connexion with the fall of the leaf, see p. 128).

The vascular tissue entering the petiole together with its associated pericycle is called a **meristele** because it is only a part of a stele. The smaller parts into which it breaks up are called **schizosteles**. The schizostele is simply a collateral bundle with a pericycle. The collateral arrangement of vascular tissue distinguishes it from the simple steles mentioned on p. 83.

§ 18. **The Lamina—bifacial type.** A section (fig. 90) through a small portion of the lamina of a bifacial leaf (p. 9) at right angles to one of the veins would show a well-marked cuticularized **epidermis** protecting the upper and lower surfaces. Between these is the ground-tissue or **mesophyll** of the leaf, with the vascular bundles running through it. Towards the upper surface the mesophyll consists of columnar or elongated cells, without intercellular spaces, and arranged at right angles to the epidermis. This is the **palisade** parenchyma. Towards the lower surface the mesophyll consists of smaller rounded or stellate cells loosely packed together so that there are numerous intercellular spaces communicating with stomata in the lower epidermis. This is the **spongy** parenchyma. The cells of both palisade and spongy mesophyll contain numerous chloroplasts. Between the two run the vascular bundles. In the diagram one of the veins has been cut transversely. It consists of **xylem** towards the upper surface, **phloem** towards the lower. Some of the smaller bundles may be cut obliquely or longitudinally. *Endodermis* and *pericycle* are present round the larger bundles (schizosteles), but are not recognizable as distinct layers except when, as frequently happens, the pericycle is lignified and the endodermal cells contain starch (starch-layer). In the mesophyll, bands of **sclerenchyma** may here and there be developed. These usually run from the vascular bundles to the epidermis on each side. Cells containing crystals or cavities containing oil are frequently found. A very definite single hypodermal layer is present in some leaves (*e.g.* holly—upper surface).

THE LEAF OF THE ANGIOSPERM.

§ 19. **Isobilateral** and **Centric** leaves. In isobilateral leaves (p. 9) there is a layer of palisade parenchyma towards both surfaces. The xylem parts of the bundles,

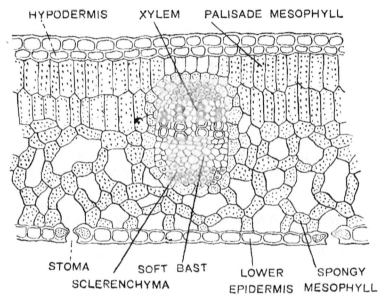

Fig. 90.—SECTION OF THE LAMINA OF A LEAF AT RIGHT ANGLES TO A SMALL VEIN.
A hypodermis is only developed in some cases.

however, are all directed towards one surface. In centric leaves (p. 9) the arrangement of the mesophyll-tissue is the same at all parts.

§ 20. **Development of the Leaf** (figs. 54, 55). The leaf originates as a small lateral protuberance at the meristematic apex of the stem. This protuberance consists of dermatogen and periblem only. The development is superficial, and therefore **exogenous** (p. 93). At first all the cells are meristematic, but later the meristematic tissue is restricted to the middle or base of the developing leaf, and growth is therefore *intercalary*. Finally, when the full number of cells has been produced, the meristem dies out. At this stage the young leaf is still very small and folded up in the bud with the other leaves. The expansion and increase in size of the leaf, when the bud unfolds, is due simply to the growth of the individual cells, not to the formation of new cells. During the earlier development procambial strands make their appearance (developed here in periblem-tissue), which sooner or later become connected with the procambial strands of the stem. These differentiate into the vascular bundles.

§ 21. **Leaf-Apex—Endings of the Veins.** It will be evident from the foregoing paragraph that the fully formed leaf has no apical growing-point such as the stem has, and further, therefore, that the terminations of the vascular bundles or veins must be different. Frequently the veinlets have no definite endings, but form anastomoses or fusions with neighbouring veinlets. Sometimes they end blindly in the mesophyll. Where the ending is definite, the vascular tissue is gradually lost. The larger xylem vessels and phloem elements disappear. The remaining small xylem elements are of the nature of tracheides. In a few cases these pass gradually into a mass of small-celled glandular tissue (*epithem-tissue*), with which are usually associated a number of water-pores (p. 52).

§ 22. **Fall of Leaves.** There are certain important processes connected with leaf-fall (**phylloptosis**). A layer of cork is formed across the base of the petiole. It is formed by the meristematic activity of the living cells (phellogen) lying immediately internal to it. This meristematic property is taken on, not only by cells of the ground-tissue of the petiole, but also by cells in the vascular bundles, so that the cork-layer is formed right across the petiole, and joins on to the cork-layer (in the case of woody dicotyledons) formed in the stem. The fall of the leaf is directly due to the disorganization of the layer of cells (**absciss-layer**) lying just outside the cork-layer. Thus the surface exposed when the leaf falls is protected by the cork-layer. The cork-layer constricts the wood-vessels in the vascular bundles, so that when these are broken across there is no exudation of sap. Preparation for this cork-formation begins even at the time when the leaf is unfolding from the bud condition; it is not completed, however, till just before the leaf falls. Leaf-fall seems to be traceable ultimately to interruption or injurious modification of the process of transpiration, brought about by excessive changes in temperature. Thus leaves fall naturally in this country towards the approach of winter; but a long period of very hot, dry weather may bring about the same result. That it is a natural process, due to vital activity, is shown by the fact that the leaves of a dead branch do not fall. There is a similar formation of cork at the base of prickles.

§ 23. We may conveniently close the chapter with a general definition of a leaf. A leaf may be defined as any natural exogenously developed outgrowth of a stem, differing from the stem itself in structure, occupying a definite position in development, and variously modified in different regions according to the functions it has to perform. A leaf-structure, however it may have been specialized, is distinguished from a stem or root by its mode of origin.

CHAPTER VII.

NUTRITION AND GROWTH.

§ 1. In Chapter I. (§ 11), we indicated, in a very general way, the physiological processes connected with Nutrition and Growth, and, here and there, we have made scattered references to the functions of various tissues and organs. We have now to give special consideration to these processes as they are exhibited in the higher plants. Although considered with special reference to the Angiosperm, they are essentially the same in all green plants showing differentiation into root, stem, and leaf. Before reading this chapter, the student should again refer to pp. 11–14.

§ 2. **Importance of Water.** That protoplasm is the essential living substance by which all these physiological processes are carried on, has already been sufficiently impressed on the student. We must again allude, however, to the important part played by water. Water forms the great bulk of a plant, in a few cases, as much as ninety per cent. of the total substance. All the organic substance is permeated with water. Water is one of the important forms in which essential chemical elements (Hydrogen and Oxygen) enter the plant. Besides this, it is the medium for the solution, absorption, and transit of all other food materials; the medium by which these are brought into intimate relation with the living substance.

§ 3. **The Food-Materials of a Green Plant.** If we make a chemical analysis of a plant—an analysis of the gases given off, and the residue or *ash* left behind on burning the plant—we recognize the following chemical elements:—carbon, oxygen, hydrogen, nitrogen, sulphur, phosphorus, calcium, potassium, magnesium, iron, sodium, silicon, and chlorine, with, frequently, traces of manganese, iodine, etc. Of these, only the first six enter into the actual composition of the living substance of the plant. It is evident that all these elements found in the plant must pass in in the food-materials absorbed, *i.e.* the food-materials consist of, or contain, these elements. We have already indicated (p. 12) that the food-materials absorbed by a green plant are of the nature of simple inorganic compounds, and that they are taken in in solution.

All the carbon used by the plant in the processes of assimilation is derived from the CO_2 of the atmosphere. This CO_2 is absorbed by the aerial *green* parts of the plant (chiefly leaves) in the presence of light. All the other elements are derived from the water and dissolved mineral substances ("salts") absorbed by the root in the process of **root-absorption**. Dissolved CO_2, or carbon taken in by the root in the form of carbonates, is not made use of in the anabolic processes. The necessary oxygen* and hydrogen are derived chiefly from water, partly from salts containing these elements. Nitrogen is absorbed in the form of nitrates (for exceptions, see p. 151), sulphur in sulphates, phosphorus in phosphates, chlorine in chlorides, silicon in silicates; iron, potassium, calcium, and magnesium, form the metallic bases of these salts. If substances absorbed are made use of in metabolism, the absorption continues. The amount absorbed depends on the amount assimilated. It has been determined by experiment that, for most green plants, the *essential elements*, *i.e.* the elements absolutely necessary for *healthy* growth, are carbon, oxygen, hydrogen, nitrogen, sulphur, phosphorus, calcium, potassium, magnesium, and iron. The others are non-essential; or, at all events, only essential for certain plants.

That carbon is essential is shown by the fact that a green plant cannot be grown in an atmosphere deprived of CO_2. That the others

* Oxygen of respiration is taken in over the whole surface of the plant—see p. 144.

are essential has been determined by the method of **water culture.** A number of plants of the same species are grown in glass jars, their roots immersed in a nutrient solution of inorganic salts. It is found, with most plants, that there is healthy growth only if the solution contains these elements in proper form and degree of concentration. Various unhealthy symptoms appear if one or more of them are absent. Thus, if iron is left out there is no development of chlorophyll. A satisfactory nutrient solution could be made, for example, by using the following substances : calcium nitrate, potassium nitrate. magnesium sulphate, potassium phosphate, ferrous chloride (or ferric phosphate). The solution must be very dilute.

§ 4. **Root-Absorption.** The soil may be regarded, in a general way, as an aggregation of organic and inorganic particles. Each particle is surrounded by a delicate film of water. This water, which is called the **hygroscopic** water, is firmly adherent to the particles, just as water adheres to or wets the surface of a glass. It is present even in the driest soil. Between the particles are spaces or interstices into which, normally, the atmospheric air penetrates; but in very damp soils these are filled with water—the **free water** of the soil. This free water is injurious to plants (except those specially adapted, *e.g.* bog or water-plants), probably because it interferes with the proper respiration of the root. It is the object of drainage to remove it.

A portion of the inorganic substance of the soil is dissolved in the hygroscopic water, and it is this water, with its dissolved salts, which is absorbed by the roots. Normally the organic substance is not absorbed, but owing to decomposition effected in it by micro-organisms and leading to the formation of simpler compounds, it serves to replenish the store of inorganic substance, and more especially nitrates, necessary to the plant.

The inorganic solutions are absorbed by a process of **osmosis.** This may be regarded as a simple physical process, modified in the case of **root-absorption** by the vital activity of the protoplasm. It may be defined as diffusion *through a permeable but non-porous membrane*. Thus, if we place in a vessel of water a bladder filled with a strong solution of a substance having an attraction for water (an osmotically active substance, *e.g.* sugar), a *large* amount of the water will diffuse (pass by osmosis) into the bladder (*endosmosis*), while a *small* amount of the solution will diffuse out (*exosmosis*). The weaker fluid diffuses faster and this continues until the same concentration is acquired, when it is equally rapid in both directions, and hence apparently ceases. It should be noticed, also, that, in the above example, a considerable pressure will be set up inside the bladder, owing to rapid endosmosis. This has an important application is connection with root absorption.

The root-hairs are the absorbing organs. The particles of the soil closely adhere to them. There is little or no absorption by the surface of the root itself. The wall of the root-hair, *together with the primordial utricle*, is the osmotic membrane; outside this is the hygroscopic water, with various salts in *very weak* solution; inside is the cell-sap, a *comparatively* strong solution, containing many organic compounds with a strong affinity for water (probably the most important of these in this connection are the organic acids). The primordial utricle exercises an important influence on the process; it prevents the entrance of certain substances which could readily diffuse through the wall of the hair alone; only very weak solutions can pass through the primordial utricle; similarly, also, it retains the water in the vacuole *even at a high pressure*, thus bringing about a very tense or *turgid* condition of the cell (cf. the distension of the bladder).

Thus in root-absorption a very large amount of hygroscopic water with dissolved salts passes into the root-hair, while only a minute quantity of cell sap diffuses out—a quantity just sufficient to moisten the tips of the root-hairs. Endosmosis is greatly in excess of exosmosis.

§ 5. **Salts insoluble in pure water** and required by the plant are brought into solution in various ways. There is always CO_2 present in the soil. Certain salts insoluble in pure water are soluble in water containing CO_2. Again, the acid sap, which moistens the tips of the root-hairs, helps in the solution of such salts. It probably brings about decomposition of the insoluble salts, and leads to the formation of soluble salts. The various chemical processes always going on in the soil no doubt also induce decompositions which bring into soluble form various substances required by plants.

§ 6. Course of Absorbed Solutions. The absorbed solutions pass *by osmosis* from the root-hairs into the cells of the cortical tissue of the root. Owing to the excess of endosmosis, seconded by the influence of the primordial utricles, a considerable pressure is set up in the cortical tissue—the cortical cells become extremely turgid. The solutions to a certain extent diffuse by osmosis through the parenchymatous tissue of the plant, but the greater proportion passes into the xylem tissue of the root, and is given off to the parenchyma at a higher level. The passage into the xylem tissue is not effected by a process of osmosis, for at first the xylem elements (vessels) are empty, so that an essential condition of osmosis is wanting. It is brought about simply by the hydrostatic pressure set up as explained in the surrounding cortical tissue. It is probable that, when the limit of turgidity is reached (*i.e.* when the pressure exceeds a certain point), a molecular change takes place in the protoplasm (primordial utricle), and the watery solutions are expelled, owing to the collapse of the cell, with considerable force. In this way the solutions, following the course of least resistance, are forced into the wood vessels. They pass up through the xylem of root and stem, and out into the xylem of the leaf-veins. There they finally diffuse into the mesophyll tissue of the leaf, where elaboration chiefly takes place. Formerly it was thought that the solutions passed upwards through the *walls* only of the wood elements; it is now known that they also pass through the cavities; and at certain periods, the vessels contain abundance of water. The student must not imagine, however, that there is a continuous current of water in the wood-vessels. The water probably forms short columns, separated by bubbles of gas or air. The current of water with dissolved salts passing upwards from root to leaves is called the **transpiration current.** It conveys food-materials to the leaves for elaboration, and makes good the loss of water **due to** transpiration.

§ 7. Root-Pressure.

We have explained that, owing to the great excess of endosmosis a considerable pressure is set up in the cells of the cortical parenchyma of the root; that this is increased by the action of the primordial utricle; and that when the cells collapse the water is forced into the wood elements. After collapse the cells recover their condition of turgidity and again collapse. In this way we can imagine that a rhythmical pumping of water into the wood elements is going on. Now, this pressure existing in the root, and which we may regard as a force driving the water into the wood elements, and upwards, is called **root-pressure**.

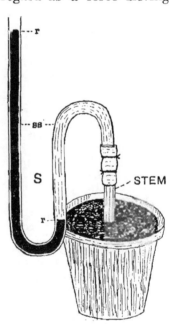

Fig. 91.—APPARATUS FOR MEASURING ROOT-PRESSURE.
(See text.)

That this pressure does exist in the root, and does tend to drive the water upwards, can be shown by a simple experiment. If the stem of a very young tree (*e.g.* the vine) be cut in the active growing period (in the spring) about a foot from the ground, there is an abundant exudation of watery sap from the vessels at the cut surface. This phenomenon is called "*bleeding*," and its manifestation continues for a considerable time. If a long piece of glass tubing be firmly attached to the cut end of the stem, the water will be found to rise to a considerable height. By using suitable apparatus the force can be measured. Thus in fig. 91, s is an S-shaped tube closely attached to the cut stem. It is filled with mercury to the level *ss*. Root-pressure forces water into s, and displaces the mercury to the levels *rr*. The force of the root-pressure is measured by the displacement. Root-pressure varies at different periods of the growing season, and also at different times of the day. It is most active in spring, and is affected by various external conditions, such as temperature, etc.

§ 8. Transpiration.

A large amount of the water absorbed by the roots, and carried to the aerial parts by the transpiration-current, is given off from the aerial surface in the form of *water-vapour*. This process is known as **transpiration**. The water vapour collects in the intercellular spaces of the parenchymatous ground tissue, and passes off through the stomata. Very little is given off from the general epidermal surface, as it is rendered almost impermeable by the development of a cuticle. The process, however, is not simply one of evaporation. It is regulated by the vital activity of the plant. That this is so is evident from the fact that usually more water is evaporated from the surface of a dead leaf than from that of a living leaf. We must remember, also that the stomata can regulate the amount of transpiration (see p. 51).

Transpiration varies according to external conditions. It is more active when the air is dry and hot than when it is moist and cold. This is not merely due to the fact that a hot dry atmosphere is favourable to evaporation, but also because it increases root-absorption. Transpiration is greater in bright sunlight, because the latter increases assimilation, and promotes osmotic activity. There is a rush of water to the assimilating cells. With this is correlated the opening of the stomata. The guard-cells always contain chloroplasts, and we now have an explanation of this. The guard-cells become turgid when there is increased assimilation, owing to osmotic activity. *When the guard-cells become turgid, the stoma opens*; when they collapse, it closes. The reason of this is found in the peculiar arrangement of thickening on the walls of the guard-cells (p. 51). The thickening is such, that the cells can expand only in a particular direction. When turgid, they bulge away from each other, becoming more **convex** on the side away from the pore, *concave* towards the pore.

Experiment. The fact of transpiration is readily demonstrated thus : —Take a small plant growing in a pot. Cover the upper part of the pot and the soil with an indiarubber membrane, tied firmly round the base of the stem. This prevents evaporation from the soil. Invert a glass jar over the upper part of the plant (stem and leaves). Owing to transpiration, moisture will collect on the sides of the jar, and run down in the form of drops. The amount of transpiration can be determined by placing a weighed quantity of calcium chloride inside the vessel. This absorbs the water, and increases in weight. The difference in weight represents the amount of water transpired. (Here, however, as the chloride removes all the water from the air, the conditions are not quite the same as if the plant were grown in the open air).

NUTRITION AND GROWTH.

§ 9. Causes of the Ascent of Water. The elucidation of the cause of the ascent of water in tall trees in opposition to the action of gravity has been, and still is, one of the problems of botanical physiology. It cannot even yet be said that they have been definitely or fully determined. The first theories were vitalistic, vaguely ascribing the ascent to the vital activity of the living protoplasmic substance. These, by themselves, were little more than a confession of ignorance. Various physical causes have since then come under discussion, such as root-pressure, transpiration, capillarity in the wood-vessels, the varying pressure in the gases contained in the wood elements, and so on. At the present time, the general tendency seems to be to regard **root-pressure** and **transpiration** as the chief factors in the ascent, the others having little or no influence on the process. A sufficient explanation has been given of the part played by root-pressure. Transpiration causes a great diminution of pressure in the upper parts of trees, and the pressure is equalized by the upward passage of water from the lower parts. Thus transpiration might be loosely regarded as a force sucking up the water from below. On physical grounds, however, these forces, by themselves, are inadequate to account for the ascent of water in high trees, and, with regard to root-pressure, it has been proved that poisonous solutions, *i.e.* solutions which would kill the living protoplasm, can be absorbed and *ascend* through the xylem, although, as has been explained, the vital activity of the protoplasm probably plays an important part in connection with root-pressure.

§ 10. Carbon-Assimilation (Photosynthesis). Carbon-dioxide is absorbed by the plant over its green aerial surface—chiefly the leaves—in the presence of light. It probably passes into the intercellular spaces by means of the stomata. It was formerly thought that gases (in solution) could pass through the cuticle. Recent experiments, however, have discredited this *cuticular absorption theory*. The CO_2 diffuses from the intercellular spaces into the parenchymatous cells—the mesophyll cells in the case of leaves. It should be carefully noticed that it passes through the walls, not as a gas, but in solution. It is dissolved in the cell-sap which permeates the walls of the cells. In the interior of these cells chemical processes, leading to the elaboration of organic compounds, take place. This elaboration goes on chiefly in the mesophyll cells of foliage leaves, although also to some extent in green herbaceous stems. Under the influence of chlorophyll and light the water (H_2O) absorbed by the root and the carbon dioxide (CO_2) are built up into carbohydrates. The first carbohydrate formed is probably **formic aldehyde** (CH_2O), and oxygen (O_2) is given off in the process. This may be represented according to the equation :—

$$CO_2 + H_2O = CH_2O + O_2.$$

But it is possible that there is an intermediate step in the process, thus :—

$$CO_2 + H_2O = (CO + O) + (H_2 + O)$$
$$= CH_2O + O_2.$$

According to the former view all the oxygen given off is derived from the CO_2; according to the latter, one half only from the CO_2, the other half from the water. From this formic aldehyde, which is a simple soluble carbohydrate, more complex soluble carbohydrates of the nature of sugars may be formed by a process of compounding (polymerization). Thus we may imagine the formation of grape-sugar ($C_6H_{12}O_6$) :—

$$6CH_2O = C_6H_{12}O_6.$$

Some recent chemical experiments, however, seem to show that, in many plants, cane-sugar ($C_{12}H_{22}O_{11}$) is the first sugar formed.

Thus, in carbon-assimilation, CO_2 is taken into the plant and, along with water, undergoes a chemical change; oxygen is given off, and the carbon made use of in the elaboration of organic substances, the first of which are carbohydrates.

The surplus amount of sugar formed in the leaves is stored up as starch in the chloroplasts (p. 34). The rest is transferred from the seat of formation to other parts of the plant and made use of in ways to be described presently.

Formerly it was thought that the starch appearing in the chloroplasts was *directly* built up from CO_2 and water, *i.e.* that the carbohydrate starch was the first carbohydrate formed in the process. We may still regard starch as the first *visible* product giving evidence of carbon-assimilation: but we can no longer regard it as being formed directly. It is simply **a** temporary *storage* of surplus carbohydrate.

NUTRITION AND GROWTH.

§ 11. Conditions of Carbon-Assimilation. It is evident there must be a supply of water and CO_2. Heat, also, is necessary; this is a general condition of plant-metabolism. Metabolism and growth can take place only within certain limits of temperature. Heat is a source of energy and a necessary condition for the initiation and continuation of all vital processes in plants. The more special factors in carbon-assimilation are light and the presence of chlorophyll. We have now to consider more fully the part played by these.

§ 12. Light. In our laboratories we can effect the decomposition of CO_2 and of water only by the expenditure of a large amount of energy—heat energy in the former case, electrical energy in the latter. The building up of complex substances containing a very small proportion of oxygen, also means expenditure of energy. The green plant carries on both these processes at ordinary temperatures. Light being an essential factor, we come to the conclusion that it is so because it is the chief source of energy. The energy used is stored up in potential form in these compounds.

When a beam of sunlight is passed through a prism it is spread out into a band, called the spectrum (which can be received on a screen), consisting of many different colours, passing gradually from one to the other. This is because sunlight consists of many different kinds of rays differing in refrangibility, *i.e.* as regards the angle through which they are bent on passing through a prism. At one end of the spectrum are *red* rays, which gradually pass through *orange and yellow*, to *blue and violet* rays at the other end. Now all these rays are not equally active in the process of carbon-assimilation. It has been determined, by direct experiment, that chiefly red rays, and, to some extent, also the blue and violet rays, are concerned in the process.*

§ 13. Chlorophyll. This is the green colouring matter of plants. It is a very complex substance, consisting of carbon, oxygen, hydrogen, nitrogen and, according to

* Plants may be grown under artificial light, *e.g.* the electric light.

some, also phosphorus. Its exact composition has not yet been definitely ascertained. It contains no iron. It is sometimes stated that chlorophyll is not a simple pigment, but a mixture of two pigments—xanthophyll, a yellow colouring matter, and cyanophyll (phyllocyanin), a bluish-green colouring matter. There is no doubt now, however, that these are decomposition products of chlorophyll. Chlorophyll is very easily decomposed under the influence of bright light if oxygen is present, and these decomposition pigments are always present in chlorophyll as extracted from plants.

Two conditions, in addition to a suitable temperature, are necessary for the *formation* or development of chlorophyll :—(a) the presence of light,* (b) a supply of iron in the food. If a plant is grown in darkness, it assumes a pale yellowish sickly appearance. This is due to the fact that a yellow colouring matter—**etiolin**—is developed in the corpuscles instead of chlorophyll. Such a plant is said to be *etiolated*. Of course, a plant grown in darkness must have some reserve store of food material to draw upon, as, *e.g.*, a potato-tuber developing in darkness. Many examples of etiolated plants will readily occur to the mind of the student, *e.g.* celery, grass covered by a roller or a board. Typical etiolated plants present many other peculiarities. Thus the internodes become very much elongated or "drawn out." For this reason the plants are called "**drawn plants.**" This has an important biological significance. In this way there is a chance of shoots reaching the light, as, *e.g.*, in seedlings smothered by other plants. In etiolated plants, also, the leaves remain small and scaly, there is an enormous development of soft succulent parenchyma and a meagre formation of lignified tissue. Large leaves would be useless in darkness; we might say, therefore, that the plant devotes all its energy to the formation of long internodes which may be of use to it.

A yellowish, sickly condition is also established if there is no iron in the food, the plastids being colourless or

* In exceptional cases chlorophyll is developed in darkness (cotyledons of ferns and of a few seeds).

containing etiolin. This condition, due to the want of iron, is called the **chlorotic** condition. It is to be carefully distinguished from the etiolated condition due to the absence of light. As soon as the plant is supplied with a *weak* solution of an iron salt, even if it is only applied to the leaves, chlorophyll is developed. Thus iron is necessary to the formation of chlorophyll, although it does not enter into its composition.

Chlorophyll can be extracted by means of alcohol, chloroform, etc. A solution of chlorophyll is fluorescent—it is green by transmitted light, red by reflected light. When a solution of chlorophyll is placed in the path of a beam of light, which is then passed through a prism, the spectrum is modified. Certain dark bands (absorption-bands) appear in the spectrum, more especially in the red and blue regions. This is because these particular rays have been absorbed by the chlorophyll. This leads us to the function discharged by chlorophyll. We have seen that the red and blue rays are especially active in carbon-assimilation. We therefore conclude that chlorophyll is a colouring matter which, by absorbing certain rays of light, supplies the living protoplasm with the energy necessary for carrying on the chemical processes connected with carbon-assimilation.

§ 14. **Experiments:**—(1) Grow a plant with its upper parts (stem and leaves) in a completely closed glass vessel, exposed to light. After a time, if the air in the vessel be examined it will contain relatively more O_2 and less CO_2. This is not the case if the experiment is made in darkness.

(2) Place a portion of a small water plant—*e.g.* the American water-weed—in a glass tube in water, exposed to bright light. Bubbles of gas are given off; if this gas be collected it will be found to be O_2. The gas is not given off in darkness. Nor is it evolved if the water was previously boiled, because the boiled water would contain no CO_2.

§ 15. **Formation of Nitrogenous Substance.** The elaboration of nitrogenous substance is not so clearly comprehended as that of carbohydrates. It has not been ascertained if light is or is not necessary to the process. The necessary supply of energy may be liberated during the katabolic changes presently to be described. According to some, from the water and nitrates absorbed by the root, ammonia (NH_3) is first formed, and this, combining with soluble carbohydrate, gives rise to a series of complex soluble nitrogenous substances called **amides**, *e.g.* $N(CH_3)_3$. However the elaboration be effected, there seems to be no doubt that these amides are, sooner or later, produced as the result of synthetic processes.

§ 16. **Transference and Destiny of the Elaborated Compounds.** Thus we have explained, as far as possible, the building-up of soluble carbohydrates (sugars) and soluble nitrogenous compounds (amides) in the assimilating cells—chiefly the mesophyll cells of leaves. With the exception of what is made use of in the assimilating cells, the sugars and amides are transferred to the various parts of the plant. *All* living cells contain these soluble carbohydrates and nitrogenous compounds conveyed to them in the cell-sap. The living protoplasm makes use of these as food substance. Together with sulphur they are built up first of all into more complex (*proteid*) substances, and, finally, into protoplasm. This final elaboration is most active where rapid growth is taking place, *i.e.* at growing points. Many of these soluble compounds, however, are made use of in the formation of insoluble storage compounds. The formation of starch in the chloroplasts, to which we have already referred, is only one example of this. Storage products may be formed in any living cell, but the formation is specially abundant in particular tissues or organs, *e.g.* medullary rays of trees, seeds, bulbs, corms, rhizomes, etc. The formation and uses of these products will be explained later.

In the process of transference from the assimilating cells, the sugars and amides may, to a certain extent, pass from cell to cell by simple diffusion. But a more rapid transference is effected through the phloem-tissue. In this way they are quickly conveyed to regions where rapid growth is going on, or where a storage of food-material is taking place. Formerly it was thought this rapid transference was effected through the sieve-tubes only. It is now certain that the phloem-parenchyma (and the neighbouring parenchyma of the ground-tissue) is chiefly concerned in the transference of the carbohydrate material. This also *may* apply to the nitrogenous substance. According to some, there is a formation of proteid or albuminous substance in the phloem-tissue, and they regard the sieve-tubes as being temporary store-places of these products. We have seen that the sieve-tubes do contain such substance.

One of the chief forms in which sulphur enters the plant is calcium sulphate. The sulphur is liberated, and thus enabled to combine with the organic food substance, by the action of an organic acid. This acid appears in many cases to be oxalic acid. The calcium of the sulphate combines with the oxalic acid to form calcium oxalate. This is probably one source of the calcium oxalate so abundantly present in the tissues of plants.

§ 17. **Use of the Metallic Elements, etc.**—So far, in connexion with these processes of elaboration, we have had no occasion to mention the metallic elements of the food material, with the exception of iron. Potassium, calcium, magnesium, and iron, do not enter into the composition of the living substance or of the cell-wall, yet they are essential elements (p. 131). Now we have seen that iron, although it does not enter into the composition of chlorophyll, is necessary for its formation. This gives us a hint as to the use of the other elements. It would seem that potassium in the same way is a necessary condition for the formation of carbohydrates, and that calcium and magnesium are necessary for the proper distribution of carbohydrates. In this connection also it may be mentioned that phosphorus is necessary in the same way to the formation of proteid substance, while it seems to be always present as a constituent of nuclear substance.

§ 18. **Katabolic Processes.** We have hitherto been dealing with the anabolic processes of metabolism (p. 13), the processes by which simple compounds are *built up* into complex organic compounds. The final result of synthesis is the elaboration of living protoplasmic substance. We have now to consider the katabolic processes (p. 13) in which the complex and unstable substance protoplasm undergoes decomposition and breaks down into simpler and more stable compounds. These processes are intimately associated with growth, and are as essential as the anabolic processes. All growth takes place *from* and *in* the living substance protoplasm. While the anabolic processes are concerned with the nourishment and building up of living protoplasm, the katabolic processes lead to the formation of various substances which are absolutely necessary either for carrying on the anabolic processes or for building up the tissues, and at the same time set free energy which is expended largely in connection with growth. As explained on p. 13 the decomposition of the protoplasmic substance is a slow process of oxidation, and the substances produced are either plastic substances, secretions, or excretions. Oxygen passes into the plant in **respiration** and is absorbed by the protoplasm. As a result the protoplasm undergoes decomposition. These katabolic changes are stimulated or influenced in various ways by the action of external agencies, such as temperature, light, gravity, etc. The general influence of environment will be discussed in the next chapter.

§ 19. **Respiration.** In this process, oxygen passes in *over the whole surface of the plant*. There are no special respiratory organs in plants, but the absorption of oxygen is most rapid in the regions or organs where the katabolic processes are most active, *e.g.* leaves, growing points, germinating seeds. Carbon-dioxide is given off as an invariable excreted product arising from the decomposition of protoplasmic substance. In aerial parts protected by a firm cuticle or a layer of cork, the oxygen enters by means of stomata or lenticels. It passes through the cell-walls into the interior of the cells in solution. The process of respiration is masked during the day, owing to the activity of carbon-assimilation.

The student must carefully distinguish between respiration

NUTRITION AND GROWTH.

and carbon-assimilation. The following table indicates the chief points of distinction :—

Respiration.	Carbon-Assimilation.
(a) A *breathing* process associated with katabolism.	(a) A *feeding* process associated with anabolism.
(b) Takes place over whole surface.	(b) Only in green aerial parts.
(c) O_2 passes in, CO_2 given off.	(c) CO_2 passes in, O_2 given off.
(d) Independent of light and chlorophyll.	(d) Dependent on light and chlorophyll.
(e) Plant loses weight.	(e) Plant gains in weight.

Experiments :—Respiration can be readily demonstrated by germinating seeds in a closed chamber. There being no chlorophyll, there is no carbon-assimilation. It will be found that the CO_2 increases in amount, while the O_2 decreases. If air, robbed of its CO_2 (by passing it through a solution of caustic potash), be drawn into the chamber, the exhaled CO_2 may be absorbed by means of caustic potash, and thus measured. As the CO_2 exhaled is as a rule nearly equal in volume to the oxygen absorbed, this would give a measurement of respiration. A similar experiment could be made on green plants *in darkness*, and also on small green twigs in the presence of light, but, in the latter case, the result would not be strictly accurate, as part of the CO_2 exhaled would be immediately made use of in carbon-assimilation.

§ 20. **Plastic Substances.** Some of these we have seen are formed anabolically, *e.g.* sugars, amides, proteids. The most important formed katabolically are cellulose, starch, oil, and very probably proteid grains (aleurone grains). Cellulose is formed in all cells where the cell-wall is undergoing extension or thickening. The others are storage food-substances—as also is cellulose, *e.g.* in many seeds.

We must now say something more with regard to the formation of starch. It is believed by many that it is formed by a *direct* conversion of sugar effected by plastids. The equation would be :—

$$C_6H_{12}O_6 - H_2O = C_6H_{10}O_5.$$

The view which is adopted here, and which seems to be steadily gaining ground, is that starch is not formed directly from sugar, but *indirectly* and *katabolically*. The process is carried on by leucoplasts, or chloroplasts. The plastids, from the soluble carbohydrates (and amides, etc.)

presented to them, *build up their protoplasmic substance.* The starch formed by the plastid is produced katabolically by the decomposition of the living substance of the plastid. In the chloroplasts the storage of starch is temporary. The starch which is formed there during the day disappears at night. In tubers, seeds, etc., the storage in connection with leucoplasts is more enduring.

While starch and cellulose, along with inulin and cane sugar in some cases (the formation of these also may be katabolic), are the chief forms in which carbohydrate is stored up, aleurone grains (and proteid crystalloids) represent the chief storage of nitrogenous substance. Oil belongs to the group of the hydrocarbons, substances containing no nitrogen and much less oxygen than the carbohydrates.

§ 21. **Secretions and Excretions.** The most important secretions are colouring matters (chlorophyll, the colouring matters of flowers, etc.), organic acids, ferments (see p. 31). The student should carefully notice the uses of these, as reference is made to them throughout the book. The chief excreted substances formed katabolically are CO_2, resins and gums, alkaloids, etc. (p. 31). It should be noticed that, in plants, there are no special excretory organs. Still, many of these excreted substances are got rid of, *e.g.* in falling leaves, the scaling of bark, etc. We find that the leaves of deciduous trees become filled with such substances towards the approach of winter. So also the bark. Falling to the ground, these substances are decomposed and brought into forms in which they may again be absorbed by plants. There is no clear line of distinction between secreted and excreted substances. Thus, certain substances (*e.g.* ethereal oils) although they are placed amongst the excreted substances, are no doubt of some indirect use, while the significance of the formation of others at present regarded as excreta is unknown.

§ 22. **Storage-Material—Ferments.** The various storage forms of food-material are, sooner or later, made use of. First of all, however, they must be brought into a soluble diffusible form. This is effected by the action of certain soluble nitrogenous bodies of the nature of unorganized ferments or enzymes, which, as we have seen, are formed katabolically from the protoplasm. These ferments are substances which have the power of inducing important chemical changes without themselves being changed. These changes are essentially similar in nature to those of digestion in animals. Certain insoluble substances are converted into forms in which they can diffuse through the organism and serve as food to the protoplasm. Many of these ferments have been extracted from plants, and there are doubtless many more which have not yet been extracted. Some of these, known as diastatic ferments, act on starch, and convert it into sugar. Others, known as proteolytic ferments, act on proteids and convert them into soluble peptones, or reduce them to simpler forms (amides). There are ferments which convert fats into glycerine and fatty acids which then undergo further alteration. Also ferments acting on cellulose, inulin, cane-sugar, etc. The soluble diffusible substances produced are often identical with the soluble substances formed anabolically (sugars and amides), and are used in the same way by the protoplasm in building up its substance. The action seems to be one of hydration. This taking-up of water is usually accompanied by decomposition, so that ferment-action comes under the head of katabolism. In nearly all cases, however, there is no absorption of oxygen or exhalation of CO_2.

§ 23. **Energy.** We have seen that energy is derived by the plant partly from heat, chiefly from light. The energy absorbed is stored up in potential form in the complex organic substances formed. In the katabolic processes there is a liberation of energy. The liberated energy is for the most part made use of again in building up the living substance of the plant, so that we may say that the greater part of the energy which enters the plant is accumulated in potential form in the complex substances formed. But a certain amount is dissipated or given off in various ways. A certain amount of energy in potential form is, for example, lost to the plant in the various complex excreted substances which are got rid of. When the katabolic processes are very active, as in the unfolding of many massive inflorescences, or the germination of large numbers of seeds, a dissipation of energy in the form of heat can be recognized ; a distinct rise in temperature is noticed. Again, some plants exhibit movement; growth itself may be regarded as a slow form of movement;

the protoplasm of some cells shows an irregular streaming movement along the primordial utricle and protoplasmic strands (circulation of protoplasm), or a regular movement round the inner surface of the cell-wall (rotation); all these forms of movement mean a loss of energy to the plant.

§ 24. **Growth** takes place as the visible result of all these metabolic processes. The conditions necessary for growth are a supply of food-material, energy, and oxygen, a suitable temperature and a condition of turgidity in growing cells. We have to take into account not only the formation of new cells, but also the growth of individual cells. In the anabolic processes there is the building up of the living substance, and this is accompanied by the storing up of energy. In the katabolic processes there is the formation, from the protoplasm, of certain substances necessary for building up the tissues, or for carrying on the various metabolic processes; the digestion of stored food-material; and the liberation of energy which is made use of in metabolism. The formation of organic substance and the storage of energy is greatly in excess of the loss of substance and the expenditure of energy. In other words (see p. 14), there is in plants, usually, a constant increase in the total amount, *i.e.* growth.

By *growth*, however, is meant not only an increase in size, but also a *permanent* change in form, owing to the formation of new substance. In many cases we can recognize a *temporary* increase in size without growth taking place, as, for example, when cells become turgid. And we may even have the formation of new substance, and yet no growth in the strict sense. For example, in a living cell new substances may be formed, and new particles of cellulose be laid down in the cell-wall (thickening of cell-wall) without the cell increasing in size or altering its form.

§ 25. **Properties of Growing-Points.** (1) It should be noticed that at growing-points the growth is greatest, not at the point where there is most rapid division of cells, but at some little distance from this point. Thus, in an apical meristem, the formation of new cells is most abundant at the apex, but the growth and increase in size of the cells takes place chiefly at some distance behind the apex. The consideration of a bud will give ample illustration of this. In the bud the internodes are extremely short. Rapid elongation takes place when the bud begins to unfold. In some cases, growth may continue for a long time in the internodes, although they are far removed from the apical meristem. Similarly, the rapid growth of the leaf is effected when unfolding, although all the cells of the leaf are present in the bud-condition.

(2) The elongation of the growing point (stem or root) is not in a straight line. As it elongates the growing point may move from side to side in a zig-zag course, or describe a spiral. The reason is that growth is not equal all round the growing point. If the growth is more rapid, first on one side and then on the other, the zig-zag movement is produced. The spiral movement is due to a wave of more rapid growth passing round the growing apex. This movement, of whatever kind, is called **nutation**. It is well seen and is of importance in twining stems.

(3) The cells at growing points are always **turgid**. There is a rapid osmosis of nutritive substances into the cells. This is always the case where metabolic processes are actively going on. The metabolic processes disturb the general equilibrium as regards the distribution of nutritive substance, and at the same time lead to the formation of osmotically active substances. Hence water with nutritive substances in solution are drawn into these cells from neighbouring cells. Turgidity is an essential condition of growth; turgidity brings about temporary changes in form which are rendered permanent by the formation of new substance.

(4) Considering this condition of turgidity, it is evident that there must be considerable pressure (or tension) in the tissues at growing points. When the epidermis is being formed as a firm and rather inextensible membrane it will be stretched to some extent by the growing cells within, and these again will be compressed owing to the inextensible nature of the epidermis. In other words, the pith, for example, tends to expand, but is prevented by the

outer tissues. Thus there is a considerable transverse tension in a stem. This is easily demonstrated by a simple experiment. Cut the growing apical region of a stem longitudinally into two halves; the two halves bend away from each other. This is due to the expansion of the pith, which will be found to present a convex surface. Again, remove a complete ring of outer tissue from a stem; on trying to replace it, it will be found that it does not go quite round owing to the expansion of the inner tissue. In stems and roots there are also longitudinal pressures or tensions, but these are not nearly so considerable as those in the transverse direction.

Phenomena Connected with Pressure. When cells become turgid they will expand in directions along which the pressure is least, and if growth is taking place these changes in form will become permanent. It is evident, then, that the question of pressure or tension in growing members has an important bearing on the study of cell-forms. It is because the pressure is relatively small in the longitudinal direction that the cells of stems or roots are usually elongated in this direction. This elongation is best seen in the elements of wood and sclerenchyma; the thickening of the walls of these elongated elements serves to meet the strains to which they are subjected. It naturally follows also that the rapid growth of a member (*e.g.* stem or root) in one direction is correlated with this relative difference in pressure.

The formation of intercellular spaces is evidently due to differences in pressure and tension in the growing cells. And several other phenomena are explained in this connection, *e.g.* the difference between spring and autumn wood, and the formation of annual rings in secondary wood. During the summer, when the cambium is active, there is a gradual increase in the transverse pressure in the stem; therefore, the autumn wood consists of narrower and more strongly thickened elements. During the winter this cambium is inactive, and the pressure is relieved. Thus there is the formation of wider elements in the following spring.

Occasionally, owing to this pressure in the secondary wood, the walls of some of the parenchymatous cells of the wood are forced through the pits into the cavities of the wood-vessels. The portion of the cell which bulges into the vessel is cut off by a wall; cell-division takes place in it, and a mass of parenchyma is formed *inside the vessel*. These masses are called **thyloses.**

§ 26. **Other Modes of Nutrition.** (1) The *free* Nitrogen of the atmosphere, although abundant, is not made use of by the green plant. There is, however, one important order of flowering plants, the Leguminosæ (the pea, bean, clover order), in which the nitrogen of the atmosphere is *indirectly* used. It was for a long time recognized that leguminous plants would readily grow in a soil containing little or no *combined* nitrogen, and that as a matter of fact the soil was often richer in nitrogen after a leguminous crop had been grown. These facts, which were at first extremely puzzling, have now been explained. Numerous small swellings or tubercles are found on the roots of these plants. When the tubercles are examined they are seen to be filled with very small unicellular bodies called **bacteroids**. By some these are regarded as of the nature of bacteria; by others as the " spores " of a fungus. The " bacterium " or fungus is always present in the soil and infects the roots of leguminous plants, thus producing the tubercles. These " bacteroids " apparently have the power of making use of the free nitrogen of the atmosphere and bringing it into combination. These compounds of nitrogen are then absorbed by the roots. Here we doubtless have an example of **symbiosis**, *i.e.* a union or association of two organisms in a common life, both deriving benefit. It is probable that, while the leguminous plant gets the benefit of the nitrogenous compounds formed, the fungus feeds on elaborated compounds (containing carbon as well as nitrogen) formed by the green plant. This symbiotic union must be clearly distinguished from parasitism (p. 14) where one organism lives *at the expense of* the other.

(2) A somewhat similar association is found in the roots of many of our forest trees, more especially members of the order Cupuliferæ, *e.g.* the beech. The roots of these trees are infested by a fungus. The threads (hyphæ) of the fungus form a dense sheath or investment at the region where the root-hairs should be. The free hyphæ projecting from the surface no doubt discharge the function of root-hairs, absorbing water with various inorganic and probably, also, organic compounds in solution. The tree gets the benefit of these while the fungus makes use of organic compounds, containing carbon, elaborated by the tree. A symbiotic association of this kind is called a **mycorhiza.**

(3) **Parasites and saprophytes** (p. 14) are not uncommon amongst the flowering plants. We can only refer to one or two of the commoner examples. Some orchids, *e.g.* the Bird's Nest Orchid, are saprophytic. They live (by means of a mycorhiza) on the decaying organic substance (humus) in woods, and have little or no chlorophyll. The carnivorous plants are examples of partial saprophytes.

The mistletoe is a partial parasite. By means of its suckers it obtains *inorganic* solutions from the wood of its host (apple, oak, etc.). But it is green and carries on the processes of elaboration in the ordinary way. The dodder is a complete parasite, living on clover, furze, heather, etc., although the *young seedling* has a root fixed in the soil. Some other common plants (*e.g.* the eyebright and cowwheat) are partial parasites. They have chlorophyll and ordinary roots, but from these roots are developed suckers (parasitic roots or haustoria), which fasten on to the roots of other plants and take up watery solutions.

CHAPTER VIII.

THE PLANT AND ITS ENVIRONMENT.

§ 1. **Irritability** is one of the fundamental properties of protoplasm; in other words, it belongs to the nature of living protoplasmic substance that it is capable of receiving impressions from, *i.e.* of being stimulated by, various external influences and of making certain responses to these stimuli. The consideration of this fundamental property opens up the whole question of the plant's relation to its environment. It is this property which brings the organism "into touch" with its surroundings. On it ultimately depends the harmony which is everywhere exhibited between a plant and its environment. The mature organs of plants, in many cases, make responses to the action of external stimuli, but such responses are best exhibited by the *growing* organs. In growth the vital activity of the protoplasm is constantly subjected to, and modified by, the stimulating influence of external agencies. The subject is a very wide one. In this chapter we can indicate only some of the more important influences on plant life.

§ 2. **Light.** The protoplasm of a green plant is in a healthy condition—a condition of *tone*—only if sufficiently exposed to light. The protoplasm loses its irritability if kept for some time in darkness, and a pathological condition is induced. It is the stimulating influence of light, however, which is quite distinct from this *tonic* influence, that we have to consider.

Variation in the intensity of light acts as a stimulus on plant members (mature or growing). This is spoken of as the **paratonic** influence of light. It is well shown by the action of light on the chloroplasts in the palisade cells of leaves. In diffuse light the chloroplasts place themselves along the outer and inner walls of the cells and are therefore as freely exposed as possible to the light. This arrangement, or rather, this movement by the chloroplasts, is called **epistrophe**. In bright light the chloroplasts place themselves along the lateral walls of the cells and are therefore more or less screened. This is **apostrophe**. The biological import of this will be evident if the student remembers that too bright light induces decomposition of chlorophyll. Again, movement is in many plants induced by the alternation of day and night. Many leaves and flowers which are freely expanded during the day close up at night. These are spoken of as **nyctitropic** movements ("sleep-movements.") We have examples in the leaves of the sensitive plant and wood-sorrel. With regard to growing members the general (paratonic) influence of light seems to be to retard the rate of growth in elongated members. We have seen a *negative* example of this in "drawn" plants (p. 140). Grown in darkness the internodes of plants become excessively elongated.

But light also exercises an influence on the *direction of growth*. This influence seems to depend, not on the varying intensity of light, but on the direction of the incident rays. Axial members tend to place their long axes parallel to the incident rays. This may be effected in two ways. The apex of the growing member may grow either *towards*, or *away from*, the light. Here we are considering the phenomena of **heliotropism**. Heliotropism may be defined as the response made by a member, as regards the direction of its growth, to the stimulating influence of light. There is **positive heliotropism** if the member turns towards the light, **negative heliotropism**, or *apheliotropism*, if the member turns away from the light. Most stems are positively heliotropic; roots are negatively heliotropic. A good example of heliotropism is seen when a plant is grown in a window. It will be noticed that, unless the plant be constantly turned round, the stem bends over towards the light. This was formerly ascribed to the retarding action of light. It was thought that the bending was due simply to the shaded side growing faster. No doubt the convex side of the stem does exhibit more rapid growth, but the explanation given is inadequate, seeing that it fails to account for the phenomena of negative heliotropism. All we can say is that these members, under the stimulating action of light, respond by tending to place their long axes parallel to the incident rays. We cannot here enter on the deeper question as to why this should be so.

The behaviour of bifacial leaves and other dorsiventral organs is different. They respond by tending to place their surfaces at right angles to the incident rays. This is called **diaheliotropism**.

With regard to the biological significance of the phenomena of heliotropism there is no difficulty. The stem, by bending over towards the light, supports the leaves in the most favourable position for receiving the light. This is seconded by the diaheliotropism of the leaf. The root, by being negatively heliotropic, has the best chance of reaching the soil.

§ 3. **Gravity.** The force of gravity also has a *stimulating* influence on the protoplasm of the growing point of stem or root. The response made is such that the root tends to grow down in the direction of the force, the stem to grow up in the opposite direction. This response made by growing members, as regards the direction of their growth, to the *stimulating* influence of gravity, is called **geotropism**. Roots are positively geotropic, stems are negatively geotropic—markedly so in twining plants.

That these opposite tendencies of root and stem are to be ascribed to gravity has been determined by experiment. There is a machine called the **clinostat** consisting essentially of a wheel mounted on a horizontal axis, and therefore rotating vertically. A plant is attached *horizontally* to the rim, and the wheel rotated *very slowly*. A little reflection will show that the normal influence of gravity is eliminated, as each side of the axis is, in turn, directed downwards. Stem and root grow in the directions in which they are placed. Another experiment is to attach a plant to a wheel which rotates *rapidly* and *horizontally*. Here another force—"centrifugal force"—somewhat similar to gravity, comes into play. If the opposite tendencies of root and stem are to be ascribed to gravity, we should expect similar tendencies to be exhibited under the action of "centrifugal force." This appears to be the case, for the root bends obliquely outwards, the stem obliquely inwards.

Lateral roots are not strongly geotropic. They tend to grow out from the parent root.

§ 4. **Hydrotropism.** The tip of the root is sensitive to moisture and responds by curving towards it. In other words the root is positively hydrotropic.

§ 5. **Contact.** It can often be observed that mechanical contact stimulates the growth of many plant organs. It is best exhibited by tendrils, root-tips, and one or two twining stems (*e.g.* Dodder). When a tendril touches some object it becomes concave at the point of contact. This is due to the stimulus being transmitted by the protoplasm to the opposite side of the tendril, there producing increased turgidity and growth of cells. More of the tendril is thus brought into contact with the object, and if the latter forms a suitable support, the process is continued and the tendril twines round it. When a growing root encounters some obstacle, *e.g.* a stone, its growth is so stimulated that it becomes convex at the point of contact and thus is turned away from the obstacle.

In the Barberry the stamens are sensitive at the base and spring up when touched by an insect. When the leaf of the Sensitive Plant is touched the leaflets close up and the whole leaf hangs down (the normal night-position—see p. 153). Here the stimulus is transmitted by the protoplasm to the pulvinus (p. 110), and drooping is due to a loss of turgidity in the cells of that region.

It is probable that in all cases the stimuli produce their effects by inducing a molecular change in the protoplasmic substance, and thus bringing about an altered condition of turgidity. In many cases the effects produced are out of all proportion to the stimulus, as in the case of the leaves of the Sensitive Plant. We can explain this only by concluding that a considerable molecular change may be set up by a stimulus, and that the protoplasm has the power of transmitting the stimulus to a considerable distance from the point of origin—thus discharging to some extent a rudimentary *nervous* function. In this connexion the continuity of protoplasm from cell to cell should be remembered.

§ 6. **Heat,** like light, has a *tonic* influence on plant growth. Vital activity only goes on within certain limits of temperature. Speaking generally, the range of temperature is from 0° C. to 50° C.

But variations in temperature also have a *stimulating* influence—cf. the paratonic influence of light.

§ 7. Adaptation. It cannot be doubted that the constant operation of external conditions has been one of the most important factors in giving rise, in the course of ages, to the wonderful adaptation to environment, which is everywhere exhibited by plants as a whole, as well as by individual plant members. Into this larger question, however, we cannot at present enter (see Chapter XVIII.). We shall here refer simply to one or two more general cases.

Consider the adaptation of an ordinary bifacial foliage leaf. Its flattened form gives it a large surface and enables it to absorb the necessary supply of carbon-dioxide. Its position is the most favourable for catching the rays of sunlight. It is covered by a protective membrane, the epidermis, which prevents excessive evaporation. Both palisade and spongy mesophyll cells, containing numerous chloroplasts, are adapted for assimilation. The palisade tissue is adapted to protect the plastids from the effects of too bright light. The spongy tissue, with numerous spaces communicating with stomata on the under surface, is specially fitted for carrying on the processes of respiration and transpiration. The spreading veins convey watery solutions to all parts of the mesophyll, and at the same time serve in the best way to support the leaf tissue. This supporting and strengthening function is also discharged by ribs or masses of sclerenchyma developed in the mesophyll.

Take again the case of a water-plant. Since absorption goes on over the whole submerged surface, the epidermis is not cuticularised. Further, as there is no transpiration, no stomata are developed (except on floating leaves, p. 52). Correlated with these features root-hairs and frequently even roots are absent, and the tracheal tissue is poorly developed. The xylem in aquatic stems (see *Myriophyllum*, p. 83) is usually central where it best meets the slight pulling strain due to the water. Support being given by the water there is little or no sclerenchyma. The leaves may be ribbon-like or much divided according as the plants grow in moving or in still water; and, as the chloroplasts need no protection, they are present in the epidermal cells and palisade tissue is absent. Finally the presence of large air-spaces facilitates the movement and diffusion of gases.

Xerophytes are plants which show special adaptations for economising their water supply. They are typically met with in dry hot sandy regions where the supply of water is uncertain and the conditions favour excessive transpiration; but low barometric pressure at high altitudes, and exposure to high winds tend to increase transpiration, while low temperature, excess of salt or humus in the soil, etc., reduce root-absorption, and thus there may be necessity for economy even in cases where water is abundant. Hence xerophytic adaptation is exhibited in varying degrees by plants in very different habitats, *e.g.* rock-plants (*lithophytes*), shore-plants (*halophytes*), plants growing on mountains at high altitudes (*alpine plants*), etc. Provision may be made for storage of water by the development of succulent stems or leaves, and for reduction of transpiration by such devices as the crowding of leaves, reduction of leaf surface, formation of a thick cuticle, coverings of hair, the protection of stomata in cavities, etc. (see Pinus, p. 298, and Appendix under Euphorbiaceæ, Ericaceæ, Chenopodiaceæ).

Plants which are adapted to live on other plants, but are not parasitic, are called **Epiphytes**. They usually develop clinging roots, which are organs of adhesion, and other roots for obtaining food-material. As their water supply is precarious they often show xerophytic characters. A ready method of seed-dispersal is an evident necessity. Epiphytes abound in tropical woods and include many orchids (p. 239).

§ 8. **Germination of a Seed.** In conclusion we give a summary of the phenomena presented in the germination of a seed, as they illustrate some of the more important processes with which we have been dealing in this and the preceding chapters. The conditions necessary for germination are:—moisture, access of air, a suitable temperature. The seed absorbs a large amount of water and begins to swell. Given a suitable temperature, chemical changes are initiated inside the seed. Ferments are produced and the digestion of storage food-material begins. Oxygen is absorbed and katabolic processes are active. The living protoplasm builds up its substance at the expense of the soluble products of digestion, and rapid growth takes place. The seed-coat is ruptured. The primary root grows out through the micropyle; being negatively heliotropic and positively geotropic it grows down into the soil, and produces branches and root-hairs. The plumule escapes from the seed, owing to the elongation of either the hypocotyl or the stalks of the cotyledons, and, being positively heliotropic and negatively geotropic, grows upwards, and comes above ground. In the presence of light, chlorophyll is developed and carbon-assimilation begins.

CHAPTER IX.

STRUCTURE OF THE FLOWER.

§ 1. **General.** The flower may be regarded as a leafy shoot highly specialized in adaptation to the performance of reproductive functions. The function of a flower is essentially to produce seed and fruit, and the various parts (stem and leaf organs) have been specially adapted to the performance of that function. It is necessary at the outset to impress this fact on the student. Stem and leaf organs in both vegetative shoot and flower have the same *morphological* value; their *physiological* value only is different. The only essentially new structures (morphologically) to be considered are certain reproductive bodies—pollen grains and ovules—which are borne on these specialized stem or leaf organs, and which are concerned in the formation of the seed.

The axis (stem portion) of the flower usually shows two regions—the **pedicel**, and the **thalamus** (also called the torus, or receptacle). The pedicel is, popularly, the stalk of the flower. It may be present or absent. If present, the flower is *pedicellate*; if absent, *sessile*. The thalamus, or torus is the portion of the axis to which the floral leaves are attached. In *typical* flowers there are four sets or series of floral leaves. To the outside are the **sepals**; collectively, they constitute the **calyx**. Internal to these are the **petals**, constituting the **corolla.** Then the **stamens** forming the **andrœcium ;** and finally, in the centre of the flower are the **carpels**, forming the **gynæceum** or **pistil**. The buttercup is a very convenient type for making a first acquaintance with these structures (fig. 92). In other flowers, owing to fusion

STRUCTURE OF THE FLOWER. 159

of the carpels, the gynæceum often shows three distinct regions—a hollow basal region, the **ovary**, and the **style and stigma** (fig. 100 and cf. fig. 92). This will be fully explained later.

The following facts support the above view of the morphological character of the flower:—(*a*) The flower, like an ordinary foliage-shoot, arises as a bud, very often in the axil of a leaf (bract).

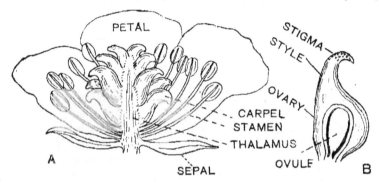

Fig. 92.—FLOWER OF BUTTERCUP.
A, Longitudinal vertical section of whole flower; B, Longitudinal section of a single carpel.

(*b*) The thalamus has the general structure of a stem, and the sepals and petals in their structure and development resemble leaves.
(*c*) While in most cases the stamens and carpels, having been highly specialized, are quite unlike leaves, there are certain conditions in which they become distinctly leaf-like. Thus, in many cultivated flowers, *e.g.* the rose, the stamens are transformed into petals; in the double cherry the gynæceum is represented by a tuft of small green leaves. In the water-lily there is a gradual transition between petals and stamens.

§ 2. **Inflorescence.** The floral, or reproductive region of the plant is usually distinctly marked off from the foliage, or vegetative region, and is known as the **inflorescence.** Sometimes the main vegetative-axis of the plant passes gradually into a single terminal flower, *e.g.* tulip and wood-anemone. Here the flower is said to be **solitary and terminal.** In other cases, the flowers are developed singly in the axils of ordinary foliage-leaves, and are called **solitary and axillary.** These are very simple types of inflorescence. Usually the flowers are aggregated on a more or less complex branch-system. According to the nature of the branching, and other points, many different kinds of such inflorescence are recognized (*e.g.* fig. 93). These will be specially considered later.

The main axis (primary axis) of the inflorescence, together with any secondary axes which may be developed (apart from the pedicels of the flowers), is called the **peduncle**. This term is applied instead of pedicel to the stalks of solitary terminal, and solitary axillary flowers. If the peduncle is an unbranched leafless axis which arises from the midst of radical leaves and bears flowers at its apex, it is called a **scape**, *e.g.* the cowslip.

§ 3. **Bracts**, etc. (fig. 93). When the flower arises as a lateral bud, the axis on which it is borne is called the **mother-axis**. This may or may not be the primary axis of the inflorescence. The side of the flower which is towards the mother-axis (or towards the growing point of the mother-axis) is said to be **posterior**; the side away from the mother-axis is **anterior**. In a solitary terminal flower it is evident that these terms are not applicable. If the flower arises in the axil of a leaf-structure, this leaf-structure is called a **bract**. Though this may be given as the strict meaning of bract, it is found convenient, in practice, to apply the term to any more or less specialized leaf-structure in the region of the inflorescence, other than the floral leaves themselves. These bracts or *hypsophylls*, present great variety of form and colour. When present, the flower is *bracteate*; when absent, *ebracteate*. The bracts may be ordinary foliage leaves, as in solitary axillary flowers, or more or less resemble them, though differing from the ordinary foliage leaves of the plant. Frequently they are small, green, and scale-like. In many plants they are reduced to small, tooth-like structures. When they

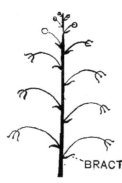

Fig. 93.—A SIMPLE INFLORESCENCE (RACEME). The main axis (peduncle) is here the mother-axis. The stalk of the flower is the pedicel.

STRUCTURE OF THE FLOWER.

are not green, but coloured like the petals of a flower, they are said to be *petaloid*. In *many* flowers, the flower-stalk bears small outgrowths of the nature of reduced leaves. These are called **bracteoles**. When present, there are *usually* two in Dicotyledons, placed laterally, and one in Monocotyledons, situated on the posterior side.

§ 4. **Perianth, or Floral Envelopes.** The outer series of floral leaves, distinct from stamens and carpels, constitute the perianth of the flower. In the great majority of flowers, the perianth consists of two series, clearly distinguished as calyx and corolla. Sometimes calyx and corolla more or less resemble each other, the sepals and petals having much the same form and colour. And not unfrequently, although two series or whorls are present, they are so closely inserted on the thalamus, and so fused together that they look like a single series. In this case the whole structure is described simply as a perianth; the terms calyx and corolla, sepals and petals are not used (*e.g.* Narcissus, lily of the valley, and many other Monocotyledons). On the other hand, the perianth may be represented by a single series or whorl. As the series of the perianth are not essential, but only accessory to the production of seed, they are frequently referred to as the *non-essential organs* of the flower. If one or both series be absent the flower is said to be *incomplete*.

If both series of the perianth are wanting, the flower is *achlamydeous*; if one only is present, *monochlamydeous*; if both are present, *dichlamydeous*. In some flowers (*e.g.* daisy, and many other Compositæ) it is recognized, by comparison with allied types, that the calyx is aborted; in such flowers the remaining series must be described as the corolla, not as the perianth. So also in cases where the corolla is aborted (*e.g.* Clematis, Anemone, and many other Ranunculaceæ), the remaining series, although petaloid, must be described as the calyx. The term perianth, however, should be used if the monochlamydeous condition is primitive, *i.e.* if it is an original or ancestral character, and not due to abortion of a second series (*e.g.* stinging nettle, oak, elm).

§ 5. **The Essential Organs.** The andrœcium and the gynæceum or pistil, because they bear the reproductive bodies necessary for the production of seed, are called the essential organs. If both are present *in the same flower* (the rule in Angiosperms) the flower is *hermaphrodite*, (symbol ☿), *bisexual* or *monoclinous*. When they are borne on different flowers, as is sometimes found, the flowers are *imperfect*, *unisexual* or *diclinous*. The unisexual flowers bearing the stamens are male (♂) or *staminate;* those bearing carpels, female (♀) or *pistillate*. If staminate and pistillate flowers are borne on the same plant (*e.g.* hazel), the plant is *monœcious*; if on different plants (*e.g.* willow and some species of campion), *diœcious*. A plant is *polygamous* if it bears staminate, pistillate, and hermaphrodite flowers (*e.g.* ash). Flowers in which both stamens and pistil have been aborted are neuter (*e.g.* ray florets of cornflower and sunflower).

§ 6. **Floral Phyllotaxis.** In most flowers the series of floral leaves are arranged in whorls, and the phyllotaxis is *cyclic*. Sometimes, however, all the floral leaves are in a spiral (*e.g.* Cactus) and the flower is said to be *acyclic*. If some of the series are cyclically arranged, others spirally, the flowers are *hemicyclic*. In the buttercup, for example, the calyx and corolla are whorled, while the stamens and carpels are spiral.

§ 7. **Number of Parts.** We may regard typical flowers as having four definite series or whorls of floral leaves—calyx, corolla, andrœcium, and gynæceum—with the same number in each series. But additional whorls may be developed in any one of these series, so that the number of floral leaves in that particular series is a multiple of the original number. This is most frequently seen in the andrœcium. On the other hand, the number in any one series may be reduced by abortion or disappearance of one or more of the parts in that series. This is very commonly seen in the gynæceum, which is the most variable part of the flower. The following examples will illustrate these points:—The violet has five sepals, five petals, five stamens, three carpels; the pea has five sepals, five petals, ten stamens, one carpel; the wall-flower has four sepals in two whorls, four petals in one whorl, six stamens in an outer whorl of two and an inner whorl of four, two carpels; many flowers have a large number of stamens in several

STRUCTURE OF THE FLOWER. 163

whorls (*e.g.* the cherry). It should be noticed that owing to the abbreviation of the thalamus and other causes, it is often difficult to distinguish the separate whorls, *e.g.* the two whorls of sepals in the wall-flower, the two whorls of stamens in the pea.

Neglecting the reduction of parts met with in particular series, and more especially in the gynæceum, we find that in Dicotyledons the series of floral leaves are, *as a rule*, arranged in twos, fours, or fives, or multiples of these numbers. In other words, the arrangement is *dimerous*, *tetramerous*, or *pentamerous*, rarely trimerous. The *trimerous* arrangement, *i.e.* in threes or multiples of three, is characteristic of Monocotyledons.

§ 8. **Alternation of Parts.** The general rule is that the leaves of the different series alternate in position with each other—the petals alternate with the sepals, the stamens with the petals, etc. If there are several whorls of stamens, these whorls alternate with each other. But there are exceptions. In spiral flowers, the parts are sometimes superposed. In cyclic flowers the departure from regular alternation arises from various causes. In the primrose order, for example, there are five sepals, five petals, five stamens, and the stamens are opposite to the petals (*antipetalous*); this is due to the suppression of an outer whorl of five stamens. Sometimes, where there are two alternating whorls of stamens, the outer whorl is opposite the petals. This (the *obdiplostemonous condition*) is due to displacement of the two whorls, which we can easily understand, remembering the abbreviation of the thalamus and the close proximity of the whorls. The carpels, owing to reduction of parts in the gynæceum, frequently have no definite position in relation to the parts of the other series.

§ 9. **Regular and Irregular Flowers.** In regular flowers the parts in each series have the same size and form, *i.e.* the sepals resemble each other, so also the petals, etc. Irregular flowers are those in which some of the floral leaves in any one series have a different shape or size from the others—for example, the petals of pea or violet.

§ 10. **Floral Symmetry** (see p. 9). Flowers may be radially symmetrical or *actinomorphic, isobilateral, zygomorphic*, or *asymmetrical*. Zygomorphy is very frequently due to irregularity, and this is the sense in which the term is used as a rule in descriptive botany; the plane of symmetry is, in most cases, antero-posterior or median (figs. 115, 116), *i.e.* it is the plane passing through the anterior and posterior sides of the flower, *e.g.* pea, violet (fig. 118), etc. Asymmetrical flowers are usually spiral, *e.g.* Cactus.

§ 11. **The Thalamus—Insertion of Floral Leaves.** The thalamus is nearly always short or abbreviated. Only occasionally is it elongated between the whorls of floral leaves. An elongation of this kind between calyx and corolla is called an **anthophore** (*e.g.* Lychnis); between the corolla and stamens, a **gonophore** or *androphore*; between the stamens and pistil, a **gynophore**. The form of the thalamus varies considerably. It may be convex and more or less dilated, or flattened, or hollow and cup-shaped. The insertion of floral leaves varies according to the form of the thalamus. In many flowers the thalamus is more or less convex, like the head of a nail. The gynæceum is developed at the apex of the thalamus; the stamens, petals, and sepals, are inserted, *in order*, on the side of the thalamus below the gynæceum. This is the **hypogynous** arrangement (fig. 94, A).

Suppose now that the thalamus is not convex, but forms a flattened circular disc. The apex of the thalamus is, of course, in the middle of the disc, and the flattened form is due to the sides of the thalamus having grown up to the same level. The gynæceum is developed in the middle of the disc, and the sepals, petals, and stamens, round the rim or margin. They are not *underneath* the gynæceum, but *round about* it. Hence this is called a **perigynous** arrangement (fig. 94, C). Sometimes the carpels are borne on a conical protuberance in the middle of the disc; this would represent a continued growth of the apex (*e.g.* strawberry or raspberry, fig. 94, B). It is with the perigynous condition that the student will

STRUCTURE OF THE FLOWER. 165

experience most difficulty; there are so many degrees of it. The thalamus may not be flat, but hollowed out, and more or less cup-like. This is due to the sides of the thalamus continuing to grow above the apex, which lies at the bottom of the cup (× fig. 94, D). The carpels (gynæceum) are developed in the cup; the sepals, petals, and stamens, from the rim of the cup. This also is a perigynous condition. It should be particularly noticed that the cup was formerly regarded as part of the calyx and called the **calyx-tube**. This term is still retained, but the student must be careful to observe that it is the thalamus. A still more extreme form of perigyny is found in the wild rose (fig. 94, E). Here there is a very deep cup. Finally, in the **epigynous** condition (fig. 94, F) the thalamus forms a deep cup as in the extreme forms of perigyny; but the carpels, as well as the sepals, petals, and stamens, are developed from the rim of the calyx-tube and simply close in the cavity. The cavity

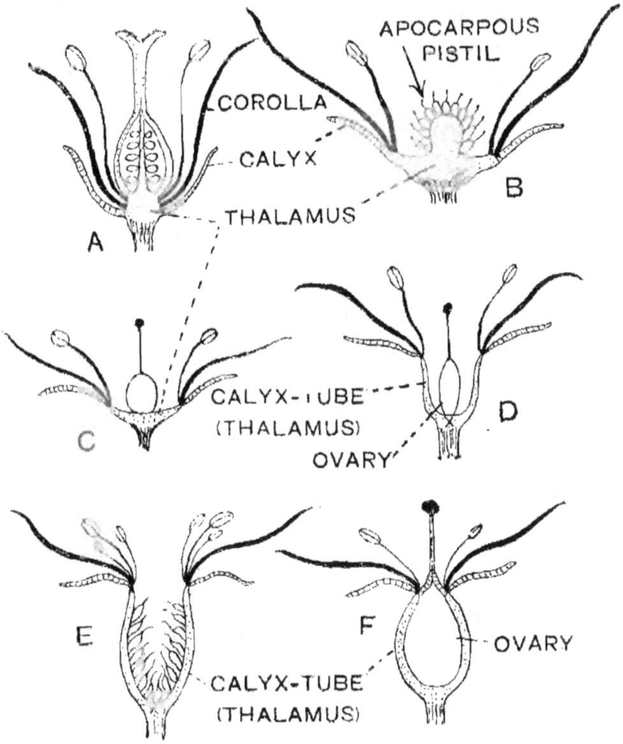

Fig. 94. THALAMUS AND INSERTION OF FLORAL LEAVES. A, Hypogynous; B—E, Perigynous; F, Epigynous. (Diagrammatic vertical sections.)

of the calyx-tube therefore becomes the cavity of the ovary (p. 159) and its wall forms the greater part of the wall of the ovary. Or, regarding it in another way, we may suppose that the carpels, developed as in the extreme forms of perigyny, are from the first adherent to the calyx tube which is for this reason considered as part of the ovary. Thus in epigynous flowers the sepals, petals, and stamens, are inserted *on* the gynæceum.* In the perigynous condition the calyx-tube remains distinct from the ovary. (Numerous examples of hypogynous, perigynous, and epigynous flowers are given in Chapter XIII.)

§ 12. **The Calyx** forms in most flowers a protective structure, serving to keep all the other parts of the flower together. If the sepals are free, *i.e.* not coherent or fused

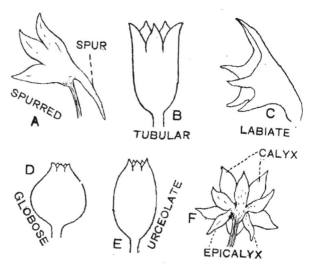

Fig. 95.—Forms of Calyx.

laterally, the calyx is **polysepalous** (*e.g.* buttercup or wallflower). If there is lateral fusion, however slight it may be, the calyx is **gamosepalous** (*e.g.* primrose) In a polysepalous calyx the outline of the individual sepal is

* The student should notice that, strictly speaking, in epigyny the gynæceum is not formed simply from carpels.

described in the same terms as are used for ordinary foliage leaves. The number of sepals in a gamosepalous calyx is usually indicated by divisions or teeth (*e.g.* fig. 95, B). If the divisions pass almost to the base of the calyx, it may be described according to their number as 3-, 4-, 5-*partite*; if about half way down, 3-, 4-, 5-*fid*; if the divisions are small, 3-, 4-, 5-*toothed*. But there are special descriptive terms with which the student must be familiar :—

The sepals are *spurred* if they are prolonged downwards into a tubular process, *e.g.* garden Nasturtium (fig. 95, A); *saccate* if pouched or dilated at the base, *e.g.* many Cruciferæ (p. 226). The calyx is *galeate* or *hooded* if one or more of the sepals form a helmet-shaped structure, arching over the other parts of the flower, *e.g.* the monkshood, fig. 96. The gamosepalous calyx is *tubular* (fig. 95, B) if regular and elongated with nearly parallel sides; *campanulate*, if regular and more or less bell-shaped (fig. 99, E); *infundibuliform* or funnel-shaped, if it widens gradually from a narrow base (fig. 99, B); *urceolate* (fig. 95, E), if expanded in the middle and narrowing towards base and apex; *globose*, if shorter and almost globular (fig. 95, D); *bilabiate* (fig. 95, C), if irregular and drawn out on each side so as to form two lips (*e.g.* calamint).

Fig. 96.—FLOWER OF MONKSHOOD. CALYX GALEATE AND PETALS REDUCED. (One sepal removed.)

In many flowers the sepals are extremely small, or aborted altogether (many Umbelliferæ and Compositæ). In many Compositæ (dandelion, thistle, cornflower) a rudimentary calyx is present in the form of a series of fine hairs developed on top of the ovary outside the corolla. This is called a **pappus** (fig. 99, A). The calyx is *petaloid* if the sepals are not green, but coloured like petals. It is *caducous* if the sepals fall off just before the flower opens (*e.g.* poppy); *deciduous* if after the flower has opened; *persistent* if they remain till the fruit is formed (pea, violet). In all hypogynous and perigynous arrangements the calyx is described as *inferior*; in the epigynous flower the calyx is described as *superior*.

In some flowers, *e.g.* the strawberry, the sepals are stipulate. The stipules fuse in pairs between the sepals, and produce an outer series of small sepal-like structures, forming what seems to be an outer calyx. This is known as the **epicalyx** (fig. 95, F). An epicalyx may also be produced by the aggregation of bracts or bracteoles beneath the calyx, *e.g.* the mallow and sweet-william.

§ 13. **The Corolla** is, in most cases, an attractive structure; its function is *usually* to attract insects to the flower in connection with the process of pollination. It may be *polypetalous* or *gamopetalous* (cf. calyx), *regular* or *irregular*, and as it to a large extent determines the symmetry of the flower, the terms *zygomorphic*, *actinomorphic*, are applied to it. According to the insertion of petals the corolla is described as hypogynous, perigynous, or epigynous. In a polypetalous corolla the outlines of the individual petals are described in the same terms as are used for the foliage leaf, and, as in the calyx, the gamopetalous corolla may be described as 3-, 4-, 5-partite, -fid, or -toothed. The following are some of the more special terms:—

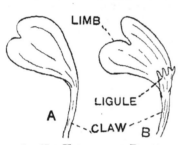

Fig. 97.—Unguiculate Petals. B also ligulate.

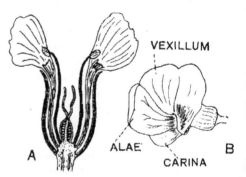

Fig. 98.—A, Vertical section of a Caryophyllaceous Flower; B, Papilionaceous Flower.

The petals are *unguiculate* or clawed (fig. 97) if they show a distinction into a stalk-like basal portion, the *claw*, and an expanded upper part, the *limb* (*e.g.* wall-flower); *ligulate* if ligules are developed at junction of claw and limb (*e.g.* pink—fig. 97

STRUCTURE OF THE FLOWER.

B); one or more of the petals may be *spurred* (*e.g.* violet) ; the petals are *fimbriated*, if they bear a fringe of hair-like processes (*e.g.*, mignonette). The following special terms are applied to **polypetalous corollas** :—*Cruciform* where the corolla consists of four unguiculate petals, arranged crosswise, *i.e.* in the diagonal planes of the flower (*e.g.* wall-flower and Cruciferæ generally, see fig. 103, A) ; *rosaceous* (fig. 94, B, C) if it consists of five spreading petals, not clawed, and attached perigynously (Rosaceæ) ; *caryophyllaceous* (fig. 98, A) if it

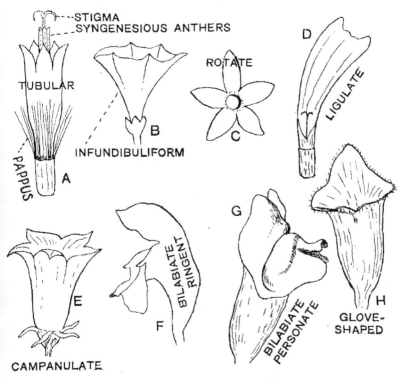

Fig. 99. —FORMS OF GAMOPETALOUS COROLLA.
A, Tubular hermaphrodite, D, Ligulate pistillate, florets of a Composite.

consists of five clawed petals, with spreading limbs attached hypogynously to the thalamus inside a slender tubular calyx (pinks and many Caryophyllaceæ) ; *papilionaceous* (from the supposed resemblance to a butterfly) if it consists of five petals, one large—the vexillum or standard, two lateral—alæ or wings, and two fused to form a boat-shaped structure, the carina or keel (pea and British Leguminosæ generally, fig. 98, B). **Gamopetalous corollas** may be *tubular* (fig. 99, A), *campanulate* (harebell, fig. 99, E), *infundibuliform* (fig. 99, B); *urceolate* (purple heather); *globose*; *bilabiate* (for these terms see under calyx p. 167); *bilabiate and ringent* if the two lips gape apart (dead nettle, fig. 99, F); *bilabiate and personate*, if the two lips are closed up (snap-dragon, fig. 99, G); *glove-shaped*, if like the finger of a glove (*e.g.* foxglove, fig. 99, H), *hypocrateriform*, or salver-shaped, if it has a *long, slender* tube and a spreading limb (*e.g.* primrose, fig. 100); *rotate* if there is a spreading limb and a very short tube (*e.g.* forget-me-not, fig. 99, C); *ligulate* if it consists of an elongated membrane which represents the one-sided development of a short basal tubular part (*e.g.* dandelion and many Compositæ, fig. 99, D).

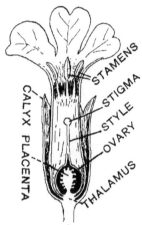

Fig. 100.—VERTICAL SECTION OF FLOWER OF PRIMROSE. COROLLA HYPOCRATERIFORM.

The petals are usually brilliantly coloured, sometimes green (sepaloid). They may be absent, *e.g.* lady's mantle and some Ranunculaceæ (Clematis, Anemone), or reduced to nectar-secreting structures, *e.g.* monkshood (fig. 96) and Christmas rose. In the latter they form a number of nectar cups situated between the calyx and stamens.

A perianth, when not distinguished into calyx and corolla, is described in much the same way; but the terms *polyphyllous* and *gamophyllous* are used to indicate the free and coherent condition, respectively, of the perianth leaves.

§ 14. **The Corona.** This is the term applied to the whole series of ligules developed on the corolla or perianth of certain flowers. In Narcissus, where the perianth is gamophyllous, the ligules are coherent, and the corona is cup-shaped.

§ 15. **Prefloration.** This has already been referred to (see p. 122). The prefloration of the perianth (or calyx and corolla) only can be studied. The ptyxis, or folding of the individual floral leaves, is described in the same terms as are used for foliage leaves (see p. 123). The æstivation of calyx or corolla (fig. 89) may be *valvate, imbricate,* or *contorted (twisted)*. *Induplicate* * æstivation is a form of valvate in which the margins of the floral leaves are folded inwards on themselves. *Quincuncial* * aestivation is a form of imbricate where there are five leaves (sepals or petals), two internal, two external, and one partly internal, partly external. *Vexillary* æstivation, characteristic of the corolla of Leguminosæ, is another form of imbricate æstivation (fig. 119). The æstivation may be recognized either by taking transverse sections of young flower-buds, or carefully removing the young floral leaves one after the other.

§ 16. **The Andrœcium.** A typical stamen (fig. 101) consists of three parts—**filament, anther,** and **connective,** The filament is the stalk of the stamen corresponding to the petiole of a foliage leaf, while the anther represents

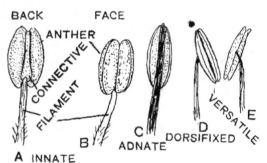

Fig. 101.—STAMENS, SHOWING INSERTION OF ANTHERS.

the highly modified lamina of the floral leaf. It usually consists of two *anther-lobes* (sometimes one), and forms a case or box in which are contained the **pollen-grains** or essential reproductive bodies. These lie in four cavities, the **pollen-sacs** (fig. 102), of which there are two in each anther-lobe. When the anther dehisces the partition between the two pollen-sacs in each lobe breaks down, so that there seems to be a single loculus in each lobe. This fusion often takes place much earlier in the development of the anther. The anther-lobes are

* Although not mentioned on p. 123 valvate *vernation* may be induplicate, and imbricate *vernation,* quincuncial

connected towards the back of the anther by a strip of tissue containing a vascular bundle. This is the *connective*. It is usually narrow, so that the anther-lobes lie close together, but may be elongated so that the lobes are widely separated, *e.g.* sage. Sometimes special appendages are developed on stamens. These generally arise as outgrowths of the connective. In the violet there is a membranous orange-coloured outgrowth on top of each anther, and, in addition to these, the two antero-lateral stamens have each

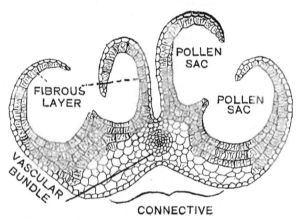

Fig. 102.—Transverse Section of Anther of Wall-flower, after Dehiscence.

a green elongated process (functioning as a nectar gland) passing down into the spur of the anterior petal (fig. 118).

Barren or rudimentary stamens are called **staminodes**. They may consist simply of filament or be represented by various peculiarly modified forms.

§ 17. **Descriptive Terms.** If the stamens are free from each other, *i.e.* not coherent, the andrœcium is **polyandrous** (diandrous, triandrous, pentandrous, etc., according to the number). If fused, the fusion may be of two kinds. (*a*) The stamens cohere by their filaments: this is the **adelphous** condition—**monadelphous**, if all fused to form a tube round the pistil, **diadelphous** if fused in two groups, *polyadelphous* if in several groups. The monadelphous condition is found for example in the mallow and some Leguminosæ (*e.g.*

broom); the diadelphous, in other Leguminosæ (*e.g.* the pea) where, of the ten stamens, nine are fused and the tenth is free; the polyadelphous, in the St. John's Wort and orange. (*b*) The stamens cohere by their anthers, the filaments being free. This is characteristic of Composites (*e.g.* daisy, dandelion, thistle, etc.), some Solanaceæ (*e.g.* the bitter-sweet and potato), etc. It is the **syngenesious** or synantherous condition (fig. 99, A).

The stamens may be hypogynous, perigynous, or epigynous; but sometimes, although they always arise on the thalamus, they become adherent to, or are carried up by, the corolla (or perianth) during its development. They appear then to be developed on the petals, and are said to be **epipetalous** (*epiphyllous*, if on a perianth). This is found in many gamopetalous, or gamophyllous, orders of Angiosperms (Compositæ, Labiatæ, etc., fig. 100). Sometimes the stamens are adherent to the gynæceum, *e.g.* orchids; this is the *gynandrous* condition (see fig. 154 and Chapter XIII., p. 240).

Where the stamens in a flower have different lengths, special terms are sometimes applied to the andrœcium. Thus, in the order Cruciferæ (wall-flower, stock, etc.), there are four long and two short stamens (fig. 103, B), and the

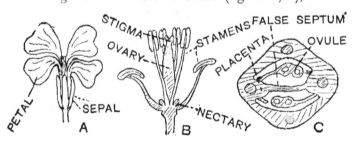

Fig. 103.—FLOWER OF A CRUCIFER.
A, Entire—cruciform corolla; B, Sepals and petals removed—tetradynamous stamens; C, Transverse section of ovary.

andrœcium is said to be **tetradynamous**. In Labiatæ (*e.g.* dead-nettle) and Scrophulariaceæ (*e.g.* foxglove), where there are two long and two short stamens, it is **didynamous**. These are the only common orders in which these terms are used.

§ 18. Insertion and Dehiscence of the Anthers.

The attachment of the anther to the filament should be noticed (fig. 101). It is *innate* or *basifixed* if the anther is fixed directly on top of the filament; *adnate* if the connective is well marked, and there is no articulation of the filament to the base of the anther, so that the filament seems to run up the back of the anther; *dorsifixed* if the filament is attached to the back of the anther and the anther is immovable; *versatile* is similar, but the anther swings on the filament.

Usually each anther-lobe bursts open (dehisces) longitudinally. According as the anther-lobes face inwards (towards the centre of the flower) or outwards, the anthers or the dehiscence is said to be *introrse* or *extrorse*. The dehiscence may be transverse, as in some Labiatæ, or, as in heaths and some Solanaceæ, it may be effected by pores at the apices of the anther-lobes. The dehiscence is brought about by the contraction of the reticulately thickened cells of the **fibrous layer,** the inner of the two layers forming the wall of the loculus (fig. 102). The line of longitudinal dehiscence lies between the two pollen-sacs of the anther-lobe.

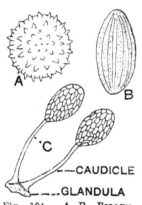

Fig. 104.— A, B, Pollen-grains (highly magnified); C, Pollinia of an Orchid (p. 240).

§ 19. The pollen,

in most plants, forms a loose, dusty powder, consisting of a large number of minute grains (fig. 104, A, B). The grains vary much in size, form, and colour, in different plants. At first (fig. 105) they are unicellular and the wall consists of two membranes or coats. The outer coat, the *exine,* is cuticularized and frequently ornamented with protuberances, spines, etc. The inner coat, the *intine,* is thin and consists of cellulose. In some plants, *e.g.* orchids, the pollen-grains are not loose, but aggregated into a single mass called a **pollinium** (fig. 104, c).

The development of the pollen-sacs and pollen-grains will be described later.

STRUCTURE OF THE FLOWER.

§ 20. **The Gynæceum or Pistil**, consisting of carpels, forms the inner essential organ of the flower. It is the part of the flower which has been most extensively and completely modified. The student, indeed, finds it difficult at first to realize that it consists of modified leaves. It is necessary, therefore, that he should read very carefully the description which follows, making sure that he fully understands the exact significance of the terms used.

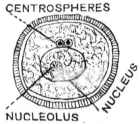

Fig. 105.—YOUNG POLLEN-GRAIN OF A LILY. (Section.)

The gynæceum may be **monocarpellary** or **polycarpellary**,* that is, it may consist of one or of several carpels. In the latter case, according to the number, it is bicarpellary, tricarpellary, etc.

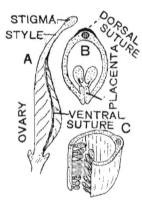

Fig. 106.—THE MONOCARPELLARY PISTIL.

A, Entire ; B, Transverse section of ovary ; C, Will indicate method of folding.

§ 21. **The Monocarpellary Pistil** (fig. 106). The student must *imagine* that during development the single carpellary leaf is folded on itself, so that its margins are brought together and become fused. In this way a cavity or space is enclosed. The slender apex of the carpel elongates and terminates in a little dilatation or knob. The hollow basal portion is the **ovary**; it contains a varying number of oval or rounded bodies, the **ovules**, which *afterwards* develop into seeds. The slender elongated process on top of the ovary is the **style**; the terminal knob is the **stigma**. Thus, a single modified carpellary leaf has formed a pistil or gynæceum showing three distinct parts—a simple ovary, style, and stigma. The student must care-

* Notice—" poly-" here has not the same significance as in polypetalous, etc.

fully distinguish the meanings of the terms carpel, gynæceum, ovary.

If we examine the ovary, we find that the ovules are *marginal, i.e.* they are developed on the fused margins of the carpel. The fused margins usually swell up to form a longitudinal ridge or cushion on the inner surface of the ovary wall. This cushion is called the **placenta**, and to it the ovules are attached. Seeing that the placenta is on the wall of the ovary, the **placentation** (= the position or arrangement of placentas in an ovary) is **parietal**. Usually, however, in the *simple* ovary the placentation is simply described as marginal. The monocarpellary pistil is easily recognized by the presence of this single placenta. The pistil of the Leguminosæ (pea, bean, etc.), is an excellent example.

§ 22. **The Polycarpellary Gynæceum.** Of this there are two conditions, according as the carpels do or do not fuse with each other. If there is no fusion, *each* carpel forms a simple ovary, style and stigma, like the single carpel of the monocarpellary pistil. This is the **apocarpous** condition (fig. 92 and cf. *polysepalous, polypetalous, polyandrous*). Here, while there is a single gynæceum or pistil in the flower, there is a number of simple ovaries. The number indicates the number of carpels. The placentation is marginal. Frequently, only one ovule is developed in each loculus (many Ranunculaceæ and Rosaceæ); if the ovule is attached to the top of the loculus it is *pendulous;* if to the bottom, *ascending.*

In the second condition all the carpels fuse together to form a single *compound* (or polymerous) ovary, and the pistil is **syncarpous** (cf. gamopetalous, gamosepalous, etc.). The fusion may or may not be complete. If complete, the ovary bears a single style and stigma (fig. 107, A), and it is only by the internal structure of the ovary that the number of carpels can be determined. If incomplete, a number of styles or stigmas are borne on the single ovary (fig. 107, B, C), owing to the apices

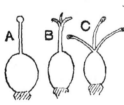

Fig. 107.—THE SYNCARPOUS PISTIL.
(To indicate different degrees of fusion.)

STRUCTURE OF THE FLOWER.

of the carpels remaining free. It is evident that the number of styles or stigmas gives the number of carpels. Thus, in Compositæ (fig. 99, A) the style is single, but there are two stigmas; hence we know that the pistil is bicarpellary.

The structure of the ovary and the placentation in the syncarpous pistil differs in different cases. The following conditions should be carefully noticed:—

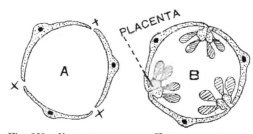

Fig. 108.—FORMATION OF THE UNILOCULAR OVARY OF A TRICARPELLARY PISTIL.
X indicates points of fusion. (Placentation parietal: Transverse section.)

(a) The carpels simply fuse by their adjacent margins (fig. 108) to form a unilocular ovary. The fused margins swell up to form placentas bearing ovules. The placentation is marginal and parietal. The number of parietal placentas indicates the number of carpels.

(b) The carpels are folded on themselves before fusing, or we might say that the fused margins run in to the middle of the ovary (fig. 109). Thus a multilocular ovary is formed, and the marginal placentas of all the carpels fuse in the centre to form a central or axile column. The placentation is marginal and **axile**.

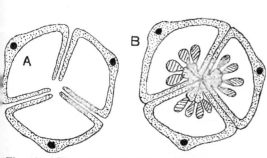

Fig. 109.—FORMATION OF A TRILOCULAR OVARY WITH AXILE PLACENTATION.
(Transverse section.)

The number of loculi, or the number of septa by which the ovary is divided, indicates the number of carpels (except where *false* septa are formed—see below).

Sometimes only one ovule is developed in each loculus. It is *suspended* if it comes off from the placenta high up and hangs down in the loculus (fig. 110); but sometimes there is no distinct axile placenta—the ovule being either *ascending* or *pendulous* (p. 176).

In the ovary of the poppy there is an intermediate condition between (*a*) and (*b*). The septa, which are covered with ovules and are therefore placentas, do not reach the middle of the ovary. The ovary is unilocular, but partially divided. The placentation is parietal.

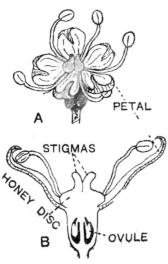

Fig. 110—UMBELLIFEROUS FLOWER. B, Vertical section.

(*c*) The carpels fuse by their adjacent margins and the ovary is unilocular as in (*a*). But the ovules are not developed on the carpellary margins. They are borne on a central axis running through the middle of the ovary. The placentation is **free-central**. In typical cases (Primulaceæ, fig. 100) the central axis is a prolongation of the thalamus into the ovary. The ovules are developed on the axis of the flower, not on the carpels. There are a few orders, however (*e.g.* Caryophyllaceæ) in which the free-central placenta is derived from an originally axile placentation by the breaking down of the septa.

Basal placentation is a modification of typical free central. Here a single ovule is inserted on the floor of the ovary. It is developed on the thalamus, which, however, is not prolonged as an axis into the ovary (*e.g.* Polygonaceæ, fig. 112, and Compositæ, fig. 111).

The ovules are, in rare cases, developed, not on the margins of the carpels, but over the whole inner surface, *e.g.* the flowering rush (apocarpous) and water-lily (syncarpous). This is called **superficial** placentation.

§ 23. **True and False Septa.** *True* septa, or dissepiments, are those which represent the inturned margins of carpellary leaves. Septa formed in any other way, *e.g.* as ingrowths from the surfaces of the carpels, are *false*. Thus, in the ovary of Cruciferæ (fig. 103, c) there are two parietal placentas; during development, a false septum grows up from the base of the ovary between the placentas.

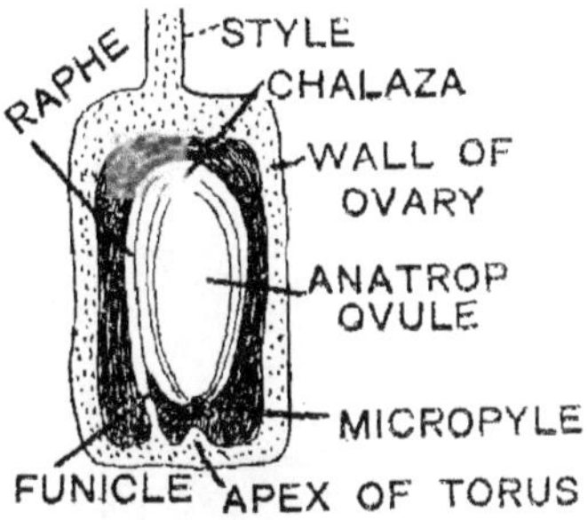

§ 24. **Superior and Inferior Ovaries.** In all hypogynous and perigynous conditions, the ovary is described as *superior*; in the epigynous condition, as *inferior*. It might seem out of place to describe the ovary as superior, and the calyx inferior, in such a perigynous condition as is shown in fig. 94, D.

Fig. 111.—INFERIOR OVARY OF A COMPOSITE.
(Longitudinal section.)

But the student must remember that the ovary here is developed at the *organic* apex of the thalamus. As already explained, the wall of the *inferior* ovary is largely formed from the thalamus (calyx-tube); the ovules may be developed from the thalamus, as in the basal ovule of Compositæ; but where the placentation is parietal, or axile, we must suppose either that the carpellary margins grow down into the cup of the thalamus, or that a polymerous ovary, such as we might find in a perigynous condition, has become completely adherent to the thalamus or calyx-tube.

§ 25. **Structure of Ovule.** If we examine a typical ovule (fig. 112), we find that it is attached by a slender stalk—the **funiculus,** or **funicle**—to the placenta. The body of the ovule consists of a mass of parenchymatous tissue, called the **nucellus,** invested by two cellular coats or **integuments.** These integuments arise from the base of the nucellus, and completely invest it, except at the apex, where a small canal or passage is left leading down to the apex of the nucellus. This passage is called the **micropyle.** In most Angiosperms there are two integuments; the outer, called the *primine,* the inner, the *secundine*; but in many (chiefly gamopetalous Dicotyledons), there is only one. The base of the nucellus, from which the integuments arise, is called the **chalaza**. The point where the body of the ovule is attached to its stalk, or funicle, is the **hilum**.

Towards the micropylar end there is a large, specially developed cell. This is the **embryo-sac**. In the embryo-sac the protoplasm is arranged as in an ordinary parenchymatous cell. There is a lining, or parietal layer and protoplasmic strands. Vacuole and cell-sap are present. In addition to this, however, several *cells* are present in the embryo-sac, formed, as will afterwards be explained (Chapter XVII.), by free cell-formation. At the micropylar end there are three cells without cell-walls constituting the egg-apparatus. The largest one is the **oosphere, ovum** or **egg-cell**; the two smaller ones are called the **synergidæ**, or **help-cells**. At the other end there are three cells with cell-walls. These are the **antipodal cells**. Embedded in the protoplasm in the middle of the embryo-sac there is a large nucleus, called the **secondary** or **definitive nucleus** of the embryo-sac.

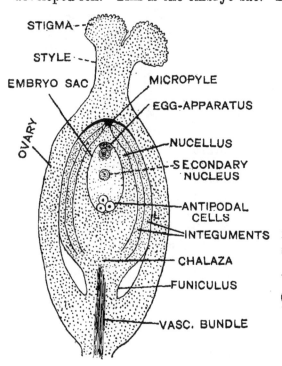

Fig. 112.—Ovary and Basal Orthotropous Ovule of Polygonum.
(Longitudinal section.)

The above gives the general structure of the fully-formed ovule, at the time when fertilization is just about to take place. The development and earlier condition of the ovule will be described later (Chap. XVII.).

§ 26. Forms of Ovule.

There are several important forms of ovule to be noticed. The typical form is the *straight* or **orthotropous** ovule (fig. 112). Here the ovule is perfectly straight, not curved or bent in any way. The chalaza and hilum lie close together, and the micropyle is at the extreme apex. In the *inverted* or **anatropous** form (fig. 113), the body of the ovule has bent over during development, and fused for some distance with the stalk or funicle. This fused portion of the funicle is called the **raphe**. In this form the micropyle and hilum lie close together, and the chalaza is towards the other end.

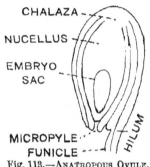

Fig. 113.—Anatropous Ovule. (Longitudinal section.)

In the **curved** or **campylotropous** form (fig. 114, B) the body is curved and bent round, so that the micropyle lies near the funicle; but there is no fusion with the funicle. Hilum, chalaza, and micropyle, all lie close together. The **amphitropous** ovule is an intermediate form in which the body of the ovule is straight, but has been twisted round, so that its long axis is at right angles to the funicle (fig. 114, A). Of these forms the anatropous is most frequently met with. Examples of the campylotropous ovule are found in many

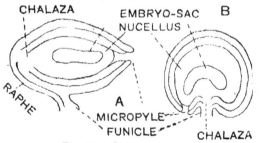

Fig. 114.—Forms of Ovule.
A, Amphitropous; B, Campylotropous.

Cruciferæ (wall-flower, etc.) and Leguminosæ (pea, bean, broom, etc.). The orthotropous ovule is less frequently found, *e.g.* Polygonum (fig. 112). The Primulaceæ and some Cruciferæ give examples of the amphitropous ovule.

§ 27. **Cohesion and Adhesion.** The student must be clear as to the meaning of these terms. *Cohesion* is fusion between members of the same series of floral leaves. Thus gamosepalous, polysepalous, polyandrous, syngenesious, apocarpous, syncarpous, are terms signifying cohesion or want of cohesion. *Adhesion* means fusion between members of different series, as when the stamens are epipetalous. The terms *hypo- peri- epi- gynous* applied to the stamens or petals, *superior* and *inferior* applied to ovary and calyx, are also terms of adhesion or want of adhesion.

§ 28. **Modification of Floral Structure.** The *primitive* typical flower, we may suppose, was regular, and showed no cohesion of parts. The countless diverse modifications now existing may be ascribed to the operation of various processes of which many examples are suggested in the foregoing pages. The chief of these are:—*abbreviation* of the floral axis; closely connected with this the *cohesion* or *adhesion* of parts, and *displacement*, as where the positions of two whorls of stamens are reversed; *branching* or *splitting* (chorisis) of parts, as in the inner whorl of stamens in the Cruciferæ, where the two pairs of stamens (fig. 103, B) are due to splitting of two originally single ones; *reduction* in the number of parts by abortion—this may be partial or incomplete, as where the petals are reduced to nectaries; *development of irregularity*, due to hypertrophy of certain parts—this is closely connected with the pollination of flowers by insects (as will afterwards be explained). The student should notice examples of these processes. At the same time he should recognize the general principle underlying all this—namely, the more or less perfect adaptation of the flower to the function which it has to carry on in relation to the conditions of its environment.

§ 29. **Vertical Sections and Floral Diagrams.** The general structure and arrangement of parts in a flower may be shown in drawings of vertical sections (figs. 94, 100,

STRUCTURE OF THE FLOWER. 183

110, B), in floral diagrams and floral formulæ. The **floral diagram** may be described as a ground-plan of the flower showing the relation of the parts to each other *and to the*

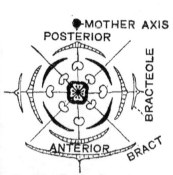

Fig. 115.—FLORAL DIAGRAM OF A REGULAR TETRAMEROUS FLOWER.

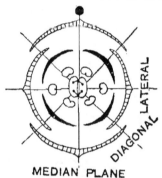

Fig. 116.—FLORAL DIAGRAM OF CRUCIFERÆ.

mother-axis (fig. 115). In making a floral diagram the student must clearly distinguish the *antero-posterior* or median, the *lateral*, and the *diagonal* planes (fig. 116). Cohesion of parts may be indicated by connecting lines (fig. 117, A), but this may be done in the floral formula which should accompany the diagram. The æstivation also may be indicated as in fig. 118. An *empirical* diagram (fig. 117, B) is one showing only the relative positions of the

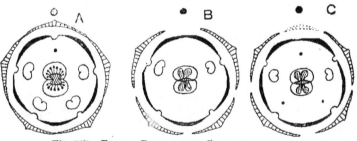

Fig. 117.—FLORAL DIAGRAMS OF SCROPHULARIACEÆ.
A, Common form; B, Empirical diagram of Speedwell; C, Theoretical diagram of Speedwell.

parts actually present. A *theoretical* one indicates, as well, by means of dots, the relative positions of parts originally present, but now lost (fig. 117, C).

The student will experience most difficulty in indicating the positions relative to the mother-axis. It will help him if he remembers that in most Dicotyledons one sepal is *posterior*. An exception is found in the order Leguminosæ (fig. 119) and there are also exceptional cases where the

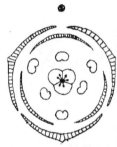

Fig. 118.—Floral Diagram of Violet.

Fig. 119.—Floral Diagram of Leguminosæ. (Monadelphous form.)

Fig. 120.—Typical Floral Diagram of Monocotyledons. (e.g. Lily.)

posterior sepal has been aborted (fig. 117, B, C). Fig. 120 shows the characteristic arrangement in typical Monocotyledons.

The **floral formula** indicates the number, cohesion, and adhesion of parts. Certain symbols must be noted. $P=$ perianth; $K=$ calyx; $C=$ corolla; $A=$ andrœcium; $G=$ gynæceum; ∞ signifies indefinite in number; brackets ( ) enclosing the number of parts indicate cohesion; $\frown$ indicates adhesion; $\downarrow=$ anti-position.* Thus:—

$$K_{(5)} \; C\overset{\frown}{_{(5)}} \downarrow A_5 \; G(\underset{\frown}{\infty})$$

reads gamosepalous calyx of five sepals, gamopetalous corolla of five petals, andrœcium of five free epipetalous stamens superposed on the petals (antipetalous), a polycarpellary syncarpous pistil of an indefinite number of carpels with a superior ovary. A line *above* the symbol for gynæceum would mean 'inferior ovary.' The floral formulæ of many common orders are given in Chapter XIII

* The antipetalous condition of stamens might be indicated by writing A_{0+5}.

CHAPTER X.

THE INFLORESCENCE.

§ 1. **The Inflorescence** is the floral region of the plant as distinguished from the vegetative. Its simplest form is the solitary terminal flower. Usually it is a more or less complex branch-system, the branches bearing flowers instead of developing into foliage shoots. Inflorescences are best classified according to the type of branching, and the special modifications in each case. Probably in all the branching is lateral, and they are either (*a*) *indefinite* or *racemose*, (*b*) *definite* or *cymose*. In the former the growing point has an indefinite power of growth; it never ends in a flower, although the actual number of lateral flowers which it produces may be few or many. In cymose inflorescences the primary axis and the successive daughter axes in turn end in flowers. It is characteristic of racemose inflorescences that the youngest flowers are always found towards the apex, or, where a cluster of flowers is formed, towards the centre (*centripetal*); while, in compact cymose inflorescences, the youngest flowers are towards the outside, *i.e.* away from the centre (*centrifugal*). This is why the terms centripetal and centrifugal are sometimes used for the two kinds of inflorescence.

§ 2. **Simple Racemose Inflorescences.** Of these, according to special modifications in each case, we recognize four chief types :—

(*a*) **The typical raceme** (fig. 93). Here the mother-axis (peduncle) is elongated, and the flowers are stalked. Examples are found in the lily of the valley, foxglove, and hyacinth. Similar to this in essential characters is the **corymb**, which may be regarded as a modification of the

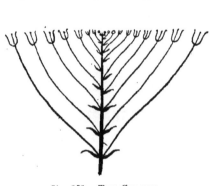

Fig. 121.—The Corymb.

Fig. 122.—A Typical Spike.

typical raceme. The mother-axis is relatively shorter, and, owing to the elongation of the lower pedicels, all the flowers come to one level (fig. 121). Good examples are found in many Cruciferæ (*e.g.* candytuft). Inflorescences intermediate in character between the corymb and typical raceme are described as corymbose racemes (*e.g.* wall-flower).

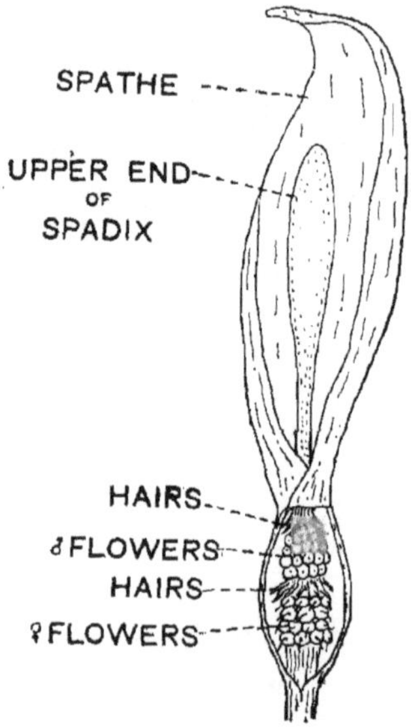

Fig. 124.—A Catkin.

Fig. 123.—Spadix of Arum. Part of lower end of spathe removed to expose the flowers.

(*b*) **The spike** is a racemose inflorescence in which the mother-axis is elongated, and the flowers are *sessile* (fig. 122), *e.g.* Spotted Orchid and Plantago. There are one or two special forms of the spike. The **spadix** is a massive fleshy spike, bearing usually unisexual flowers. It is protected by a large enveloping leaf, sometimes green, more usually petaloid, known as a *spathe*.* The arum lily (fig. 123) is a good example. The **catkin** or *amentum* (fig. 124) is a long, more or less pendulous, deciduous spike, bearing unisexual flowers. It is found in many nut-bearing and other trees, *e.g.* birch, hazel, poplar.

Fig. 125.—THE SIMPLE UMBEL.

(*c*) **The umbel** (fig. 125) is a racemose inflorescence in which the flowers are stalked, but, owing to the abbreviation of the mother-axis, are all given off at one level. There is an indefinite growing point which throws off a large number of lateral flowers, but does not give rise to an elongated mother-axis. We may suppose this to have been derived from the raceme or corymb by compression, all the daughter-axes being brought

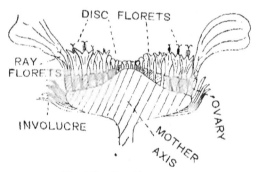

Fig. 126.—THE CAPITULUM.
(Vertical section.)

to one level, just as the flower itself may be described as a compressed shoot, in which, owing to the shortening of the internodes, all the whorls of floral leaves lie close together.

(*d*) **The capitulum** (fig. 126) is a racemose inflorescence in which the flowers are sessile, and crowded together on a reduced or abbreviated mother-axis (peduncle). We may

* The spathe is usually called a bract—the term bract being used in the wider sense (see p. 160).

suppose it to have been derived from the spike in the same way as the umbel from the raceme. The mother-axis is called the *disc* or *receptacle*.* It is sometimes flat, more frequently dilated and convex. Examples are found chiefly in the Compositæ (daisy, dandelion, etc.). The student must clearly recognize that the heads of the daisy, dandelion, etc., are not single flowers, but inflorescences bearing a large number of sessile flowers. The capitulum is invested by a number of small, scaly, overlapping leaves (barren bracts). The protective investment thus formed is called the **involucre**.

§ 3. **Cymose Inflorescences.** These are either *uniparous*, *biparous*, or *multiparous* (see p. 64). In the uniparous forms each successive axis ends in a flower after producing one daughter-axis. They may be *helicoid* or *scorpioid*, are *sympodial*, and sometimes resemble typical racemes (see fig. 39, B, D). Uniparous cymes resembling racemes can be distinguished by the fact that the bracts, if they are present, are on the opposite side of the sympodial axis from the leaves. If bracts are not developed they are not easily distinguished. In the *biparous* cyme each axis ends in a flower after producing two daughter-axes. It is also called the **dichasium** or *false dichotomy* (fig. 127). Typical examples are found in many Caryophyllaceæ. Sometimes the daughter-axes

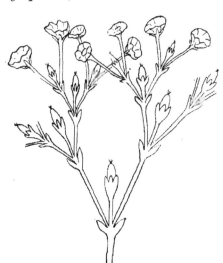

Fig. 127.—DICHASIUM OR BIPAROUS CYME.

* The term receptacle is ambiguous, being also applied to the thalamus. The student must clearly distinguish the two applications of the term.

are not given off at the same level, *e.g.* some buttercups, Christmas rose, etc. In the *multiparous* cyme, a whorl of daughter-axes is given off before the mother-axis ends in a flower. Here a cymose umbel is formed, clearly distinguished from the racemose or typical umbel by the fact that the oldest flower is in the middle.

§ 4. **Compound and Mixed Inflorescences.** Many inflorescences have not the simple characters above described, and present considerable difficulty to the student. In practical work—and it should be remembered that book knowledge is worth little or nothing here—he should begin by carefully recognizing the simple forms before proceeding to the analysis of the more complex.

Many inflorescences are **compound**, *e.g.* a raceme of racemes, a spike of spikelets (rye grass), an umbel of umbels.

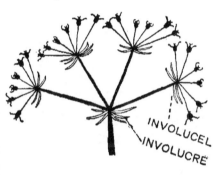

Fig. 128.—THE COMPOUND UMBEL.

The *panicle* is a compound irregularly branched raceme. The *compound umbel* (fig. 128) is frequently found in the order Umbelliferæ; here the bracts at the base of the chief branches constitute the *involucre*—the smaller bracts at the base of each secondary umbel, the *involucel*. The inflorescence in the elder is a compound multiparous cyme, in which some of the branches given off are larger than others.

Many inflorescences are **mixed**. We may, for example, have a raceme of spikes, a raceme of capitula, a spike of capitula, a raceme of cymes, etc. The *panicle of spikelets* is a common form in many grasses (*e.g.* the oat); in the horse-chestnut there is a raceme of short cymes and the inflorescence is called a *thyrsus*. In the lilac the inflorescence is of the same nature, but the branching is much more copious. Some times the name *racemose* or *panicled thyrsoid cyme* is applied to such an inflorescence.

§ 5. **Special Forms.** There are many inflorescences which, owing to abbreviation of axes or special crowding of the flowers, do not so readily yield to careful analysis. In the hawthorn, for example, the inflorescence might be mistaken for a typical corymb. Examination will show, however, that the lateral axes borne on the main axis are really cymes. It is a *corymbose cyme*. In the cultivated geranium* and many species of Narcissus the inflorescence, at first sight, appears to be an umbel. But it will be found that the young flowers are not by any means aggregated towards the centre, and that the flowers are arranged in a number of groups. These are really cymose clusters. We may speak of the whole inflorescence as an *umbellate cymose head*. These are found in many plants. In Narcissus the inflorescence is protected by a membranous *spathe*. In the dead nettle, and many other members of the Labiatæ, the leaves are opposite and decussate, and at each node there seems to be a whorl of flowers. These *apparent* whorls are called **verticillasters**. Careful analysis shows that there is an inflorescence in the axil of each leaf; it is a dichasium of scorpioid cymes, *i.e.* a biparous cyme which passes on either side into a uniparous form by suppression of one of the branches at each branching. It is difficult to recognize this because the axes have been reduced and the flowers are sessile. It is easily recognized in many Labiatæ where the flowers have short stalks. In fig. 129 the axis which ends in flower 1 gives rise to two daughter-

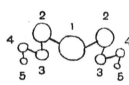

Fig. 129.—Diagram indicating the relation of flowers in half of a verticillaster.

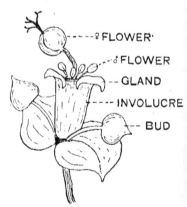

Fig. 130.—Cyathium of Euphorbia.

* Which is really a Pelargonium.

axes, 2, ending in flowers. Each of these gives rise to an axis, 3, and so on. In the sweet-william and some other plants there is a copiously branched biparous cyme, in which the axes are short and all the flowers crowded together. This clustered form of cyme is called a *fascicle*.

The **cyathium** (fig. 130) is a peculiar inflorescence found in Euphorbia (the spurge). There is a cup-shaped involucre the margin of which bears a number of crescent-shaped glandular scales. Inside the cup there are a number of stamens; also a gynæceum borne on a stalk. The whole structure looks like a single flower; but each stamen is really a male or staminate flower, and the gynæceum with its stalk is a female or pistillate flower. This is borne out by the fact that each stamen is articulated to a stalk and has a scaly bract at its base.

CHAPTER XI.

REPRODUCTION AND LIFE-HISTORY OF THE ANGIOSPERM.

§ 1. Hitherto we have been dealing with the individual plant—its structure, nutrition, and growth. We have now to see how the plant reproduces its kind and perpetuates the species, for that is the ultimate object of its existence. In connexion with the reproductive processes we shall consider the general course of development—in other words, the life-history of the plant.

§ 2. **Vegetative Reproduction** (p. 15). The essential feature in all vegetative reproduction is that the more or less specialized part which is separated off from the vegetative region of the parent *directly* develops into a new plant *resembling the parent*. The part thus separated off has different forms in different cases, but in all cases in the higher plants it either consists of a bud, more or less specialized, or bears one or more buds. The bud, being either buried in the soil, or in contact with the soil, sends down roots from its base, while above ground it develops into a shoot. The development, at first, is dependent on stored food-material.

Copious vegetative reproduction is found amongst the Angiosperms, and it takes many forms. Sometimes specialized buds are developed for this purpose, *e.g.* bulbils (p. 68), while in many plants, if a shoot happens to be buried in the soil, the ordinary buds in the axils of the leaves send down roots, and develop into shoots which become separated. Vegetative reproduction by means of stolons, runners,

REPRODUCTION AND LIFE-HISTORY OF THE ANGIOSPERM. 193

suckers, tubers, bulbs, corms, rhizomes, etc., has already been referred to in Chapter IV. As a rule, vegetative reproduction does not lead to any very wide distribution of a plant.

§ 3. **Reproduction by Seed.** This is by far the more important method, tending, as it does, not only to maintain the vigour and vitality of the species, but also to secure its more extended distribution. We shall find that in the formation of the seed a sexual process (sexual reproduction, p. 15) takes place in the plant, similar in all essential points to that which is found in animals. The processes which in Flowering Plants lead up to the formation of seed and fruit are very complicated, and we must now consider them in detail.

§ 4. **Pollination.** In order that seed may be produced, the pollen grains must be transferred from the anthers to the stigma. The meaning of this will be explained in what follows. This transference of the pollen grains is called **pollination**. It may be effected in various ways, and thus we have different kinds of pollination. There may be (*a*) **self-pollination**; or (*b*) **cross-pollination**.* In the former, the pollen grains simply fall, or are transferred in some way, from the anthers to the stigma or stigmas *of the same flower*. In the latter, they are carried in various ways to the stigma or stigmas *of other flowers*, either on the same plant, or on different plants of the same species. As the great majority of Angiosperms have hermaphrodite flowers, one would perhaps naturally expect, in most cases, to find self-pollination. As a matter of fact, however, cross-pollination is by far the commoner method, although self-pollination takes place much more frequently than is usually supposed. This being so, we must believe that there is some considerable advantage accruing to the species from cross-pollination. It has been determined, by direct observation and experiment, that stronger seed is often formed from cross-pollinated flowers. This is more especially the case where

* The terms self-fertilization (**Autogamy**) and cross-fertilization (**Allogamy**) are commonly used. The student, however, must carefully distinguish between pollination and fertilization.

Bot.

the pollen is transferred, not simply from one flower to another on the same plant, but from one plant to another. Now we may correlate this with the known facts of heredity. In the formation of seed, a sexual process takes place, and we know that in sexual reproduction the offspring inherits characters from both sides. In self-pollination there is simply the mixing of practically similar characters, while in cross-pollination there is a mixing, to a greater or less extent, of dissimilar characters. This we know by observation to be of advantage to a plant or animal.

Cross-pollination may be effected by various agencies. Thus, the pollen may be transferred by means of the wind, water, or animals, and the flowers are said to be **anemophilous, hydrophilous,** or **zoophilous** respectively. Grasses, meadow rue, nettle, are good examples of anemophilous flowers. Hydrophilous flowers are found in some water plants. While we have examples of flowers being pollinated by means of such animals as slugs, snails, humming-birds, etc., the animals thus employed are, in nearly all cases, insects (flies, moths, bees, etc.). Such flowers are said to be **entomophilous.** The great majority of Angiospermous flowers are such. Here we recognize an intimate interrelation existing between the plant and animal kingdoms.

§ 5. **Contrivances and Conditions favouring Cross-pollination.** There are in flowers many forms, conditions, and mechanisms, the significance of which becomes clear only on the view that cross-fertilization is advantageous, and that the flowers have become adapted to it. Such mechanisms, etc., either entirely prevent self-pollination, or tend to do so, and are distinct adaptations for cross-pollination. Neglecting cases where the position of anthers and stigmas prohibits autogamy, we shall consider some of the more striking conditions. In many plants cross-pollination is absolutely necessary if seed is to be produced, owing to the flowers being *diclinous* or unisexual. We have this condition in its extreme form where the staminate and pistillate flowers are on different plants (willow). **Dichogamy** is a condition in which, though the flowers are hermaphrodite, the anthers and stigmas come to maturity at different times.

and which when completely developed entirely prevents self-pollination. There are two kinds of dichogamy:— (*a*) **protandry**, where the anthers ripen first; here the pollen grains are transferred to an older flower: (*b*) **protogyny**, where the stigma ripens first; here the pollen grains are transferred to a younger flower. Protandrous flowers are much more common than protogynous. Examples of the former are found in Compositæ, Labiatæ, etc.; of the latter in Arum, sweet vernal grass, Plantago, Paeony, etc. Most wind-pollinated flowers are protogynous.

Anemophilous, entomophilous, and hydrophilous flowers have each special characters of their own, so that as a rule we can distinguish them at a glance. In anemophilous flowers the pollen is produced in great abundance, as much of it must be wasted; the flowers are small and inconspicuous; there is no honey or perfume; and frequently the stigmas are branched and feathery, to catch the pollen grains. The greatest variety, however, is shown in entomophilous flowers; there is no difficulty in recognizing these as being the most highly specialized. As a rule they have large, conspicuous, or highly-coloured corollas, or are arranged in conspicuous inflorescences; they usually secrete honey and give out perfume. The pollen is not produced in any great abundance, as the provision for its transference is more perfect. The bright corollas, the perfume and honey, serve to attract insects. To understand this, the student must remember that insects do not visit flowers for nothing. They go in search of food. In return for the service rendered by insects the flowers sacrifice part of their nutritive substance in providing food to the insects (honey and pollen), and make a further sacrifice of energy in developing certain structures (corollas) which will attract the insects. By instinct insects associate these attractive structures with the presence of a supply of food-material.

Many entomophilous flowers are further characterized by the presence of ingenious mechanical devices, which guide and control the movements of the insect and turn them to the best account. Thus, in many cases, the corolla is so modified that the insect must alight on the flower or enter in a special way (*e.g.* Labiatæ, Leguminosæ); the same result may be attained by the secretion of nectar into special receptacles or spurs (*e.g.* violet); often the insect, on entering a flower, pushes against special processes or outgrowths which move the stamens and bring the anthers in contact with its body (sage); or the stamens may be jerked, and the pollen scattered over the insect's body. The general result of all these devices is that the insect receives the pollen on a special part of its body, and when it enters another flower the pollen is deposited on the stigma. In many protandrous flowers this is secured by the style bending over so that the stigma is in the position formerly occupied by the stamens. A very special, but at the same time very simple, arrangement for making the best use of the insects is the condition known as **heterostyly**. It is seen in the primrose. Here there are two types of flower borne on different plants. One kind has long stamens (with anthers in the throat of the corolla tube) and a short style; the other has a long style and short stamens; thus in the two types the positions of anthers and stigma are simply reversed. Evidently pollination will be most readily effected by transference between these two forms and not between two flowers of the same form; and experiment has proved that the best seed is produced when this is the case. In the primrose there are two kinds of flower; this is the dimorphic form of heterostyly. In Lythrum there are three, *i.e.* flowers with long, short, and medium stamens; this is the trimorphic form.

§ 6. **Special Examples of Floral Mechanism.** (*a*) In *Viola tricolor* (pansy) the pollen is shed on the surface of the anterior petal, which serves as a landing place for the insect and is prolonged downward into a spur receiving the honey secreted into it by the nectaries developed on the antero-lateral stamens (fig. 118). The stigma is rounded and hollow. It has a little flap or valve opening downwards (fig. 131). The insect entering the flower opens the valve and deposits pollen received from another flower. Pushing down into the spurred petal, the insect receives a fresh supply of pollen; but as it leaves the flower it closes the valve and self-pollination is prevented. (*b*) In many leguminous flowers

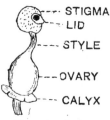

Fig. 131.—Pistil of Pansy.
(From Green.)

(fig. 98, B) honey is secreted at the bottom of the tube of stamens. The essential organs are enclosed in the keel, and thus protected. Leguminous flowers are usually visited by bees. The bee alights on the alæ. These are articulated with the keel. When the bee, seeking for the honey, pushes down the alæ, the keel also is depressed and the essential organs liberated. There are modifications in detail, but generally the stigma comes out first, and, if there is pollen already on the insect's body, cross-pollination is effected. Self-pollination may also take place.

§ 7. **Special arrangements for Self-Pollination.** Many annual plants cannot afford to undertake the risks and sacrifices attendant on cross-pollination and are commonly self-pollinated (*e.g.* Groundsel, Chickweed). They have small flowers, often without honey or smell, and are either **homogamous**, that is, their anthers and stigmas mature at the same time, or so slightly dichogamous that self-pollination is secure. Even in flowers evidently adapted for cross-pollination there is commonly the possibility of self-pollination as a final resource. Many of them are distinctly dichogamous, but not completely so, there being usually a short period during which self-pollination becomes possible. To effect this there are sometimes special contrivances such as the curling back of the stigmas to reach the pollen (*e.g.* Compositæ, Campanulaceæ). A very special adaptation for self-pollination is the production of **cleistogamous flowers**. These are closed flowers produced late in the year by certain plants which had previously produced entomophilous flowers, *e.g.* the sweet violet, the wood-sorrel, *Lamium amplexicaule* (one of the dead-nettles), etc. In these plants the ordinary entomophilous flowers frequently fail to produce seed. The cleistogamous flower is inconspicuous and apetalous. The calyx never opens, and the stamens and pistil are developed in a closed case.

§ 8. **Germination of the Pollen Grain. Processes leading up to and ending in Fertilization.** At first the pollen grain is unicellular (fig. 105), but later, even before it leaves the anther, its nucleus and protoplasm divide to form two cells (fig. 132, A). One of these, the **generative cell**, is small and lies freely in the protoplasm of the larger or **vegetative cell**. Either before or after pollination it divides again into two generative cells. There are no cell-walls between these cells. Further development takes place on the stigma. The vegetative cell bursts the exine at a point where it is thin and grows out into a very slender pollen-tube (fig. 132, B). The pollen-tube grows down through the tissue of stigma and style and finally enters the ovary. Its development and growth is at first stimulated and directed by chemical substances contained in the stigma. This is an example of **chemiotaxis** or sensitiveness to chemical stimuli. Its growth through the tissue of stigma and style is effected by ferment action and closely resembles that of the hypha of a fungus. After entering the ovary the pollen-tube is guided in various ways towards an ovule, which it enters usually by the micropyle. It pierces the apex of the nucellus and comes in contact with the embryo-sac near to the oosphere and synergidæ. The generative cells together with the nucleus of the vegetative cell have by this time passed down to the apex of the pollen-tube. One generative cell only is concerned in the actual process of fertilization. It passes from the pollen tube into the embryo-sac and fuses with the oosphere. In this process the apex of the pollen tube and the wall of the embryo-sac at that particular point are both absorbed to allow the passage of the generative cell. The synergidæ assist in the process, hence their name synergidæ or "help-cells" (Gr. συν, with, εργον, work). This fusion of the protoplasm and nucleus of the generative cell with those of the oosphere constitutes fertilization in the strict sense. It is evidently a sexual union similar to that which occurs in animals. The oosphere

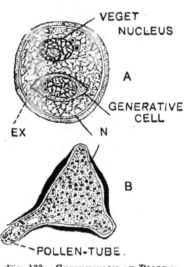

Fig. 132.—Germination of Pollen-grain.

REPRODUCTION AND LIFE-HISTORY OF THE ANGIOSPERM. 199

is the female cell or gamete; the generative cells are male gametes (see p. 15). During the process the vegetative nucleus is disorganised. The fate of the other generative cell is described on p. 304. The fertilized oosphere forms a cellulose wall and is then called the **oospore** (p. 40).

§ 9. **Modifications of the Process.** In a very few Dicotyledons, *e.g.* the hazel and birch, the pollen-tube does not enter the ovule by the micropyle, but by piercing the chalaza or base of the ovule. This is known as **chalazogamic** fertilization as distinguished from the usual or **porogamic** method, and we speak of the plants as chalazogams and porogams respectively.

Sometimes the embryo-sac protrudes into the micropyle, so that the pollen-tube comes in contact with it immediately on entering the micropyle.

§ 10. **Development of the Embryo and Endosperm.** The stimulus of fertilization induces changes in the embryo-sac and ovary, and leads to the development of the seed and fruit. The embryo is developed from the oospore. After fertilization the synergidae disappear. The oospore first of all divides into two cells, an

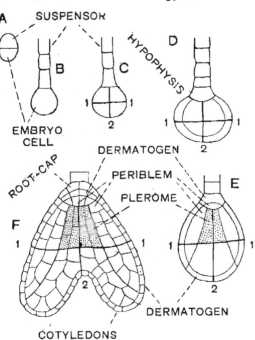

Fig. 133.—DEVELOPMENT OF DICOTYLEDONOUS EMBRYO
(*Capsella bursa-pastoris*, THE SHEPHERD'S PURSE).
A, First division of oospore. Only two of the octant-walls (1, 2) can be shown.

upper and a lower (figs. 133, 134). The upper (hypobasal) cell further divides by a series of divisions parallel to the first and gives rise to a row or filament of cells called the

suspensor. The lower (epibasal) cell is attached to the end of this and divides by three walls at right angles into eight cells (octants). Four of these are posterior (next the suspensor), four anterior. This little mass of tissue is called the **embryonal mass.** The suspensor becomes attached to the wall of the embryo-sac. While this development is going on other changes take place in the embryo-sac. The secondary nucleus begins rapid karyokinetic division, and produces a large number of small nuclei embedded in the protoplasm of the embryo-sac. The protoplasm aggregates round these nuclei to form protoplasts (p. 18) and finally cell-walls are laid down between them. Thus by a process of free cell-formation (p. 38), starting from the secondary nucleus, a tissue is formed in the embryo-sac. The cells of this tissue become laden with food-material (starch, oil, aleurone grains, etc.) built up from soluble compounds which diffuse into them from the placenta. The nutritive tissue thus formed *in* the embryo-sac is the **endosperm.**

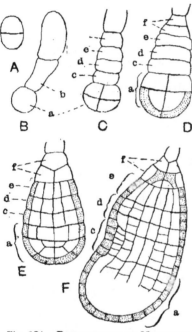

Fig. 134.—DEVELOPMENT OF MONOCOTYLEDONOUS EMBRYO.

a is the Embryonal Cell (in B) forming the embryonal mass (in C, D, E, F); it gives rise to the cotyledon. *c, d, e, f,* are formed by division of *b*, the terminal cell of suspensor. *c* gives growing-point of stem; *d, e,* give hypocotyl; *f,* the growing-point of root (cf. fig. 137).

As the embryonal mass increases in size the various parts of the embryo are gradually differentiated. In most **Dicotyledons** (fig. 133), the *terminal* plumule and the two cotyledons are derived from the four anterior octants, the hypocotyl from the posterior octants: the growing point of the radicle—in other words, the primary root—is derived from the terminal cell of the suspensor, called the **hypophysis cell**. The marking out of plerome, etc., can be readily followed in fig. 133. In the majority of **Monocotyledons**, on the other hand, a large part of the embryo is derived from the suspensor. The embryonal mass simply gives rise to a single large *terminal cotyledon*. The primary root and the primary stem, which here arises *laterally*, are developed from the cells formed by division of the terminal cell of the suspensor (fig. 134).

§ 11. **Modifications of the above Processes.** In some plants no suspensor is formed at all, and the whole of the embryo is derived from the rounded mass formed by the segmentation of the oospore. In other plants large, *massive* (not filamentous) suspensors are produced. In a few plants, more especially when the embryo sac is large, the endosperm is not produced by free cell-formation, but by ordinary cell-division of the embryo-sac, *i.e.* the secondary nucleus divides into two, and then a wall is laid down dividing the embryo sac into two cells, in each of which the process is repeated.

§ 12. **Formation of Seed and Fruit.** All this time the embryo-sac with its developing contents has been increasing in size. As this proceeds, the nucellus becomes gradually crushed and disorganized, and finally disappears. The integument or integuments of the ovule become dry and firm, and form the seed coat surrounding and protecting the embryo and endosperm. The testa is derived from the outer integument. When an inner coat, the endopleura (see p. 57), is present, it is derived from the inner integument. In all seeds at an early stage endosperm is present. If the embryo remains small, and the endosperm persists, the fully-formed seed is *albuminous* (most Monocotyledons, and many Dicotyledons). In many Dicotyledons, however, and in a few Monocotyledons (*e.g.* orchids) the embryo absorbs the endosperm tissue while the seed is ripening; in this case, the endosperm disappears, and the embryo is large. These are *exalbuminous* seeds. In a very few cases the nucellus is not completely disorganized, but, like the endosperm tissue, becomes laden with food-material. The nutritive tissue thus formed *outside* the embryo-sac, and therefore quite distinct from the endosperm, is called **perisperm** (*e.g.* water lily and pepper).

The seed, then, is the highly specialized reproductive structure which is formed in Flowering Plants by developmental changes induced in the ovule by the stimulus of fertilization. But the student has now to notice that these induced or stimulated developments are not confined to the ovule, but extend to other parts. Processes of *secondary growth* are set up in the ovary and the neighbouring parts of the flower. The entire result of these changes constitutes the **fruit** in the wide sense of the term.

The function of the fruit is to protect the seed, and to secure its proper dispersal at the right time.

§ 13. **Development of the New Plant.** A period of rest, previous to germination, usually follows the complete development of the seed. This period may be short or long. Many seeds can retain their vitality for many years, but if this period is indefinitely prolonged, the vitality is, sooner or later, lost. Thus the development of the oospore into the adult plant takes place in two stages—one inside the seed, leading to the formation of an embryo; the other when germination takes place, and the embryo develops into the adult plant (see p. 157).

§ 14. **Suspensor and Cotyledons.** We have seen that the endosperm contains a supply of food which is made use of (sooner or later) by the embryo. In nearly all cases the cotyledons are the absorbing organs. In exalbuminous seeds this absorption is effected during the ripening of the seed, and this is why the cotyledons become so large and fleshy. In albuminous seeds it is effected at germination, and explains why the cotyledons are in this case usually hypogeal; they remain in the seed to absorb the endosperm. In all cases the absorption is effected by the action of enzymes (p. 146). Only occasionally does the suspensor act as an absorbing organ. Its function is simply to push the embryo down into the endosperm. Sometimes, however, when the the suspensor is massive, it sends out processes which pierce the nucellus and integuments, bury themselves in the placenta, and take up food material (some orchids).

CHAPTER XII.

SEED AND FRUIT.

A. THE SEED.

§ 1. The general structure of the seed having been described and illustrated in pp. 55-60 it simply remains to indicate a few peculiarities which will be apt to present themselves and give rise to difficulty.

§ 2. **Form and Position of Embryo.** The structural characters of seeds are made out by means of sections (Appendix, § 11). There is little or no difficulty in this where the embryo is straight, as in the examples given in pp. 55-60. But there are many seeds in which the embryo is not straight, but curved or twisted in various ways. The cotyledons also may be *folded*. Thus a section may pass through radicle and cotyledons in various ways (*e.g.* fig. 141, c). In albuminous seeds there is an additional element of difficulty in the fact that the embryo has different relations to the endosperm. The student can overcome these difficulties only by diligent practical work. He should carefully study the directions given in the Appendix —also figs. 135-138.

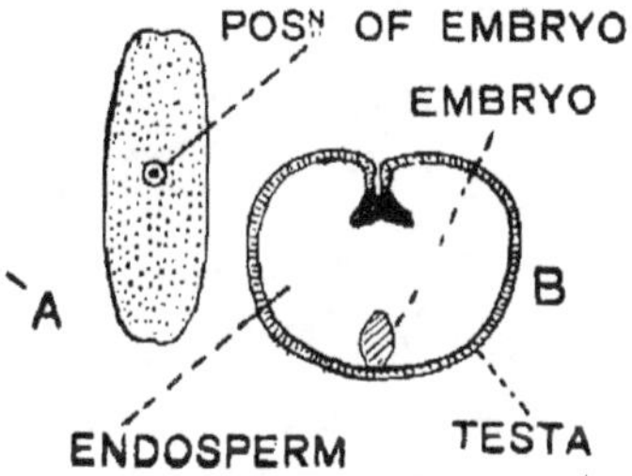

Fig. 135.—A, SEED (STONE) OF DATE; B, TRANSVERSE SECTION THROUGH EMBRYO.

The **brazil-nut** is a seed in which the shell is the testa. It is exalbuminous. The embryo consists almost entirely of a huge *hypocotyl* laden with food-material (oil and aleurone grains), the cotyledons being rudimentary and the plumule and primary root represented only by growing points.

In *most* Monocotyledons the plumule is very minute. Its position is recognized by a tiny notch at the base of the

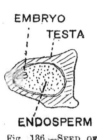

Fig. 136.—SEED OF MONKSHOOD. (Longitudinal section.)

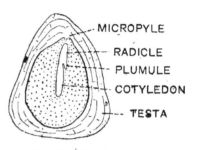

Fig. 137.—SEED OF A LILY. (Longitudinal section.)

large terminal cotyledon (fig. 137). But see also maize, oat, etc. (figs. 37, 139 c).

§ 3. **Aril and Caruncle.** Sometimes *after fertilization* an additional investment is formed on the seed called the *aril*. It may be developed from either the funicle or the micropyle, and is usually fleshy, but may have other forms. In the spindle-tree it is fleshy and micropylar in origin; in the willow and poplar it is hairy and funicular. The mace of the nutmeg is an aril developed from both micropyle and funicle. Smaller growths on the seed are called *caruncles*, e.g. the pansy, in which the caruncle is formed at the hilum, the castor-oil (fig. 138) and spurge, where it is formed at the micropyle; the tuft of hairs on the seed of the willow-herb is also of this nature. Most botanists apply the term aril to *all* growths or investments formed on the seed-coat after fertilization.

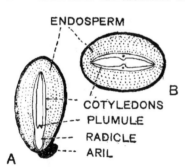

Fig. 138.—SEED OF CASTOR-OIL. A, Longitudinal section; B, Transverse The dark outline is the testa.

B. The Fruit.

§ 4. Definitions. The fruit in the wide sense is the entire result of secondary growth induced in the ovary and neighbouring parts of the flower by the stimulus of fertilization. Botanists, however, have distinguished between *true* fruits and *false*, or spurious, fruits. A *true fruit* is one which consists only of the ripened ovary and nothing else. It has been defined as a "pericarp containing seeds," the pericarp (see p. 55) being the case developed from the altered ovary. A *false fruit* is one in the formation of which other parts of the flower have a share, *e.g.* the thalamus or perianth leaves. False fruits are called **pseudocarps**. This distinction, though in some respects convenient, is essentially artificial. In the classification given below true and false fruits will frequently be found grouped together.

§ 5. Classification of Fruits. Fruits have so many forms and modifications that it is necessary to classify them. Many different classifications have from time to time been framed, and even yet there is none which may be considered quite satisfactory. Fruits may be **simple**, **aggregate**, or **composite**. A *simple fruit* is one which is formed from a *single* flower in which the pistil is monocarpellary or syncarpous, *e.g.* the pod of the pea, the capsule of the poppy. An *aggregate fruit* is one which is formed from a *single* flower in which the gynæceum is apocarpous. Here each carpel (or rather ovary) gives rise to a fruitlet, and the fruit therefore consists of an aggregation of fruitlets. A *composite fruit*, on the other hand, is formed from an *inflorescence*, not from a single flower. Here all the flowers increase in size, become aggregated together, and form a single mass. These composite fruits are called *syncarps*, and, of course, are all pseudocarps.

The **Simple Fruits** are further subdivided into **dry** and **succulent** according as the pericarp is dry and firm or more or less fleshy and juicy. The *dry simple* fruits are either **achenial, capsular,** or **schizocarpic**; the *succulent simple* fruits may be **drupaceous, baccate,** or **pomes**. The *aggregate* *fruits* are collections of one or other of these simple forms. The *syncarps* have peculiarities which distinctly mark them off from these others. We will now proceed to describe examples of these various forms.

§ 6. **Achenial Fruits.** Achenial fruits may be defined as *dry, indehiscent, one-seeded* fruits. The term indehiscent means that the pericarp does not naturally burst open to allow the seed to escape. The pericarp and testa are both

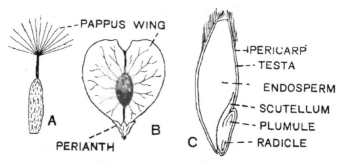

Fig. 189.—ACHENIAL FRUITS.
A, Cypsela with pappus; B, Samara of Elm; C, Caryopsis of oat.
(C, Longitudinal section—bract and bracteole removed.)

ruptured when the embryo begins to develop at germination. According to special peculiarities there are different kinds of achenial fruits :—

(*a*) The **typical achene** is formed from the superior ovary of a monocarpellary pistil in which the pericarp and testa are free from each other. It is difficult to give familiar examples of the simple typical achene; perhaps the best is that of the lady's mantle (p. 230). But, as will be shown presently, many aggregate fruits consist of collections of these.

(*b*) The **cypsela** (figs. 34, 139 A) is an inferior achenial fruit. It is formed from the inferior ovary of a pistil consisting of more than one carpel (usually two). The cypsela is the characteristic fruit of the Compositæ (sunflower, daisy). In many forms this fruit is crowned by a persistent hairy *pappus* (p. 167) which serves to disseminate the fruits. The whole structure would constitute a pseudocarp (*e.g.* the dandelion, groundsel, thistle, corn-flower). The cypsela is also found in other plants, *e.g.* the valerian, where it has a pappus.

(*c*) **The caryopsis** (figs. 37, 139 c) differs from the typical achene in that the testa and pericarp are fused together, forming one membrane. It is formed from a superior ovary. The caryopsis is the characteristic fruit of the Grasses (*e.g.* oat, maize, wheat, barley—see p. 244). In some the true fruit or caryopsis is invested by the persistent bract and bracteole, and the whole structure is therefore a pseudocarp, *e.g.* the oat.

(*d*) The **samara**, or *winged* achene, is an achenial fruit in which the pericarp is prolonged into a membrane or wing which plays an important part in the dispersal of the fruit, *e.g.* the ash (fig. 36) and elm (fig. 139, B).

(*e*) The **glans** or **nut** is an achenial fruit in which the pericarp is hard and leathery, or woody, forming a shell. It is formed from a multilocular ovary (with several ovules) in which only one loculus and one ovule develop, the others being aborted. Typical examples are found in the hazel, oak, beech, edible chestnut. In these typical examples the nuts are invested by a hard or membranous structure called the **cupule**, derived from the fusion of bracteoles developed beneath the flower. Sometimes the cupule encloses one nut, sometimes several. The whole structure would constitute a false fruit. The cup or cupule of the acorn, the membranous " husk " of the hazel-nut, are well known. In the edible chestnut two nuts are enclosed in a spiny cupule, and in the beech there are usually two triangular nuts enclosed in a cupule which is almost closed and also slightly spiny. The student must be careful to distinguish this structure from the capsules presently to be described. He should also notice that many structures, popularly called nuts because they have a hard shell, are not really nuts. Thus the "brazil-nut" is a seed (derived from a capsular fruit). The walnut is part of a drupaceous fruit (p. 212).

§ 7. **Capsular Fruits.** These are *dry, dehiscent, many-seeded* fruits. The term dehiscent means that the fruits break open naturally to allow the seeds to escape. There are different kinds of capsular fruits.

(a) The **legume**, or *pod* (fig. 76, B), is a dry, dehiscent, many-seeded (*i.e.* a capsular) fruit, formed from the ovary of a monocarpellary pistil. The pod splits open, both in front and behind, into two halves, or valves, which separate from each other (Examples: —pea, bean, broom, laburnum, and Leguminosæ generally).

Fig. 140.—Collection (Etærio) of Follicles of Monkshood.

(b) The **follicle** resembles the legume very closely. It differs only slightly in form, and in the fact that it opens on one side only, either in front or behind, but not both (fig. 140). There is no common example of the simple follicle; but many aggregate fruits, as will be explained presently, consist of aggregations of follicles.

(c) The **siliqua** is a characteristic fruit of the Cruciferæ, *e.g.* wall-flower and stock. It is developed from the ovary of a bicarpellary pistil, in which there are two parietal placentas and a *false* septum stretching between them, so that the ovary is bilocular. It is a long, cylindrical fruit, and, in dehiscing,

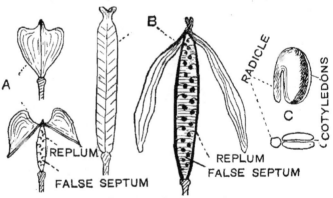

Fig. 141.—Fruit and Embryo of Cruciferæ.
A, Silicula; B, Siliqua; C, *One* form of embryo, entire and in section.
A and B show the dehiscence.

the two walls of the loculi break away from the two placentas and false septum, and hang freely suspended from the apex of the fruit (fig. 141, B). Thus the two placentas

SEED AND FRUIT. 209

are left behind, forming a two-ribbed framework called the **replum**, across which the false septum stretches. The seeds are, of course, exposed on this structure.

(*d*) **The silicula** (fig. 141, A) may be regarded as a very short, flat form of siliqua. It has the same essential structure and mode of dehiscence (*e.g.* the shepherd's purse, honesty, etc.).

(*e*) **The capsule.** All other capsular fruits which do not have the special characters of the legume, follicle, siliqua, or silicula are simply called capsules. They are formed from polycarpellary, syncarpous pistils, in which the ovary may be either unilocular or multilocular.

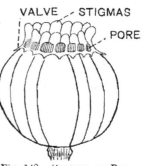

In the capsule many modes of dehiscence are met with. Sometimes the seed escapes through little holes or pores, either at base or apex—*porous dehiscence*, *e.g.* the poppy (fig. 142); sometimes by the separation of little tooth-like portions of the wall of the ovary, either at base or apex—dehiscence by teeth (*e.g.* at the apex in the pink).

Fig. 142.—CAPSULE OF POPPY. (Porous dehiscence.)

In many Primulaceæ, *e.g.* the pimpernel, there is a transverse dehiscence, leading to the separation of a lid from the apex of the capsule. Such a capsule is called **a pyxidium.** Usually, however, the dehiscence is by longitudinal slits (fig. 143). In *multilocular capsules*, with axile placentation, the dehiscence is *loculicidal* if the slits run down the middle of the carpels (*i.e.* open into the loculi),

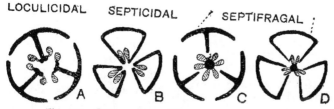

Fig. 143.—DEHISCENCE OF MULTILOCULAR CAPSULES. (Diagrammatic transverse sections.)

the *septa and placentas* breaking away in the middle (Iris); *septicidal*, if the slits run down the middle of the septa, the placentas separating in the middle (Rhododendron); *septifragal*, if the slits appear as in loculicidal or septicidal dehiscence but the septa break and the placentas and seeds are left in the middle (Datura, the thorn-apple). In *unilocular capsules* (with parietal placentation) the slits may pass down the middle of the carpels between the placentas (violet), or split the placentas (cf. loculicidal and septicidal above).

§ 8. **Schizocarpic Fruits (Schizocarps).** These are dry, many-seeded fruits, which, as they ripen, *split up* into a number of one-seeded and usually indehiscent parts resembling achenes and called **mericarps**. The best-known forms are:—

(*a*) **The lomentum** (fig. 144, A), an elongated fruit somewhat resembling a pod, marked off by a number of *transverse* constrictions into one-seeded parts, which afterwards (usually) break away from each other, *e.g.* Ornithopus, the bird's foot and some foreign Leguminosæ. The fruit of the radish is often given as an example, but is really a siliqua resembling a lomentum owing to transverse constriction,—in other words, a lomentaceous siliqua.

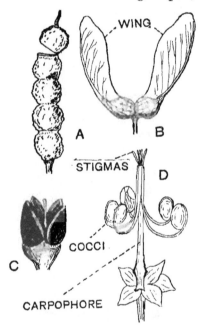

Fig. 144.—Schizocarpic Fruits.
A, Lomentum; B, Double samara of Maple; C, Carcerulus; D, Regma of Geranium.

(*b*) **The cremocarp** (fig. 145, A, B). This is the characteristic fruit of the order Umbelliferæ. It is developed from a bicarpellary pistil with a bilocular, inferior ovary in each loculus of which there is a single suspended ovule (fig. 110, B). As it ripens the cremocarp splits *longitudinally* (between the two loculi) into two mericarps, which remain for some time attached to a prolongation of the axis called the *carpophore*. Each mericarp contains a seed. These mericarps are *popularly* called seeds, *e.g.* the caraway "seed" (fig. 145, c).

(c) **The carcerulus** (fig. 144, c). This is the characteristic fruit of the orders Labiatæ and Boraginaceæ (see Chapter XIII.). In these orders the fruit is formed from a bicarpellary pistil with a *superior* ovary which becomes quadrilocular owing to the formation of two false septa (fig. 150). There are four ovules on an axile placenta. As the fruit ripens the four mericarps separate from each other towards the middle and look like four little nutlets or achenes. In the carcerulus of the mallow the *superior* ovary of the polycarpellary pistil splits into a large number of mericarps.

(d) **The regma** (fig. 144, D). This is a schizocarp which breaks up into one-seeded *dehiscent* parts, which are not called mericarps, but *cocci*, e.g. geranium and castor-oil. In the geranium the pistil is formed of five carpels fused round a long carpophore. The five styles of the carpels are also adherent to the carpophore. When ripe the cocci break away and remain suspended by their styles from the apex of the carpophore.

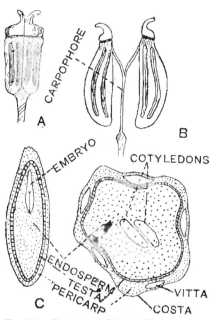

Fig. 145.—FRUIT AND SEED OF UMBELLIFERÆ.
A, B, The Cremocarp; C, Longitudinal and transverse sections of a mericarp of the Caraway.

(e) **The double samara** (fig. 144, B). This fruit is found in the maple, and consists usually of two, but often of three parts, in each of which the pericarp is prolonged into a membrane, so that the fruit resembles two or three samaras joined together. The samaras finally break away from each other.

§ 9. **Drupaceous Fruits.** We now pass from the dry to the succulent simple fruits. A typical drupe is a *true* fruit formed from a monocarpellary pistil with superior ovary, in which the pericarp is distinguished into three regions—the outermost forming a skin called the *epicarp*, the middle one soft and fleshy called the *mesocarp*, and the innermost one hard and stony (sclerenchymatous) forming a "stone," the *endocarp*, inside which there is usually one, sometimes two seeds. Thus the "stone fruits" are examples of drupes (cherry, plum, greengage, apricot, peach, etc.). The almond is a drupe with a somewhat firm instead of a succulent mesocarp. In the *commercial* form the epicarp and mesocarp have been removed. The shell of the almond is the endocarp. The following are examples of drupaceous fruits which differ from the typical drupe in various respects, but more especially in being developed from a syncarpous pistil. In the *walnut* there is an epicarp and a thin succulent mesocarp; the shell is the endocarp and consists of two valves. The cartilaginous partitions passing inwards between the cotyledons (which are covered by the testa) are ingrowths from the endocarp. The **coco-nut** is another curious drupaceous fruit. The shell is the endocarp; the edible substance is endosperm; and the brown layer covering it the testa. A minute monocotyledonous embryo is embedded in the endosperm at one end. There is a space in the middle of the endosperm filled with sap (so-called "milk"), owing to the fact that the endosperm could not fill the whole of the large embryo-sac. The fibres outside the endocarp are the remains of a mesocarp. Because of this fibrous investment the coco-nut is often called a *fibrous drupe*. Some drupaceous fruits are popularly called "berries." Thus the holly "berry" is a drupe in which the endocarp consists of a number of parts (four usually), each with a seed. There is a similar fruit in the elder.

§ 10. **Baccate Fruits, or Berries.** These are succulent fruits in which the succulent mass is more or less pulpy, and the seeds, which are usually hard, are embedded in the pulp. The berry differs from the drupe essentially in the fact that there is no stony endocarp, although epicarp, mesocarp, and endocarp may be differentiated. The endocarp, if present, is never stony. Baccate fruits may be derived from inferior ovaries (*e.g.* currant, gooseberry, melon, cucumber), or from superior (*e.g.* grape, orange). The *orange* is a multilocular superior berry with axile placentation; the outer glandular skin is the epicarp, the underlying white substance the mesocarp, and the inner membrane lining the loculi the endocarp. The juice is secreted by a large number of multicellular hairs developed from the walls of the loculi. The *gooseberry* is formed from an inferior unilocular ovary in which there are usually two, sometimes three, parietal placentas. The pulp is derived chiefly from the placentas, partly from the seed-coats. The *date* is recognized as a *berry* and *not a drupe* by the fact that the "stone" is not endocarp but seed (fig. 135). The outer skin of the date is the epicarp; the sticky mass underneath, the mesocarp. Surrounding the stone is a thin membranous endocarp. The cell-walls of the endosperm are strongly thickened, constituting a store of cellulose—hence the hardness of the seed. The banana is a berry from which, through over-cultivation, the seeds have disappeared. The pomegranate, also, is a kind of berry.

§ 11. **The pome** is the fruit found in the apple, pear, and several other Rosaceæ. We may take the apple as an example. The flower of the apple shows an extreme form of perigyny. There is a hollow, cup-like thalamus. The pistil developed inside the cup is apocarpous and consists of five carpels. As development proceeds, however, the carpels fuse together and also become adherent to the calyx-tube (thalamus). The whole fused mass develops into the fruit, *i.e.* the pome. Evidently the pome is a pseudocarp. The cartilaginous core, being formed from the pistil, is the true fruit or pericarp containing the seeds. The pome is, therefore, clearly distinguished from the drupe which is a true fruit; and the terms epicarp, mesocarp, and endocarp, sometimes applied to the outer skin, the fleshy substance, and the core respectively, are inadmissible because in the pome these are not regions of the pericarp. In the *hawthorn* there may be one or several carpels and they become stony. When there is only one the pome of the hawthorn closely resembles a drupe. Examination, however, would show the remains of the sepals, etc., at the apex of the fruit. These, of course, are not found in the drupe. The quince and medlar, also, are pomes.*

§ 12. **Aggregate Fruits.** These aggregates of simple fruitlets are called "*Etærios.*" We may have etærios of achenes, follicles, or drupes (drupels). (*a*) **Etærio of achenes.** A typical etærio of achenes is found in the buttercup (fig. 92). There the achenes are all grouped together on a slender prolongation of the thalamus. In the ordinary hedge Clematis (old man's beard) there is an etærio of achenes, which presents a feathery appearance, because the styles of the carpels are persistent, elongate, and become hairy. The *strawberry* is a pseudocarp consisting of an etærio of achenes scattered over the surface of an enlarged, fleshy thalamus. In the young strawberry flower the carpels are all crowded together on a protuberant thalamus. As the fruit develops, this thalamus expands enormously, and the *achenes* are spread out over its surface. Popularly these achenes are called seeds. The student must avoid this error. The hip of the *wild rose* is a pseudocarp consisting of an etærio of achenes enclosed in a persistent hollow thalamus or calyx-tube (see fig. 94, E). This pseudocarp of the wild rose is called a "**cynarrhodium.**" These examples will show the student that he must exercise care and discrimination in the examination of fruits. (*b*) **Etærios of follicles** (fig. 140). These are found in many Ranunculaceæ, and some Rosaceæ (see Chapter XIII.). They will present no special difficulty. (*c*) **Etærios of drupes.** Examples of these are found in the blackberry (bramble) and the raspberry, both belonging to the order Rosaceæ. Here the little rounded, succulent bodies, which have exactly the structure of a typical drupe, are inserted on a fleshy conical thalamus.

---

* The pome, being formed from a single flower with an apocarpous pistil, might be compared with some forms of aggregate fruits, *e.g.* the wild rose (§ 12). The fusion of the carpels, however, and their adhesion to the calyx-tube, leads *practically* to a syncarpous condition.

§ 13. **Composite Fruits.** These are not numerous; the best examples are the *fig, pineapple, mulberry,* and *hop.*
(*a*) **The fig.** Here the inflorescence is a peculiar hollow, pear-shaped form of capitulum, the flowers being developed inside (fig. 146, A). The female flowers produce little fruits like achenes, popularly regarded as seeds. The composite fruit formed from this inflorescence is called a **syconus.**
(*b*) **Pineapple** and **mulberry.** Here the composite fruit is called a **sorosis.** It is formed from a spike. In the *pineapple* the fleshy axis and the flowers all fuse together. The areas on the surface of the fruit represent the flowers. Seeds are rarely formed. Above the flowers the axis produces a number of leaves forming the "crown." In the *mulberry* (fig. 146, B) the perianths in the female spike become fleshy, and enclose the true fruits. The whole composite fruit resembles closely the fruit of the blackberry. They must be carefully distinguished. The fruit of the blackberry is an etærio of drupels developed from the apocarpous pistil *of a single flower.*

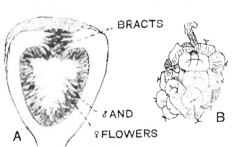

Fig. 146.—Composite Fruits.
A, Syconus of Fig (cut vertically); B, Sorosis of Mulberry.

(*c*) **The hop.** Here the composite fruit is formed from an inflorescence consisting of an axis bearing a number of membranous scales. On the upper surface of each scale, at its base, are two female flowers. The fruit is called a **strobilus.** The true fruits are achenes.

§ 14. There are many fruits which it is difficult to classify. For example, the ivy "berry" is a fleshy fruit containing several seeds; these are not enclosed in a stony endocarp, but there is a firm investment round each. The fruit, to some extent, resembles a drupe, and may be called a drupaceous berry. Again, the fruit of the horse-chestnut is more or less fleshy, but is dehiscent, and contains two seeds, thus having the characters of a capsule. It may be described as a succulent capsule. So with others,

C. The Dispersal of Seeds and Fruits.

§ 15. Its Significance. It is manifestly of advantage to the species that the seeds should be carried some distance from the parent plant. It gives the young seedlings a better chance in the struggle for existence, for they are saved, to a large extent, from the competition with each other in the matter of food, light, etc., which would naturally arise if they were crowded together round the parent. This arrangement for dispersal has an immense importance, also, in connection with the study of the distribution of plants over the earth's surface.

§ 16. Means of Dispersal. There are four ways according to one or other of which seeds are *regularly* dispersed—viz. (*a*) wind, (*b*) water, (*c*) animals, (*d*) mechanisms in the fruit.

(*a*) **Wind.** In a few plants the seeds are so small and light that when they are set free from the fruits they are freely blown about by the wind, *e.g.* orchids. In others, where the seeds are larger, the fruit opens in such a way that the seeds can be blown out only by a high wind, usually being jerked out as the plant sways in the wind; this is commonly spoken of as the "censer mechanism." Examples are seen in the monkshood (where the fruit is an etærio of follicles), the poppy (capsule with porous dehiscence), many Cruciferæ (a siliqua or silicula) and those Caryophyllaceæ in which the fruit is a capsule dehiscing by apical teeth, *e.g.* the pink. Many achenial fruits, also, are scattered in this way, *e.g.* the sunflower. Special structures are often present which enable the seeds or achenial fruits to be more readily carried by the wind. Good examples are seen in the hairy aril of the willow, the hairy caruncle of the willow-herb, the wings developed on some seeds, the wing of the samara, the pappus of many cypselas, the hairy styles of Clematis. These are sometimes spoken of as the "parachute mechanism." It is interesting to notice in what various ways these structures, which are adaptations to wind-dispersal, are developed.

(b) **Water.** This mode of dispersal is found in water-plants. A few have special arrangements for keeping the seeds afloat, e.g. the water-lily, where the seed has little air-cavities between the testa and aril.

(c) **Animals.** Seeds or fruits may be dispersed by animals, either by adhering to them or by being eaten by them. In the former case the seeds or fruits have developed certain structures by means of which they become attached to the animal. These processes are usually of the nature of hooked spines. Such seeds or fruits are called "burrs," e.g. the burdock, a Composite, where the pointed leaves of the involucre become hardened and hooked. Succulent fruits, e.g. drupes, berries, pomes, etc., are eaten by animals. The succulent character is an adaptation to this mode of dispersal. In cases where the seeds pass through the alimentary canal of the animal they must be protected. This is secured either by a resistant testa (berries), or pericarp (e.g. strawberry, wild rose), or by the seed being enclosed in a strong endocarp (drupes). Here again it is interesting to notice from what various parts the succulent mass may be developed—e.g. mesocarp in drupes, thalamus in pome, strawberry, and wild rose, perianth in mulberry, the aril on some seeds, etc. The animals concerned in this mode of dispersal are *chiefly* birds.

(d) **Fruit-Mechanisms.** Some fruits have "explosive mechanisms" by which the seeds are forcibly ejected to some distance, e.g. Touch-me-not and wood-sorrel. The "explosion" is due to tensions set up by certain parts of the pericarp or seed becoming extremely turgid.

§ 17. **Occasional Dispersal.** Seeds may *chance* to be dispersed in other ways. Some can withstand long immersion in water, and may be accidentally carried in this way. Many are carried on floating timber, many in the mud adhering to the feet of water-birds—and so on. We must, however, distinguish between these *occasional* means, and the regular modes of dispersal *to which plants have adapted themselves.*

CHAPTER XIII.

CLASSIFICATION—SYNOPSIS OF NATURAL ORDERS.

(The study of this chapter may be deferred.)

§ 1. Now that the student has made a general acquaintance with the form, structure, and physiology of the Angiosperm, he may proceed to the special study of certain of the more familiar orders and genera. He must clearly understand, however, that such a study is essentially practical. This chapter has not been written with a view to doing away with the actual examination of specimens. It is intended to serve as a stimulus and a guide to the student in conducting such practical work.

The Angiosperms form an enormous assemblage of plants, presenting such an infinite variety of vegetative and floral characters that the elementary student feels hopelessly bewildered when he first contemplates them. He is at a loss when he tries to follow the guiding principles which enable us to detect the essential resemblances due to affinity and so reduce this huge assemblage to something like order. In other words, he cannot recognize in their proper order of value those morphological characters on which classification is based. The subject cannot be fully dealt with here; all that is hoped is that the present chapter may prove a satisfactory introduction to it.

§ 2. **Species, Genera, etc.** Many plants have a very close resemblance to each other. There are many characters which they *all* exhibit. These are all characters handed down from parent to offspring, and may be called constant characters. On the other hand, there are minute *individual* differences. The individual characters exhibited by any one plant are not exactly reproduced in another, *i.e.* they are not constant. For example, every one can recognize a raspberry plant. He does so because all raspberry plants resemble each other so closely that he might easily imagine them to have been derived from the same parents. At the same time he recognizes that all raspberry plants present individual differences, *e.g.* very minute differences in the form of the leaves. Such a group of plants constitutes a **species**. A species may be defined as a group or assemblage of individual plants which resemble each other in all *constant* characters and are therefore presumably descended from a common ancestor. Again, certain species resemble each other very closely, while each possesses the peculiar constant characters which mark it off as a species. Such a group of species constitutes a **genus**. For example, the *Raspberry* and *Blackberry* are such species; together with others, they form the genus *Rubus*. We name a plant by giving both its generic and specific names—thus: *Rubus Idæus* is the *Raspberry*, *Rubus fruticosus* the *Blackberry*. In the same way, according to *wider* resemblances, the genera are grouped into **Natural Orders**, the orders into *Cohorts*, the Cohorts into *Series*, and so on to *Sub-classes*, *Classes*, *Divisions*, and, finally, *Groups*, or *Sub-kingdoms*. Many of these terms, however, are used arbitrarily. Even the application of the terms *species* and *genus* is not definitely fixed. Frequently there is greater sub-division, or less, than we have indicated, *e.g.* the sub-division of an Order into families or tribes, each containing a number of Genera. Into these details, however, we cannot enter.

§ 3. **Classification of Angiosperms.** The group *Phanerogamia* (p. 4) is divided into the two Divisions, *Gymnospermæ* and *Angiospermæ*. The division of the Angiospermæ into the two classes *Dicotyledones* and *Monocotyledones* has already been indicated and all their distinguishing characters fully dealt with.

The Dicotyledones are sub-divided into **three** sub-classes

—*Polypetalæ, Gamopetalæ,* and *Incompletæ* (Apetalæ or Monochlamydeæ)—according to the characters of the floral envelopes. These may be further divided into series according to the insertion of parts. In each series is placed a number of cohorts and orders distinguished from each other by characters relating chiefly to the structure, arrangement, and number of parts in the andrœcium and gynæceum, and the characters of seed and fruit.

The Monocotyledones are divided into three sub-classes, *Spadicifloræ, Glumifloræ, Petaloideæ,* according to the characters of the perianth, etc. There is further subdivision as in Dicotyledones.

The classification is indicated in the following table:—

CLASS DICOTYLEDONES:—leaves with reticulate venation; flowers with parts usually in twos, fours, or fives, seldom in threes; etc.

I. **Sub-class Polypetalæ:** corolla typically polypetalous.
1. **Series Thalamifloræ:**—hypogynous, sepals usually free, stamens borne directly on the thalamus.
Natural Orders:—Ranunculaceæ, Polygonaceæ, Caryophyllaceæ, Papaveraceæ, Cruciferæ, Violaceæ, Malvaceæ.
2. **Series Discifloræ:**—hypogynous, sepals free or united, thalamus usually flattened and forming a disc (above the calx) on which the petals and stamens are inserted.
Natural Orders:—Geraniaceæ, Euphorbiaceæ.
3. **Series Calycifloræ:**—perigynous or epigynous.
Natural Orders:—Leguminosæ, Rosaceæ, Crassulaceæ (perigynous); Saxifragaceæ (perigynous or epigynous); Onagraceæ, Umbelliferæ (epigynous).

II. **Sub-class Gamopetalæ:**—corolla typically gamopetalous.
1. **Series Hypogynæ.**
Natural Orders:—Ericaceæ, Primulaceæ, Oleaceæ, Convolvulaceæ, Solanaceæ, Bora-

ginaceæ, Labiatæ, Scrophulariaceæ, Plantaginaceæ.
2. **Series Epigynæ.**
Natural Orders:—Campanulaceæ, Caprifoliaceæ, Compositæ.

III. **Sub-class Incompletæ:**—monochlamydeous or achlamydeous; frequently diclinous.
Natural Orders:—Cupuliferæ—including Betulaceæ (birch and alder), Corylaceæ (hazel), and Fagaceæ (beech, oak); Salicaceæ (willow, poplar); Chenopodiaceæ, Urticaceæ, Ulmaceæ (elm). Cupuliferæ, Salicaceæ, and certain others are combined under the cohort Amentiferæ (Amentaceæ).

CLASS MONOCOTYLEDONES:—leaves with parallel venation; flowers with parts in threes; etc.
I. **Sub-class Spadicifloræ:**—inflorescence usually a spadix; flowers without a perianth.
Natural Orders:—Araceæ (arum lily), Palmaceæ (palms).
II. **Sub-class Glumifloræ:**—perianth scaly or absent; flowers invested by scaly bracts.
Natural Order:—Graminaceæ (grasses).
III. **Sub-class Petaloideæ:**—perianth usually petaloid.
1. **Series Hypogynæ.**
Natural Order:—Liliaceæ.
2. **Series Epigynæ.**
Natural Orders:—Amaryllidaceæ, Iridaceæ, Orchidaceæ.

§ 4. It should be particularly mentioned that amongst the Polypetalæ and Gamopetalæ are a number of plants *without petals*. They are placed in these sub-classes, and not in the Incompletæ, because their true affinities are shown by their other characters. The absence of the corolla is due to suppression; it is not a primitive character. Amongst the Incompletæ should be placed only plants in which we can recognise a primitive mono- or a-chlamydeous condition.

A. Dicotyledons.

§ 5. Ranunculaceæ (fig. 147).

Distinguishing characters:—Flowers polypetalous, hypogynous; stamens indefinite (∞); gynæceum apocarpous.

Most of the plants belonging to this order are herbaceous, with alternate or radical and, usually, much-divided, exstipulate, sheathing leaves. Clematis (*e.g. C. vitalba*, the well-known "traveller's joy" or "old man's beard") is a shrub with *opposite* leaves climbing by means of its petioles. The plants usually perennate by means of sympodial rhizomes. The primary root is lost, and adventitious roots are developed on the rhizome. Many of these adventitious roots form root-tubers, *e.g.* monkshood and lesser celandine (p. 97).

In *most* cases the **inflorescence** is cymose. In *Anemone* the flowering shoot bears a terminal flower. The flowers are hermaphrodite, and may be actinomorphic or zygomorphic. There is a general tendency throughout the order to the reduction of petals to nectariferous structures (*e.g.* monkshood, fig. 96); and frequently the petals are wanting altogether. Some, however, regard the nectariferous organs as representing modified stamens.

The arrangement of floral leaves may be cyclic, hemicyclic, or spiral. In *Aquilegia* (the columbine, fig. 147, A) it is perfectly cyclic; in the buttercup it is hemicyclic, the stamens and carpels being spirally arranged (fig. 147, B); in *Clematis* and *Anemone* it is spiral.

Calyx and **corolla** sometimes consist of five sepals and five petals, *e.g.* the common buttercup (fig. 92), and *Aquilegia;* but the number varies. In those forms where the petals are absent or modified the sepals are usually petaloid, and are apt to be mistaken for the corolla. The calyx and corolla present many other interesting modifications (see below). The æstivation is imbricate except in *Clematis*, where the calyx is valvate. The calyx is polysepalous.

The **stamens** are indefinite in number (∞), hypogynous, free; anthers innate. The **gynæceum** is polycarpellary, apocarpous; the number of carpels varies. There may be one or a number of anatropous ovules in each ovary ; if one it may be erect or pendulous.

The **seed** (fig. 136) is albuminous. The **fruit** varies; it depends on the number of carpels and ovules ; it may be

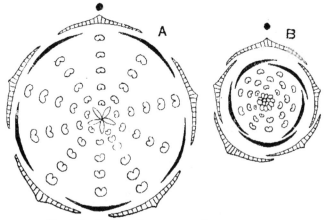

Fig. 147.—FLORAL DIAGRAMS OF RANUNCULACEÆ.
A, Aquilegia (Columbine) ; B, Species of Ranunculus.

an etærio of achenes or of follicles (fig. 140)—rarely a berry or capsule.

The following flowers may be specially noticed :—

(a) *Ranunculus* (buttercup):—actinomorphic : 5 sepals ; 5 petals, each with a pocket-shaped nectary at its base ; carpels numerous; fruit an etærio of achenes. *R. ficaria*, the lesser celandine, has 3 sepals and 8 petals.

(b) *Aquilegia* (columbine) :—actinomorphic ; 5 petaloid sepals; 5 petaloid spurred petals, secreting nectar; 5 carpels; fruit an etærio of follicles.

(c) *Aconitum* (monkshood):—zygomorphic; 5 petaloid sepals, the posterior one large and galeate ; 2 petals represented by nectariferous organs enclosed in the hood of the calyx ; carpels 2–5 ; fruit an etærio of follicles.

(d) *Delphinium* (larkspur) :—zygomorphic ; 5 petaloid sepals, the posterior one spurred ; 2 spurred petals projecting into the spurred sepal and secreting honey ; carpels 1–5 ; fruit an etærio of follicles (sometimes a single follicle).

(e) *Clematis* :—actinomorphic ; 4 or 5 petaloid sepals; petals 0 ; carpels numerous ; fruit an etærio of achenes, with persistent hairy styles serving for dissemination.

(f) *Anemone* :—actinomorphic ; 5 or more petaloid sepals ; petals 0 ; carpels numerous ; fruit an etærio of achenes. Beneath the terminal flower are three leaves which are referred to sometimes as an involucre, sometimes as an epicalyx ; in some species they lie so close to the flower that they might be mistaken for a calyx.

(g) *Thalictrum* (meadow rue) :—actinomorphic ; 4 or 5 more or less petaloid sepals; petals 0 ; fruit, etærio of achenes. Some species are protogynous and wind-pollinated.

(h) *Trollius* (globe-flower) :—actinomorphic ; numerous petaloid sepals; numerous small petals; fruit, an etærio of follicles.

(i) *Caltha* (marsh-marigold) :—actinomorphic ; usually 5 petaloid sepals ; petals 0 ; fruit, an etærio of follicles.

(j) *Pæonia* (pæony) :—actinomorphic ; 5 sepals ; 5 or more large petals ; fruit an etærio of 2–5 follicles. Flowers protogynous.

(k) *Helleborus* (Hellebore, *e.g.* Christmas rose) :—actinomorphic ; 5 greenish or petaloid sepals ; a number of petals reduced to nectar-cups ; fruit an etærio of follicles.

In the *baneberry* the fruit is a single berry ; in *Nigella* (love-in-a-mist), a capsule, the carpels being fused.

Typical floral formula :—$K_5 \ C_5 \ A\infty \ \underline{G\infty}$.

§ 6. Caryophyllaceæ.

Distinguishing characters :— Flowers polypetalous, hypogynous ; stamens definite, usually twice as many as the sepals and petals ; gynæceum of 2–5 carpels, syncarpous ; ovary unilocular ; free central placentation ; the swollen nodes and opposite leaves are characteristic.

The plants belonging to this order are mostly herbs with swollen nodes and opposite, simple, usually entire and exstipulate leaves. The **inflorescence** is cymose, typically a dichasium (fig. 127). The flowers are regular, usually hermaphrodite and pentamerous, exceptionally unisexual or tetramerous.

The **calyx** consists of 5 (or 4) sepals, polysepalous or gamosepalous, the **corolla** of 5 (or 4) petals, polypetalous. The **stamens** are hypogynous (rarely perigynous), free, 10 or 8, occasionally 5 or 4; anthers innate; usually the flower is obdiplostemonous (p. 163). **Gynæceum**, 2-5 carpels, syncarpous, with free styles (fig. 107, C); ovary unilocular, superior; ovules usually numerous, amphitropous or campylotropous, with free central placentation; sometimes septa can be recognized at the base of the ovary, showing that the placentation is derived from an originally axile placentation by the breaking down of the septa. **Seed** albuminous; embryo curved. **Fruit** usually a unilocular capsule dehiscing by teeth separating at the apex; seeds scattered by the censer mechanism (p. 216).

In the order two very distinct tribes are recognized:—(a) *Silenoideæ*, in which the calyx is gamosepalous—e.g. *Dianthus*, the pink, of which the sweet-william and carnation are cultivated forms; *Silene* and *Lychnis*, which include the various forms of campion and catchfly; (b) *Alsinoideæ*, in which the calyx is polysepalous, and the stamens *occasionally* perigynous—e.g. the chickweed, the stitchwort, the sandwort, the pearlwort, etc.

In the *Silenoideæ* the calyx is tubular and the corolla caryophyllaceous (fig. 98, A). The petals are clawed, and sometimes ligulate (*Lychnis*, fig. 97, B). The flowers are protandrous, and adapted to pollination by long-tongued insects (butterflies, etc.). Honey is secreted at the base of the stamens. The *ragged robin, corncockle*, and *red campion* are species of Lychnis. An androphore (p. 164) is present in species of this genus. *Lychnis dioica*, the red campion, is diœcious.

In the *Alsinoideæ* the flower is spreading, and adapted to pollination by short-tongued insects. The flowers are protandrous, and honey is secreted at the base of the stamens. The petals are often deeply cleft.

Typical floral formulæ:—Silenoideæ $K_{(5)} C_5 A_{5+5} G_{(2-5)}$; Alsinoideæ, $K_5 C_5 A_{5+5} G_{(2-5)}$.

§ 7 Cruciferæ (fig. 116).

Distinguishing characters:—Flowers polypetalous, hypogynous; parts in twos or fours; tetradynamous stamens; placentation and structure of ovary and fruit.

The plants belonging to this order are herbaceous, occasionally shrubby. The leaves are alternate and exstipulate. Familiar examples are :—wall-flower (*Cheiranthus*), stock (*Matthiola*), shepherd's-purse (*Capsella*), garden-cress (*Lepidium*), candytuft (*Iberis*), honesty (*Lunaria*), turnip, cauliflower, and cabbage (*Brassica*), etc. In the genus *Brassica* (and others) there are many interesting modifications of the vegetative regions and inflorescence—*e.g.* cauliflower, in which there is a fleshy inflorescence; brusselssprouts, in which miniature cabbages are borne in the leaf axils; kohl-rabi, in which there is a thickened stem; and turnip, in which there is a thickened tap-root.

The **inflorescence** is usually a raceme or corymb; there are no bracts. The flower as a whole (figs. 103, 116) is usually isobilateral, occasionally zygomorphic. The polysepalous **calyx** consists of four sepals in two whorls. The lateral sepals are more or less saccate or pouched at the base. The æstivation is imbricate as a rule. The **corolla** is polypetalous and cruciform. The petals are usually clawed. The æstivation is imbricate. The **andrœcium** consists of six hypogynous stamens in two whorls, and is *usually* tetradynamous; the two short *lateral* stamens form the outer whorl; the four inner stamens are supposed to have been produced by chorisis (p. 182). The nectaries are small green glands, situated on the thalamus at the bases of the short stamens. The **gynæceum** is bicarpellary and syncarpous; the two small stigmas are borne on a short style; the ovary is superior, and bilocular, owing to the development of a false septum between the two parietal placentas; the ovules usually numerous, amphitropous, or campylotropous. The **fruit** (fig. 141) is a siliqua (wall-flower, stock), or silicula (candytuft, honesty). The **seed** is exalbuminous; the testa is frequently mucilaginous (mustard, and garden-cress), and thus serves to fix the seed to the soil favourably for germination. The embryo is always folded (fig. 141, c).

The flowers are only slightly dichogamous, and self-fertilization is of regular occurrence. The arrangement in corymbs should be noticed. This massing of small flowers serves for attraction and more rapid and abundant pollination; it avoids the necessity of each flower producing a large corolla (cf. the umbel and capitulum). Frequently the outer petals of the outer flowers of the corymb are larger than the inner ones, so that the corollas are zygomorphic

Typical formula*:—$K_{2+2} \; C_{\times 4} \; A_{2+2^2} \; G_{\underline{(2)}}$.

§ 8. Leguminosæ (sub-order *Papilionaceæ*)—fig. 119.

Distinguishing characters:— Flowers polypetalous, perigynous, papilionaceous; stamens 10, mon- or diadelphous; gynæceum monocarpellary; fruit a legume (or lomentum).

The sub-order is a very large and important one, including herbs, shrubs, and trees of various form and adaptation in which the leaves are alternate, stipulate, and frequently compound. The French bean is an example of a twining plant; the pea, of a plant climbing by means of leaf-tendrils. Robinia (the garden acacia) has spiny stipules. Many other modifications have been dealt with generally in preceding chapters. All have tubercles on their roots connected with the absorption of nitrogen (see p. 151). Familiar examples are broom (*Cytisus scoparius*), gorse (*Ulex*), *Genista*, laburnum (*Cytisus laburnum*), the clovers (*Trifolium*), the vetches (*Vicia*), the peas and beans, etc.

The **inflorescence** is always racemose—either racemes in which sometimes, owing to the abortion of the growing point, only two or three flowers are developed, *e.g.* sweet pea (fig. 76, B), or stout spikes. The flowers are hermaphrodite, zygomorphic, papilionaceous, and perigynous. The perigynous condition is not strongly marked. The mechanism of pollination is described on p. 197.

The **calyx** is gamosepalous, five-toothed, sometimes bilabiate, inferior. The odd sepal is *anterior* (fig. 119). The **corolla** is polypetalous, papilionaceous (fig. 98, B). The **andrœcium** consists of ten stamens, perigynous, either

* × = Cruciform.

monadelphous, as in the broom, or diadelphous, as in the sweet pea; in the latter case the posterior stamen is free. The **pistil** is monocarpellary (fig. 106); the ovary, unilocular, superior; the ovules, indefinite (∞), anatropous, or campylotropous; marginal placentation. The style is elongated, and the stigma terminal. The **fruit** is a legume (fig. 76, B), occasionally a lomentum. The **seed** is large and exalbuminous.

Floral formulæ:—$K_{(5)} \; C_5 \; A_{(5+5)} \; G_{\underline{1}}.$

or, $K_{(5)} \; C_5 \; A_{(5+4)+1} \; G_{\underline{1}}.$

All the British Leguminosæ belong to the sub-order *Papilionaceæ*. But there are two other sub-orders, the Cæsalpineæ and Mimoseæ, which are tropical; the Mimoseæ include the mimosas and acacias. The characters of these are in many ways peculiar. The flowers are not papilionaceous, and the stamens often (acacia and mimosa) numerous. The acacias have spiny stipules, and many of them have phyllodes. *Mimosa pudica* is the sensitive plant. The so-called acacia of our gardens (*Robinia pseudacacia*) is not an acacia.

Many of the Leguminosæ describe sleep **movements** (p. 153).

§ 9 **Rosaceæ** (fig. 148).

Distinguishing characters:—Flowers polypetalous, perigynous, regular; stamens usually numerous; pistil apocarpous.

This is a large order of herbs, shrubs, and trees. The leaves are alternate, simple, or compound, and usually stipulate. Some forms of vegetative reproduction are worthy of notice, *e.g.* runners in the strawberry, suckers in the raspberry.

The **inflorescence** is very various, and includes both racemose and cymose forms. The flowers are regular, pentamerous (or tetramerous), usually hermaphrodite, perigynous.

The **calyx** is gamosepalous, five-sepalous; æstivation imbricate. An epicalyx is *sometimes* present, *e.g.* strawberry. The **corolla** is polypetalous and rosaceous, with usually five petals imbricate in the bud. **Stamens** two, three, or four times as many as the petals, or ∞. **Gynæceum** of 1 to ∞ carpels, apocarpous. There are 1 to 2 anatropous ovules in each carpel. **Fruit** various—a drupe, a pome, etærios of drupes, achenes, or follicles. The **seed** is exalbuminous.

The order gives a good illustration of the various forms of perigyny (see fig. 94, B–E). Considering the variety in the forms of the thalamus and the number of carpels, there is little to wonder at in the variety met with in the fruit. The following notes on a few common forms will illustrate these points :—

Fig. 148.—FLORAL DIAGRAM OF ROSE.

(*a*) In the strawberry (*Fragaria*), raspberry (*Rubus Idæus*), and blackberry (*Rubus fruticosus*), the thalamus is a flattened disc with a conical protuberance (fig. 94, B). This protuberance expands, and forms the fleshy mass of the fruit of the strawberry (p. 214), while in the two others it bears the etærio of drupels.

(*b*) In *Spiræa ulmaria*, the meadow-sweet, the fruit is an etærio of five follicles borne on a flattened thalamus.

(*c*) In the cherry, plum, apricot, almond (species of the genus *Prunus*), etc., only one carpel with two ovules is present, and the thalamus forms a cup (calyx-tube). The fruit is a drupe with usually one seed, the second ovule aborting. The calyx-tube decays and breaks away; it has no part in the formation of the fruit.

(*d*) In the wild rose (*Rosa*), the thalamus forms a deep cup (fig. 94, E). It is persistent, and encloses the etærio of achenes (see p. 214).

(*e*) In the apple, pear, mountain-ash (species of genus *Pyrus*), hawthorn (*Cratægus*), the medlar and quince, etc., the thalamus forms a deep cup, and a complete fusion takes place between it and the carpels forming a pome (p. 213).

(*f*) In the lady's mantle (*Alchemilla*) there is an epicalyx, and a calyx of four sepals, *no corolla*, four stamens alternating with the sepals, and one carpel forming a single achene. This tetramerous arrangement is also met with in a few other types of Rosaceæ. In some, the flowers are unisexual, e.g. *Poterium Sanguisorba*.

General floral formula :—$K_{(5)} \; C_5 \; A\infty \; G_{\underline{1-\infty}}$.

§ 10. Umbelliferæ.

Distinguishing characters :—Flowers polypetalous, epigynous, pentamerous ; 5 stamens ; structure of ovary and fruit.

This is a very large and important order, easily recognized by the general habit of the plants and their fruits. The plants are either herbs or shrubs with hollow (fistular) stems and alternate, amplexicaul, exstipulate leaves, which are usually much divided. Common examples are the carrot (*Daucus carota*), the chervil, fool's parsley (*Æthusa cynapium*), the caraway (*Carum carui*), pig-nut (*Conopodium denudatum*), wild celery (*Apium graveolens*), etc.

The **inflorescence** is nearly always a simple or compound umbel (figs. 125. 128). The **flowers** (fig. 110) are *usually* hermaphrodite and regular; they are epigynous and markedly protandrous. An epigynous honey-disc is present on top of the ovary, and accessible to all kinds of insects. **Pollination** is generally effected by small flies.

The **calyx** is rudimentary or absent. The **corolla** is polypetalous; the petals are usually white or yellow, and often have reflexed tips; æstivation usually imbricate. The **stamens** are five in number and epigynous. The **gynæceum** is bi-carpellary, syncarpous; the epigynous honey-disc bears two stigmas; the ovary is bilocular, with one suspended ovule in each loculus. The **fruit** is a cremocarp (fig. 145, A, B). The mericarps are usually marked by ridges (*costæ*) with intervening oil-ducts (*vittæ*). The **seed** is albuminous (fig. 145, C).

The aggregation of flowers in umbels has the same biological significance as the aggregation into corymbs (p. 227). The outer flowers of the umbel are frequently irregular and zygomorphic. Unisexual flowers sometimes occur.

Floral formula:—$K_0 \; C_5 \; A_5 \; G_{\overline{(2)}}$.

§ 11. Primulaceæ (fig. 149).

Distinguishing characters:—*Flowers gamopetalous, hypogynous, regular; stamens epipetalous, antipetalous; pistil syncarpous; free central placentation.*

The plants belonging to this order are herbs perennating by means of rhizomes or tubers. The primrose has a sympodial rhizome; the cyclamen has a large massive persistent tuber. The leaves are exstipulate; either radical, or cauline and (usually) opposite.

The **flowers** (fig. 100) are usually regular, actinomorphic, hermaphrodite, pentamerous (or sometimes tetramerous); many are heterostyled (*e.g. Primula, Glaux*).

The **calyx** is gamosepalous and persistent, 5- (or 4-) cleft. The **corolla** is gamopetalous, regular, exceptionally polypetalous or absent (Glaux, the sea milkwort). The **stamens** are 5 (or 4) in number, epipetalous and antipetalous. The **gynæceum** is polycarpellary (5 carpels), syncarpous, with a single style and a capitate stigma; the ovary is unilocular, superior; the ovules are numerous, anatropous or amphitropous; the placentation is free central. The **fruit** is a capsule, dehiscing by 5 valves (hence we infer 5 carpels), *e.g.* primrose; or a pyxidium *e.g.* pimpernel. The **seed** is albuminous.

Fig. 149.—FLORAL DIAGRAM OF PRIMULA.

Primula vulgaris is the primrose, with solitary flowers; *P. veris* the cowslip, with flowers in an umbel borne on a long, leafless scape arising from amongst the radical leaves. Both are dimorphic and heterostyled (p. 196). The length and narrowness of the tube of the hypocrateriform corolla render pollination by long-tongued insects (bees or butterflies) necessary.

The antipetalous position of the stamens is explained by supposing that an outer whorl of stamens has disappeared. This receives support from the fact that staminodes alternating with the petals are occasionally met with.

Floral formula of the primrose: —$K_{(5)} \overparen{C_{(5)} A_{0+5}} G_{(5)}.$

§ 12. **Labiatæ** (fig. 150).

Distinguishing characters:—*Flowers gamopetalous, hypogynous, zygomorphic; stamens didynamous and epipetalous; fruit a carcerulus. The square stems, opposite decussate leaves, inflorescence, and bilabiate ringent corolla are characteristic.*
This is an important order, consisting of herbs or undershrubs with square stems and opposite decussate, simple, exstipulate leaves. Most of them are land plants; a few are marsh-plants. Suckers are found in many forms (fig. 42). In most of them there are numerous epidermal glands secreting volatile oil.

The characteristic **inflorescence** is the verticillaster (p. 190). The flowers are pentamerous, with suppression of certain parts, and hermaphrodite. The **calyx** is gamosepalous, tubular or bilabiate, persistent. The **corolla** is zygomorphic, bilabiate ringent, imbricate in the bud. The **stamens**, owing to the suppression of a fifth (the posterior one), are four in number, epipetalous and didynamous. The **gynæceum** is bicarpellary and syncarpous. Early in its development a median constriction appears in the ovary and produces two false septa. The style is *gynobasic*, *i.e.* it arises from the base and comes up from between the four parts of the ovary; there is a bifid stigma. The ovary is quadrilocular (two true and two false septa) with one erect anatropous ovule in each loculus. The placentation is axile. The **fruit** is a carcerulus (p. 211). The **seed** is exalbuminous (or nearly so).

Fig. 150.—Floral Diagram of Labiatæ.

Well-known examples are the dead-nettles (*Lamium*), the sage (*Salvia*), the thyme, lavender, rosemary, the mints (*Mentha*), the balm and horehound, the ground-ivy, etc.

There is a honey disc at the base of the ovary. Most of the flowers are protandrous; but this is scarcely at all the case in the dead-nettles. The insects (mostly bees, in the case of British species) alight on the lower lip of the corolla; the upper lip protects the essential organs. When the stigmas are fully developed, they protrude beyond the stamens, so that they are touched first. There is a remarkable mechanism in *Salvia*, the sage. There are only two stamens. In each stamen there is a very long connective separating the two anther lobes, of which one is fertile, the other barren. The connectives swing on the tops of the filaments. The barren lobes of the two stamens are in close contact, and when they are touched the fertile lobes are brought down on the insect's body.

Common formula :—$K_{(5)} \overparen{C_{(5)} A_4 G_{(2)}}$.

§ 13. Scrophulariaceæ (fig. 117).

Distinguishing characters :—Flowers gamopetalous, hypogynous, more or less irregular and zygomorphic; stamens usually 4, sometimes 2, rarely 5, epipetalous; fruit usually a bilocular capsule.

The order is distinguished from Labiatæ chiefly by its fruit. Most of the plants belonging to it are herbs or under-shrubs, with alternate or opposite exstipulate leaves. The stems are cylindrical. The order includes a number of semi-parasites, e.g. the eye-bright (*Euphrasia*), cow-wheat (*Melampyrum*), the yellow-rattle (*Rhinanthus*), lousewort (*Pedicularis*)—see p. 151.

There are various forms of **inflorescence.** Racemes and spikes are common; but sometimes the lateral branches are cymose and we find racemes of cymes. Bracts and bracteoles are generally present. The flowers are typically pentamerous, but there are various modifications, due to suppression and fusion.

We may first consider the common form (fig. 117, A) as found in the snapdragon (*Antirrhinum*). The **calyx** is gamosepalous, 5-lobed, persistent. The **corolla** is gamopetalous, bilabiate and personate (fig. 99, G). The **stamens** are four in number, epipetalous and didynamous; a posterior stamen has been suppressed. The **gynæceum** is bicarpellary, syncarpous; the style single, the stigma entire or two-lobed. The ovary is bilocular, superior. The anatropous ovules are numerous and borne on a large axile dumb-bell-shaped placenta. The **fruit** is a capsule, dehiscing by pores or loculicidal. The **seeds** are numerous and albuminous.

The following modifications should be noticed. In *Scrophularia* and *Pentstemon* the posterior stamen is represented by a staminode. In the mullein (*Verbascum*) the flowers are nearly regular, and have five stamens. In *Veronica* (speedwell—fig. 117, B, C) the posterior sepal is suppressed and the two posterior petals are fused, so that there are four sepals and four petals; the corolla is rotate; there are two stamens, the two anterior ones in addition to the posterior being suppressed; the seeds are relatively few in number. In *Calceolaria* we find very much the same condition, but the corolla is inflated, bilabiate and globose, and is described as calceolate or slipper-shaped. In the toad-flax (*Linaria*) the corolla is spurred; there are four stamens. In the foxglove (*Digitalis*) the corolla is described as glove-shaped (fig. 99, H).

There is a honey-disc at the base of the ovary. The common type is pollinated by bees; those with open flowers, like speedwell and mullein, by bees and flies; *Scrophularia* by wasps.

Floral formulæ:—common type—$K_{(5)} \, \widehat{C_{(5)} \, A_4} \, G_{\underline{(2)}}$.

Speedwell—$K_{(4)} \, \widehat{C_{(4)} \, A_2} \, G_{\underline{(2)}}$.

§ 14. Compositæ (fig. 151).

Distinguishing characters:—flowers gamopetalous, epigynous; stamens 5, syngenesious, epipetalous; pistil bicarpellary; ovary unilocular with one basal ovule. The capitulate inflorescence is very characteristic, as also is the fruit.

This is the largest and most widely distributed order of flowering plants, including over ten thousand species. The plants are nearly all herbaceous; only in certain parts of the globe is the order represented by shrubs. They present great variety in their vegetative organs, as would naturally be expected, considering they are found everywhere and in all situations. Water-plants, climbing-plants, and epiphytes (p. 157) are not common in the order. There is usually a tap-root, often more or less thickened, *e.g.* dandelion. The leaves are usually radical, or alternate, less frequently opposite (sunflower), usually exstipulate.

CLASSIFICATION—SYNOPSIS OF NATURAL ORDERS.

Laticiferous vessels are found in some (*e.g.* dandelion) and oil-ducts are common in the order.

The **inflorescence** is with few exceptions a capitulum, containing numerous small flowers (florets) surrounded and protected by an involucre (fig. 126). The disc is usually flattened or convex. The flowers may be all alike (Dandelion) or may have different characters (Daisy). The youngest are in the centre. Small scaly bracts, called **paleæ**, are frequently present between the flowers (*e.g.* Sunflower), and the capitulum is described as *paleaceous*. The capitula may be arranged in racemes, panicles, spikes, etc.

Fig. 151.—FLORAL DIAGRAM OF COMPOSITÆ.

The **flowers** (fig. 99, A, D) are epigynous and usually pentamerous. They may be actinomorphic or zygomorphic; hermaphrodite, unisexual, or neuter (p. 162). The arrangement of the flowers in the capitulum is described below. The **calyx** is absent, or represented by a ring of small teeth or a pappus borne on top of the ovary. The **corolla** may be tubular, ligulate, or (in some foreign Composites) labiate; it is valvate in æstivation. The **stamens** are five in number, epipetalous and syngenesious. The **pistil** is bicarpellary, syncarpous; the style is single and slender or filiform, the stigma bifid. The ovary (fig. 111) is inferior and unilocular; there is a single anatropous ovule with basilar placentation. The **fruit** (figs. 34, 139, A) is a cypsela; the **seed** exalbuminous.

The British Compositæ are divided into two series or sub-orders:—

(1) The *Liguliflorœ*, in which the flowers in a capitulum are all ligulate and hermaphrodite. All have laticiferous vessels. Examples: dandelion (*Taraxacum officinale*), lettuce (*Lactuca sativa*), chicory (*Cichorium intybus*), endive (*Cichorium endivia*), hawkweed (*Hieracium*), sow-thistle (*Sonchus*).

(2) The *Tubuliflorœ*, in which the flowers of the disc are not ligulate, and there is no latex. Amongst these there are two common arrangements: (*a*) the flowers are all

tubular and hermaphrodite, as in the thistles; (*b*) there are ray and disc florets—those of the disc tubular and hermaphrodite, those of the ray usually ligulate and pistillate *e.g. Bellis perennis*, the Daisy (**gynomonoecious condition**). The mechanism of pollination in Compositæ is interesting. The flowers are protandrous. The pollen grains are shed into the tube of anthers. The style and stigma elongate through the anther-tube, and *gradually* brush out the pollen grains, which are carried away in large numbers by insects visiting the capitulum. The stigmas do not unfold till they have grown out of the anther-tube; and, as only the upper surface of the stigma is receptive, self-pollination is prevented. If cross-pollination does not take place, self-pollination is effected, in most cases, by the stigmatic lobes bending back and reaching the pollen. There is a ring-like nectary at the base of the style; the tubular corolla protects the nectar and pollen, and allows the honey to be reached only by fairly long-tongued insects. In some flowers, where the corolla tube is short—*e.g.* milfoil—pollination is effected by small flies. The aggregation of flowers in a capitulum secures the pollination of a large number of flowers at each insect-visit.

A few interesting forms may be specially noticed. In the sunflower (*Helianthus annuus*), the *ligulate* ray-florets are neuter, so also the large *tubular* ray-florets in the cornflower. In the groundsel (*Senecio vulgaris*), all the florets are tubular and hermaphrodite; the flowers are rarely visited by insects, and are regularly self-pollinated. In *Tussilago*, the colt's-foot, the capitulum has about forty central male flowers, surrounded by a large number of female flowers. The flowers appear in spring before the leaves. The male flowers have a style acting as a pollen brush, but no stigmas; only the male flowers secrete honey.

The capitulum of the Compositæ is a very highly specialized inflorescence, showing considerable division of labour. The involucre discharges the same function for the aggregation of flowers which a calyx does for an individual flower. There is no necessity for a protective calyx; the calyx, when present (as a pappus), has a new function in connexion with fruit dispersal. The aggregation of *small* flowers serves to attract insects. As a rule, only some of the flowers (ray-florets) develop large corollas; this seems to be done at the expense of one or both sets of essential organs, hence the frequent occurrence of pistillate or neuter florets in the ray. The mechanism of pollination is simple and very effective, and is accompanied by an arrangement for ultimate self-pollination. There is so little chance of pollination failing to take place that there is no need for more than one ovule to be developed in each flower. From a study of the inflorescence it has been generally agreed that the Compositæ represent the highest development of Angiospermous Flowering Plants. There is no reason to wonder at the vast number of species and their wide distribution.

Common floral formula :— $K_{0 \text{ or pappus}} \overparen{C_{(5)} A_{(5)}} G_{\overline{(2)}}$.

§ 15. **Salicaceæ** (fig. 152).

Distinguishing characters :—flowers in catkins, diœcious; perianth rudimentary or absent; stamens, 2–30; fruit, a capsule.

This is the order of the willows (*Salix*) and poplars (*Populus*); it contains only the two genera. They are trees or small shrubs. Some of the willows are very small. There is abundant vegetative reproduction by means of suckers. The leaves are alternate and stipulate. In the willow the terminal bud usually dies off, and the growth of the axis is continued by a lateral bud. The terminal bud persists in the poplar. The winter bud of the willow is protected by a single scale, that of the poplar by a number of scales.

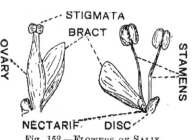

Fig. 152.—FLOWERS OF SALIX. (Male and Female.)

The habit of the trees is very varied. *Salix babylonica* has pendulous branches; it is the weeping willow; *Salix viminalis* is the osier. *Populus alba* is the white poplar; *P. nigra*, the black poplar; *P. tremula*, the aspen.

The **inflorescences** are unisexual catkins borne on different plants (*diœcious*). The catkins make their appearance in autumn on short dwarf shoots; they develop early in the following spring. The flowers are borne in the axils of bracts.

Willow. The bracts are entire. The **male flower** consists of two (rarely three) stamens, with a honey-disc at the base on the side next the axis (posterior). Occasionally the stamens are monadelphous. The **female flower** consists of a bicarpellary, syncarpous pistil, with cleft stigma. A honey-disc is present. The ovary is unilocular (superior), with two parietal placentas, and numerous anatropous ovules. The **fruit** is a loculicidal capsule. The **seed** is exalbuminous, and bears a tuft of hairs developed from the base (hairy *aril*, p. 204), and serving for seed-dispersal.

Poplar. The bracts are toothed. In both male and female flowers there is a rudimentary cup-shaped perianth. By some this is considered to be, not a perianth, but a honey-disc. In different species the stamens vary from four to thirty. The characters of the pistil, fruit and seed are the same as in the willow.

The willow is entomophilous. As its honey-secreting flowers appear early, before other flowers appear, they attract many insects, especially bees. The flowers of the poplar secrete no honey, and are anemophilous. The greater number of stamens is correlated with this.

B. Monocotyledons.

§ 16. **Liliaceæ** (fig. 120).

Distinguishing characters :—Perianth usually petaloid; flowers hypogynous; 6 stamens; trilocular ovary.

The plants are mostly herbs perennating by means of rhizomes, *e.g.* Solomon's Seal (*Polygonatum*, fig. 43), bulbs, *e.g.* lily, onion, and hyacinth (fig. 46), or corms, *e.g.* the meadow saffron (*Colchicum*, fig. 45). A few are shrubs or trees, *e.g. Dracœna* and *Yucca*, often showing secondary growth (p. 93). Some have vegetative reproduction by means of bulbils, *e.g. Lilium bulbiferum*. Some are climbing plants, *e.g. Smilax* (p. 124). *Ruscus*, the butcher's broom, is a shrub with phylloclades (fig. 50), bearing unisexual flowers. Many species are xerophytic (*e.g.* Aloe). The **inflorescence** may be racemose, or cymose. The umbellate heads found in many, *e.g.* onion and *Agapanthus*, are cymose. In the tulip there is a solitary terminal flower. The **flowers** are actinomorphic, usually hermaphrodite, hypogynous. The **perianth** consists typically of six parts in two whorls, and may be polyphyllous or gamophyllous. There are six **stamens** in two whorls, hypogynous or epiphyllous, with usually introrse anthers. The **pistil** is tricarpellary, syncarpous; the ovary trilocular, superior; the ovules indefinite, anatropous; the placentation axile. The **fruit** is usually a loculicidal or septicidal capsule, occasionally a berry (*e.g. Asparagus officinalis*, the asparagus, and *Convallaria majalis*, the lily of the valley). The **seed** (fig. 137) is albuminous.

Pollination. The flowers are entomophilous. Honey is secreted by nectaries situated in the wall of the ovary between the carpels (*e.g. Scilla nutans*, the bluebell or wild hyacinth) or at the bases of the perianth leaves (*e.g. Lilium*).

Common formula :— $K_3 \, C_3 \, A_{3+3} \, G_{(3)}$.

or $P_{3+3} \, A_{3+3} \, G_{(3)}$.

§ 17. Orchidaceæ (figs. 153—155).

Distinguishing characters :—Perianth petaloid, zygomorphic; flowers epigynous, gynandrous; stamens reduced in number; ovary unilocular.

This is a very large and interesting order, comprising only herbs perennating by means of rhizomes, tubers (fig. 46), etc. They are of very diverse habit, including land plants, epiphytes (p. 157), saprophytes, etc. Many of the land and epiphytic forms are adapted to xerophytic conditions (p. 157), storing up water and reserves either in thickened internodes called *pseudo-bulbs* or in fleshy leaves. The epiphytic orchids which abound in the tropics are especially interesting. The epiphytic adaptation is shown in many ways. They support themselves by means of clinging adventitious roots (cf. ivy) on which are developed absorbing roots which penetrate into the humus which collects between the clinging roots and the support. Then finally there are the aerial roots, each of which has at its apex a sheath of tracheides (the *velamen*), serving to collect water which trickles down the roots. The production of abundant small seeds easily carried by the wind is also an adaptation to epiphytic conditions.

The **inflorescence** is racemose, most frequently a spike. The flowers (figs. 153, 154) may be compared with the typical monocotyledonous form, the many striking differences met with being due to suppression, adhesion, and hypertrophy of certain parts. The **perianth** is petaloid, zygomorphic, and consists of six segments in two whorls. The posterior segment (petal) of the inner whorl is always more strongly developed than the others; it forms the *labellum*. Owing to the twisting of the inferior ovary, the labellum comes to be anterior (the *resupinate* condition), and serves as the landing stage for the insect. The labellum is sometimes spurred. The **androecium** usually consists of one stamen and two staminodes, but in some forms there are two stamens and one staminode. The stamens are fused with a prolongation of the axis of the flower called the **gynostemium** (the *gynandrous* condition, p. 173), which also bears the three stigmas on its apex. In the majority of Orchidaceæ the pollen grains are united into *pollinia*. The **pistil** is tricarpellary and syncarpous; the ovary inferior and unilocular. The ovules do not develop till after pollination; they are anatropous, and borne on three parietal placentas. The **fruit** is a capsule containing an enormous number of very small, light seeds. The **seed**

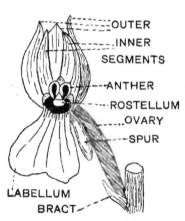

Fig 153.—Flower of *ORCHIS*.

is exalbuminous, and its embryo is not differentiated into plumule, radicle, and cotyledon.

There are two divisions of Orchidaceæ,—the *Monandræ*, *e.g.* Orchis, in which there is one fertile stamen (fig. 155, B), and the *Diandræ*, *e.g.* Cypripedium, in which there are two (fig. 155, A). The former is by far the larger division. We will describe Orchis and Cypripedium as types of these:

Orchis, *e.g. O. mascula*, the purple orchis, and *O. maculata*, the spotted orchis. The single fertile stamen is the anterior one of the outer whorl; the two staminodes are the anterior ones of the inner whorl.

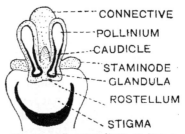

Fig. 154.—CENTRAL PART OF FLOWER OF *ORCHIS*.
Perianth segments removed.

Pollinia are present. One of the stigmatic surfaces is incapable of being pollinated, and develops into a projecting structure called the **rostellum**. The pollen grains are held together by delicate threads which run together at the base of the anther lobe to form a mucilaginous cord called the **caudicle**, which is attached below to a sticky substance called the glandula (fig. 104, c). The two functional stigmatic surfaces are below the rostellum, the two anther lobes above it, one on each side; the two caudicles lie in a little pouch one on each side of the rostellum. These various structures are borne on the gynostemium.

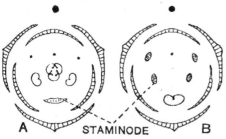

Fig. 155.—FLORAL DIAGRAMS OF ORCHIDACEÆ.
A, Cypripedium; B, Orchis.

Cypripedium The stamens are the anterior ones of the inner whorl, *i.e.* those represented by staminodes in Orchis. The staminode is the anterior one of the outer whorl, *i.e.* the fertile one

16

in Orchis. There is no rostellum. The three stigmata are fused in one. The pollen grains are not aggregated into pollinia, but are sticky.

Pollination. The flowers are entomophilous. No honey is *secreted*; the insects have to pierce the tissues to get it The contrivances in Orchidaceæ are innumerable, in many cases extraordinary. We may consider the native orchids as an example. The insect alights on the labellum, and seeks in the spur of the labellum for the honey. The back of the insect comes in contact with the rostellum, and pushes aside the pouch containing the caudicles, which thereupon become fixed. While the insect is piercing the spur, the mucilaginous substance (glandula) of the caudicle "sets," and when the insect leaves the flower the pollinia are dragged out. At first they stand erect on the back of the insect, but very gradually they are lowered, owing to hygroscopic properties which their stalks possess, so that when the insect enters another flower they touch the stigmas.

Floral formulæ[*]:—*Orchis*, $K_3 \; C_3 \; A_{1+\dagger 2} \; G_{\overline{(3)}}$

Cypripedium, $K_3 \; C_3 \; A_{2+\dagger 1} \; G_{\overline{(3)}}$.

§ 18. **Gramineæ** (figs. 156—158).

Distinguishing characters : — *Perianth scaly or absent; flowers hypogynous, enclosed in scaly bract and bracteole; pistil monocarpellary; fruit a caryopsis : the leaves are ligulate, and the leaf-sheath is split.*

This is a very large and easily recognized order, including the familiar cereals and grasses generally. Nearly all the plants are herbaceous. with hollow internodes and jointed nodes. The maize (Zea), and a few others have solid internodes, and the bamboos often grow to a great height. The leaves are alternate, and are usually arranged in two opposite series or orthostichies (distichous arrangement—divergence $\frac{1}{2}$). The leaves have no petiole, but long sheaths which are split on the side opposite the lamina (fig. 74, E). The lamina is linear, and bears a ligule at the base. Many are annual; on the other hand, many are perennial and

---

[*] † = staminode.

CLASSIFICATION—SYNOPSIS OF NATURAL ORDERS. 243

either have a rhizome or develop a tufted habit by copious branching at the base.

The **inflorescence** is more or less complex. The flowers are arranged in spikelets, but the spikelets are grouped together in various ways to form compound inflorescences. In the wheat (*Triticum*) and perennial rye-grass (*Lolium perenne*) the spikelets are arranged on a main axis forming a compound spike. In many other species the spikelets are borne on numerous branches given off from the main axis; in these forms the inflorescence is a panicle of spikelets which may be loose, *e.g.* the oat (*Avena sativa*), or close and cylindrical, owing to the shortness of the branches, *e.g.* the fox-tail grass (*Alopecurus*) and Timothy-grass (*Phleum pratense*).

The spikelet (fig. 156) consists of a slender axis bearing a number of scales in two rows. The two basal scales, one on each side, are barren, *i.e.* have no flowers; they are called the **glumes**. The other scales are bracts with flowers in their axils; they are called the lower or *outer paleæ* or flowering glumes. The lower pale sometimes bears a long process called the arista or awn. The number of flowers in a spikelet varies; there may be only one perfect flower; sometimes one or more of the flowers are rudimentary, *e.g.* the oat. The axis of the flower bears a scaly bracteole called the *upper* or *inner palea*; it is opposite the bract. The flower (fig. 157) lies between the upper and lower paleæ. It has usually three hypogynous **stamens** corresponding to the outer whorl of the typical monocotyledonous flower; sometimes only two. The stamens have long slender filaments,

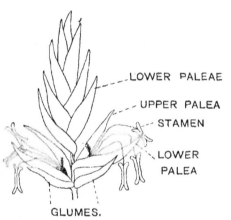

Fig. 156.—TYPICAL SPIKELET OF A GRASS.

and the anthers are versatile and extrorse. The **gynæceum** is monocarpellary, and usually bears two feathery stigmas. The presence of two stigmas gave rise to the view that the pistil is bicarpellary, but the stigmas in Gramineæ are a development of the carpellary margins, not of the apex. The monocarpellary condition is clearly shown by the presence of a fold on the posterior side of the ovary. The ovary is unilocular and superior, and contains one erect anatropous ovule. The **fruit** is a caryopsis (figs. 37, 139, c). The **seed** is albuminous. At the base of the ovary on the anterior side (next the bract) are two little scales called **lodicules**. These together may represent another bracteole. Some consider that they represent two parts of a perianth (petals) otherwise completely suppressed (fig. 158).

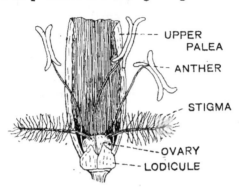

Fig. 157.—TYPICAL FLOWER OF A GRASS. Lower palea removed.

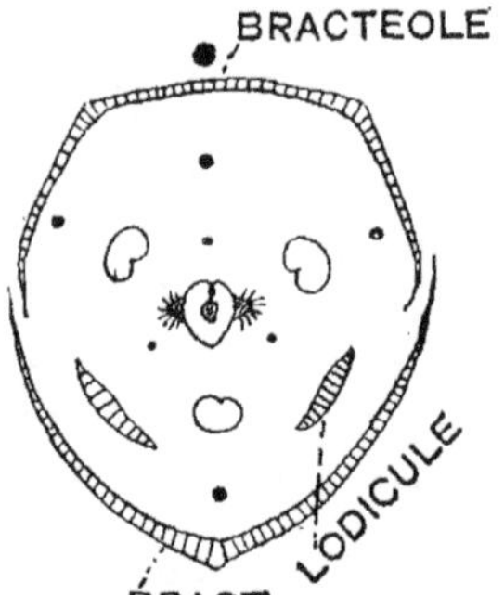

Fig. 158.—FLORAL DIAGRAM OF A GRASS. (Cf. with figs. 120, 155.)

Pollination. The flowers are anemophilous and commonly protogynous. At the time of flowering the lodicules swell up and force the bract and bracteole apart. The filaments of the stamens elongate and the anthers hang out. The pollen grains blown about by the wind are caught by the feathery stigmas.

Familiar grasses are:—maize (*Zea mais*), in which ♂ flowers are borne in a loose apical panicle and ♀ flowers on a stout lateral fleshy spike (spadix) sheathed in leaves, oat (*Avena sativa*), barley (*Hordeum*), wheat (*Triticum*), perennial rye-grass

CLASSIFICATION—SYNOPSIS OF NATURAL ORDERS.   245

(*Lolium perenne*), vernal-grass (*Anthoxanthum odoratum*), meadow-grass (*Poa*), brome-grass (*Bromus*), honey-grass (*Holcus*), etc.

Floral formula :— $K_0 \, C_0 \, A_{3+0} \, G_{\underline{1}}$, or if the lodicules be considered as representing perianth, $K_0 \, C_{0+2} \, A_{3+0} \, G_{\underline{1}}$.

Questions on the Angiosperm.

(Many of these were set at London University Examinations.)

1. Give an account of the external conditions requisite for the germination of a seed and the growth of the embryo, and explain how these conditions operate.

2. What is a sieve-tube? What is its structure? What is its position in a Dicotyledon? What is its probable function?

3. Describe the structure and development of an ordinary foliage-leaf of any Phanerogam, and point out how it is suited to its functions.

4. What is *cell-sap*? Where does it occur and what does it contain? What purpose does it serve in the economy of the plant?

5. Give a diagram of a longitudinal median section through the growing-point of any root continued backward to the tissues which have assumed their permanent character. Also diagrams of *transverse* sections showing :—

(1) The mode of first occurrence of the vascular elements.

(2) The origin of lateral roots.

Give explanatory references in each case.

6. What are cambial tissues? Where, and under what conditions, and whether as transitory or permanent, do such appear in (1) a Dicotyledon and (2) a Monocotyledon?

7. What are the *reserves* available for new growths in plants before they are in a position to absorb and elaborate food independently of such reserves? In what organs and under what forms do they occur?

8. Give a concise summary of the various contrivances favouring the cross-fertilization of flowers.

9. What is meant by transpiration? Describe simple experiments by which its occurrence and nature may be demonstrated.

10. Enumerate the principal morphological types of leaf, and their respective functions in the plant economy.

11. Describe the appearance and structure of a typical stoma. What is its position? What is its function? How does it act?

12. What is a *stele*? Describe the tissues of the stele of a typical dicotyledonous stem.

13. What is meant by carbon-assimilation in plants? What conditions are necessary for the process to go on? What interchange of gases takes place between the plant and the atmosphere during the process? What is the first *visible* product? Where does it appear?

14. Describe the structure of a vascular bundle. How do vascular bundles originate?

15. Draw a series of diagrams representing successive stages in the development of a dicotyledonous embryo from its origin. Name the corresponding parts in the several diagrams.

16. How are new cells formed? Describe the phenomena observed in ordinary cell-division.

17. What is growth? What are the conditions necessary for growth? Give an account of the properties exhibited by growing-points.

18. What is a fruit? Give a description of the fruits of apple, strawberry, orange, fig.

19. What is "*secondary growth*"? Describe the process of secondary growth in a typical dicotyledonous stem.

20. What do you understand by the respiration of plants? How would you demonstrate it?

21. Explain clearly the biological significance of:—(*a*) brightly coloured, (*b*) irregular, (*c*) regular, and (*d*) inconspicuous flowers.

CLASSIFICATION—SYNOPSIS OF NATURAL ORDERS. 247

22. Give a general account of the structure, origin, occurrence and functions of plastids.

23. What is heliotropism? How would you account for the phenomena presented? Give examples to indicate the biological significance of heliotropism.

24. Describe various ways in which seeds may be dispersed.

25. Draw up a table, indicating *characteristic* differences between typical Monocotyledons and Dicotyledons, under the following heads:—
(a) seed, (b) root, (c) stem, (d) leaf, (e) flower.

26. How does the nutrition of an animal differ from that of a green plant, in respect of (1) the substances assimilated, (2) the mode in which they are assimilated?

27. Describe in general terms the mode in which the manufacture of albuminoids is effected by the plant. From what sources and in what form do plants obtain their nitrogen?

28. Describe in detail the structure of a typical ovule just before fertilization takes place.

29. Give a short account of the characters of the flower, fruit, and seed of the Umbelliferæ.

30. Give an account of the process of root-absorption. By what means are roots enabled to absorb substances which are insoluble in water? Enumerate the chemical elements, contained in the substances absorbed, which are essential to the nutrition of green plants.

31. Describe various methods in which plants may reproduce themselves vegetatively.

32. Give an account of the structure, position and function of nectaries.

33. What is root-pressure? How would you demonstrate and measure it?

34. Give an account of the way in which the processes of metabolism in plants lead to the storage of starch in their seeds.

35. What is an inflorescence? Describe the characteristic inflorescences of Compositæ and Umbelliferæ.

PART III.—VASCULAR CRYPTOGAMS AND FLOWERING PLANTS.

CHAPTER XIV.

STRUCTURE AND LIFE-HISTORY OF THE FERN.

§ 1. **General Characters.** The Ferns are by far the most important group of the Vascular Cryptogams. They are for the most part shade- and moisture-loving plants,* and grow abundantly in woods, hedges, and on hill-sides. The fern-plant shows a well-marked differentiation into root, stem, and leaf. The stem has various forms—*e.g.* in the Tree-fern it is aerial, erect, and unbranched; but in most cases it is a *rhizome*, growing either horizontally or obliquely upwards through the soil. The roots are fibrous and *adventitious*, being developed from the surface of the rhizome or from the leaf-bases. The leaves are always large and highly developed. The lamina is sometimes entire (*e.g.* the Hart's-tongue Fern), but is usually much divided or branched. To illustrate the structure and life-history of the group, we shall consider more especially *Aspidium filix-mas*, the Male Shield Fern. Frequent reference, however, will be made to *Pteris aquilina*, the common bracken.

§ 2. **Rhizome, Leaf, and Root.** The rhizome in Aspidium (fig. 159) is almost erect, its apex just reaching the surface. It is a stout structure, and its surface is covered with the numerous withered bases of what were formerly foliage leaves. There is usually no lateral branching, but adventitious buds are developed on the bases of the leaves, and these may separate to form new plants. Lateral

---

* A few ferns are xerophilous, *i.e.* they prefer dry exposed situations.

branching occurs in many ferns, but it is rarely axillary. As the rhizome grows in front, it gradually decays and dies off behind; in this way the adventitious buds or the lateral branches become separated, and form independent rhizomes (vegetative reproduction). The leaf or *frond*,* is large, compound, and much branched.

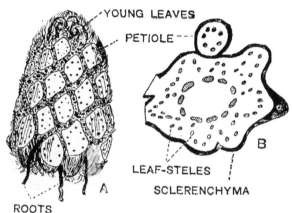

Fig. 159. RHIZOME OF ASPIDIUM.
A, Upper part, from which the older leaves have been cut off at the base and most of the roots removed; B, diagrammatic transverse section.

The dark coloured base of the petiole is continued upwards as the *rachis*, which bears the green, flattened pinnæ and pinnules. A rosette of leaves unfolds each year, but each leaf takes two years to develop. Practically only the petiole is formed during the first year of its growth. All the young leaves and the bases of the old leaves are covered with numerous brown ramenta (p. 52), which are characteristic of ferns in general. The *ptyxis* of the leaf is circinate (p. 123); this also is characteristic; each leaf is rolled on itself like a crosier from the apex to the base. The venation is described as *furcate* or divergent. One main vein enters each pinnule and gives off branches, which bifurcate and end near the margin without anastomosing.

The **fibrous adventitious roots** are developed chiefly from the bases of the leaves.

The dorsiventral rhizome in the **bracken** is an elongated straggling structure which grows horizontally through the soil and branches at intervals. The branching is really lateral, but simulates dichotomy.

* The name applied to the large leaves of ferns.

As in Aspidium, adventitious buds are developed at the bases of the petioles. Only one leaf is unfolded each year on each branch of the rhizome. It unfolds in the spring of the third year after beginning its development. The leaves are separated by long internodes. The rachis is branched. In some ferns adventitious buds are developed on the upper surface of the lamina. They may separate, strike root and form new plants.

§ 3. **Structure of the Rhizome.** The rhizome of Aspidium and of *most* ferns is **polystelic** (p. 83). Instead of having a single stele, as is the case in the stems of most Flowering Plants, it has a number of steles, which can be traced upwards to the apex, where they arise from a number of plerome strands. Fig. 159, B, represents a transverse section of the rhizome. It shows a series of steles arranged in a ring. The ground tissue is chiefly parenchymatous, but there is a hypodermal band of sclerenchyma. In the ground tissue outside the ring of steles there are a number of small steles passing out to the leaves. The central ground tissue being extrastelar should not be called the pith (see definition of pith, p. 81). Fig. 160 shows a portion of the vascular system isolated. The steles fuse at intervals and form a cylindrical network surrounding the central ground tissue. The meshes of the network correspond to the insertion of the close-set leaves, and are therefore called the *foliar gaps*. The steles passing out to the leaves are given off as branches from the edges of the foliar gaps. There is no secondary growth.

Fig. 160.—Part of the Vascular System of Aspidium Dissected out.

§ 4. **Structure of the Stele.** Fig. 161 shows the structure of the stele. The outline as seen in transverse section is more or less oval or elliptical. In the middle is a mass of wood or **xylem** consisting chiefly of long slender scalariform tracheides (fig. 162) and small-celled *xylem parenchyma* (conjunctive parenchyma) containing starch. The stele, according to its size, may have one, two, or three small **protoxylem** groups. These consist of small spiral

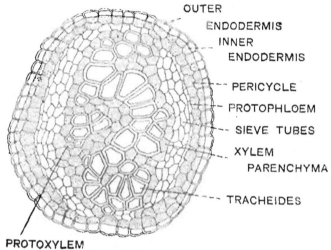

Fig. 161.—STELE OF ASPIDIUM. Transverse section.

tracheides. Frequently one is found at each end of the xylem. The xylem is *surrounded* by the **phloem**. This consists of a layer of sieve-tubes with associated phloem parenchyma (conjunctive parenchyma), and, outside this, a narrow irregular layer of small fibrous cells, the **protophloem**. In longitudinal sections the sieve-tubes are seen to consist of narrow elongated pointed cells with albuminous contents. The sieve-tubes contain no starch; numerous *sieve-plates* are present on their lateral walls only; there are *no companion cells*. These points are characteristic of ferns. Outside the protophloem are the **pericycle** and **bundle-sheath**.

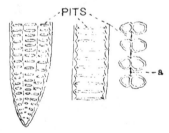

Fig. 162.—SCALARIFORM TRACHEIDES OF FERNS.
A, B, Small portions of tracheide in surface-view; C, portion of the wall in longitudinal section; a = torus of bordered pit.

Typically each of these consists of a single layer (fig. 164); but in Aspidium and other ferns, the bundle-sheath is double round the greater part of the stele. The cells of the pericycle and inner layer of the bundle-sheath contain starch and are called the *phloem-sheath*. The outer layer of the bundle-sheath consists of thickened cells and presents the characters typical of an endodermis. Where the bundle-sheath is a single layer the pericycle constitutes the phloem-sheath. It will be noticed that there is very little intra-stelar or conjunctive ground-tissue; it consists of the xylem-parenchyma, phloem-parenchyma, and pericycle. The bundle-sheath, of course, belongs to the surrounding extra-stelar tissue.

§ 5. **Monostely and Polystely.** It has already been explained (p. 83) that such simple steles were formerly regarded (and are still spoken of) as "cauline concentric" bundles. With regard to the monostelic condition in most Flowering Plants, and the polystelic condition in most ferns, it may be said that in Flowering Plants the necessary amount of vascular tissue is secured by the enormous development of a single stele, while in most ferns it is secured by the branching of steles. In this connexion it should be noticed that the stem of the very young fern is monostelic; in some ferns the monostelic condition persists, but in the great majority the original concentric stele branches as the stem grows in size. There are, therefore, many steles, but each is small, and shows little or no expansion. In each the development of vascular tissue and intra-stelar ground-tissue is slight. We can recognize the number of bundles in each stele by the number of protoxylem groups.

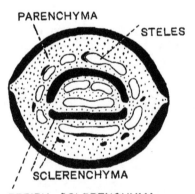

Fig. 163.—RHIZOME OF PTERIS.
(Diagrammatic transverse section.)

STRUCTURE AND LIFE-HISTORY OF THE FERN. 253

§ 6. In the **bracken fern** (figs. 163, 164) there are two series of steles, between which lie two stout bands of sclerenchyma. The outer steles are more numerous, but smaller. The hypodermal sclerenchyma does not form a continuous band; it is interrupted on each side of the rhizome, and at these points the parenchymatous ground tissue reaches

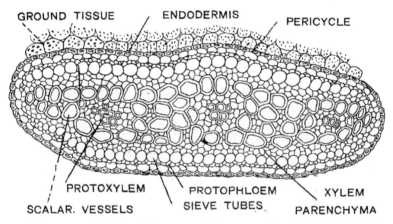

Fig. 164.—STELE OF THE RHIZOME OF PTERIS
(Transverse section: From Green—altered).

the epidermis. It is in this way that provision is made for the transference of oxygen to the more deeply situated tissues. A regular cylindrical network of steles is usually not found in ferns with dorsiventral rhizomes (bracken, etc.). The leaves are few in number, and the foliar gaps are very much elongated. Thus the steles form long, irregular strands, which fuse only at considerable intervals. In the bracken the pitted ends of the large wide tracheides are perforated so that they communicate with each other. For this reason the wood-elements are called *vessels*. This is exceptional amongst Vascular Cryptogams.

§ 7. **Structure of the Root.** The root of ferns is *monostelic*, and the stele is *diarch*. The pericycle and endodermis are single layers of thin-walled cells. In the older parts of the root the cortical tissue immediately outside the endodermis is usually strongly lignified, and forms a stout strengthening sheath. The outer cortical tissue is parenchymatous. The outermost layer is the piliferous layer, or epiblema. There is no secondary growth.

§ 8. Development of Roots.

Here an important difference has to be noticed as compared with the development of the roots of Flowering Plants. Lateral branch-roots, while they are *endogenous* in origin, are not developed from the pericycle, but from the endodermis. They arise opposite the protoxylem. These root-producing cells of the endodermis are called **rhizogenic cells**. In the same way the adventitious roots developed from the rhizome or petiole take their origin in rhizogenic cells of the endodermis investing a stele.

§ 9. Apex of Rhizome and Root (fig. 165).

At the apex of the rhizome there is, as in Flowering Plants, a mass of meristematic tissue. An important difference, however, must be noticed. In the Fern there is at the extreme apex one very large distinct cell from which all the tissues are produced. This is the **apical cell**. There is no such single cell in Flowering Plants. In the rhizomes of most ferns (e.g. Aspidium), this cell is bounded by four walls—three flat walls meeting in a point below, and a curved wall closing in the cell on top. The cell, therefore, is tetrahedral in form, its apex being directed inwards. Segments are cut off, *in succession*, parallel to the flat walls. After the formation of each segment, the apical cell increases to its original size. The segments are indicated in the figure. There are no segments cut off parallel to the curved wall in the rhizome. In ferns with distinctly dorsiventral rhizomes (*e.g.* bracken), there is a two-sided

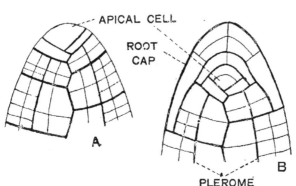

Fig. 165.—APEX OF RHIZOME AND ROOT OF FERN.
A, Rhizome; B, Root. (Diagrammatic longitudinal section.)

instead of a three-sided apical cell, and there are, therefore, only two instead of three series of segments.

The segments cut off undergo division, and thus the tissues of the rhizome are produced. The first division is into inner and outer halves (fig. 165, A). All the plerome strands make their appearance in the tissue formed from the inner halves of the segments just behind the extreme apex. As already explained, they differentiate into the steles. The rest of the tissue gives rise to extra-stelar ground tissue. The layer immediately surrounding *each* plerome strand forms an endodermis. It is evident that there is no distinct dermatogen layer; the outermost layer of tissue is specialized to form the protective external covering, the "epidermis" of the rhizome.

In the **root**, also, there is a single apical cell. It lies just behind the root-cap. In all cases it is three-sided. The segments cut off parallel to the flat walls divide in the same way as in the rhizome; but the inner halves of the segments constitute a single plerome, differentiating into a single stele. The outer halves may be called the periblem. Segments are also cut off parallel to the curved wall, and give rise to the tissue of the root-cap. These segments represent the **dermatogen**, and the root-cap is as usual to be regarded as a many-layered epidermis. The tissue of the root-cap does not persist behind the apex—hence, the piliferous layer is the outermost layer of cortical tissue. The endodermis represents the innermost layer of cortical tissue.

§ 10. **Structure and Development of the Leaf.** The leaf is developed from a single superficial cell of the growing point (exogenous). This cell persists at the apex of the leaf as a two-sided apical cell until an adult condition is reached. One or more branch-*steles* enter the petiole from the rhizome (several in Pteris and Aspidium, figs. 159 B, 160). The steles branch out into the pinnæ, but in the pinnules usually break up into schizosteles (p. 126). The mesophyll of the leaf-blade shows a differentiation into palisade and spongy layers, although not so distinctly as in the bifacial leaves of Flowering Plants. The ordinary epidermal cells have chloroplasts. Stomata are confined to the lower surface of the leaf.

§ 11. **Sporangia and Spores.** Early in the summer a number of structures called **sori** appear on the under surface of the pinnules of the leaves. These are at first of a light green colour, but when older they become dark brown (fig. 166). They are developed immediately over the veins. If a young sorus be carefully removed and examined under the low power, it will be found to consist of a collection of very small stalked bodies called **sporangia** which are covered over and protected by a horse-shoe-shaped scale called the **indusium**. The sporangia and indusium are both developed on a little cushion of tissue, the **placenta**, formed immediately over a vein. The relative position of these various parts is clearly shown in fig. 167, which represents a transverse

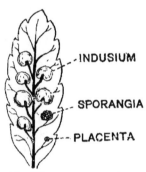

Fig. 166.—PINNULE OF ASPIDIUM BEARING SORI.

The indusium has been removed from one sorus, the indusium and sporangia from another.

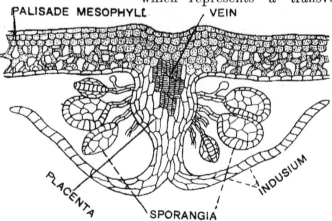

Fig. 167.—SPORANGIA OF ASPIDIUM.
Transverse section through a pinnule and sorus.

section of a pinnule passing through a sorus. In some

STRUCTURE AND LIFE-HISTORY OF THE FERN. 257

ferns in which the sori are arranged as in Aspidium there is no indusium. A sorus, therefore, may be described as a collection of sporangia developed on a placenta, either with or without an indusium. The fully formed sporangium (fig. 168) is a small structure consisting of a tiny capsule borne on a slender multicellular stalk. The stalk in Aspidium often bears a little glandular cell (fig. 167), the function of which is doubtful. The capsule is biconvex, and its wall consists of a single layer of cells. The cells are small and thin-walled except round the edge of the capsule, where they are large, specially thickened, and cuticularized. This specialized layer, which, however, is incomplete on one side, is called the **annulus**. Inside the capsule lies a loose powdery substance which on examination is found to consist of extremely small reproductive bodies called **spores**. Typically sixty-four of these are produced inside each capsule. A sporangium, therefore it a reproductive organ containing spores. The spore, which is of a brown colour and irregular or somewhat triangular in shape (fig. 168), is a single cell (unicellular)

Fig. 168.—SPORANGIUM AND SPORE OF FERN.

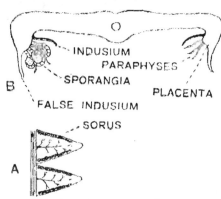

Fig. 169.—SPORANGIA OF PTERIS.
A, Two fertile pinnules; B, Transverse section of pinnule.

consisting of protoplasm and nucleus invested by a wall which is differentiated into two layers or coats. The inner layer, called the *endosporium*, is thin, and consists of cellulose; the outer, called the *exosporium*, is thickened and cuticularized.

Pteris (fig. 169). The sporangia and spores have the same structure and appearance as in Aspidium, except that no glandular cell is developed on the stalk of the sporangium. The sporangia, however, are differently arranged. Instead of being grouped together in small sori, they are developed in a continuous series on a placenta running along the under margin of the pinnule. In other words, there is a continuous **linear sorus**. On the inner side of the placenta in the common bracken (not in all species of Pteris) there is borne a delicate membrane of a yellow colour, representing a true membranous indusium. The margin of the pinnule also, by bending over, serves to protect the sporangia, and it is termed a *false indusium*. Between the sporangia are epidermal outgrowths (hairs) called *paraphyses*.

§ 12. **Development of the Sporangium** (fig. 170). The sporangium is developed from a single epidermal cell of the placenta, and may therefore be regarded as a very highly specialized epidermal outgrowth. The cell in question grows out and forms a little protuberance, which is cut off by a wall. It is then divided transversely into two cells (A). The lower cell, by further longitudinal and transverse divisions, develops into the stalk. The upper cell produces the capsule. First of all (B) a series of outer cells is divided off from a large central tetrahedral cell by four walls resembling those bounding the apical cell of the rhizome, *i.e.* three flat and one curved (only two of the flat walls can be shown in the figure). The outer cells by further divisions at right angles to the surface, form the single-layered wall of the sporangium. The tetrahedral cell is called the **archesporium**. It undergoes division by walls formed parallel to the first set (C). The outer cells so cut off undergo further division, and are called **tapetal cells**. The remaining central cell is the *archesporium proper*; by repeated cell-division it produces, typically, sixteen **spore-mother-cells** (D). These

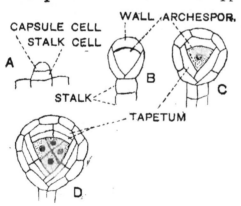

Fig. 170.—Development of the Sporangium of the Fern.

separate from each other, and, owing to the disorganization of the tapetal cells, float freely in a fluid filling the cavity enclosed by the sporangium wall. The nucleus of each mother-cell divides karyokinetically into two, and these two divide again, so that four nuclei are formed. Then cell-walls are laid down between the nuclei, and thus the mother-cell is divided into four cells, called the "*special mother-cells*," in each of which a spore is formed by rejuvenescence (p. 39). The protoplasmic contents of each special mother-cell form a cell-wall which is differentiated into exosporium and endosporium. The walls of the special mother-cells are disorganized and the spores lie free in the cavity of the sporangium. The developing spores are partly nourished by the fluid formed by the breaking down of the tapetal cells, and food-materials are conveyed to them through the stalk of the sporangium. During development some of the cells of the wall of the sporangium are specialized to form the annulus. The archesporium may be defined as the meristematic cell or cells (single cell in the Fern) which are found in a developing sporangium and give rise to the spores.

§ 13. **The Sporophyll.** A leaf bearing sporangia is called a sporophyll. In most ferns the sporophylls more or less resemble ordinary foliage leaves (Aspidium and Pteris). They are simply leaves which function both as vegetative and reproductive organs. In some ferns, however, the sporophylls differ considerably from the foliage leaves, *i.e.* they are more or less specialized (e.g. *Osmunda regalis*, the Royal Fern); but there is never a specialized reproductive *shoot* distinct from the ordinary vegetative one (see p. 7).

§ 14. **Germination of the Spore** (fig. 171). When the sporangium is ripe, the cells of the annulus become dry, contract, and thus put a strain on the thin part of the edge of the capsule. The latter bursts open at this point, the *stomium*, and the spores are set free. If a spore falls in a suitable soil, it germinates. For germination a supply of oxygen and sufficient warmth and moisture are required, and when these conditions are fulfilled the exosporium is ruptured, and the endosporium grows out into a short tube. From this, a colourless hair resembling a root-hair, and called a root-hair, arises, and passes down into the soil. The tube (germ-tube) elongates, and forms at first a short filament divided into cells by a series of parallel transverse divisions (fig. 171, A). After that, divisions are formed in the other two planes, and a small green flat

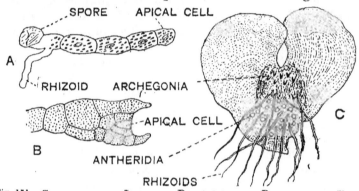

Fig. 171.—Germination of Spore and Development of Prothallus of Fern.

plate of tissue is produced. This, in early stages, grows by means of a two-sided apical cell (B), but later by a group of meristematic cells. Owing to the more rapid growth of the marginal cells the plate eventually becomes more or less heart-shaped. The structure thus developed is called the **prothallus** or *prothallium* (c).

§ 15. **The prothallus** is a very small flat plate of tissue, measuring only one-fourth or one-third of an inch across. It consists of rounded parenchymatous cells, containing numerous chloroplasts. Towards the margin it consists of a single layer, but in the central region it is thickened, owing to the division of its cells parallel to the surface. This thickened region is called the *cushion*. Long, brown unicellular **root-hairs** (also called *rhizoids*) are developed from the cells of the under surface, and pass down into the soil. It will be recognized that the prothallus is an *independent plant*. By means of its chlorophyll it can assimilate the carbon dioxide of the atmosphere, and by the aid of its root-hairs it absorbs nutritive salts from the soil It is a distinct self-sustaining plant, whose vegetative body is a thallus (p. 6). Owing to its thinness gases can penetrate to all parts with relative ease, and hence no stomata are developed upon it.

§ 16. **The Sexual Reproductive Organs** (fig. 171, c) are produced on the *under* surface of the prothallus—the **antheridia**, or *male sexual organs*, on the posterior region, the **archegonia**, or *female sexual organs*, on the cushion in the anterior region near the notch of the heart-shaped prothallus. The antheridia are developed first. The *antheridium* (fig. 172, c) is a spherical capsule, the wall of which consists of a single layer of cells containing chloroplasts.* Inside are a number of small cells called **spermatocytes**, or *spermatozoid mother cells*, each of which gives rise to a male sexual cell, or gamete, the **spermatozoid** (or *antherozoid*). The *archegonium* (fig. 173, E) is a flask-shaped organ, consisting of two parts; (*a*) a swollen basal portion, the *venter*, completely sunk in the tissue of the prothallus; (*b*) a more slender portion called the *neck*, projecting freely from the surface. In the cavity of the venter lies a rounded nucleated mass of protoplasm, with nucleus and nucleolus. It has no cell-wall.

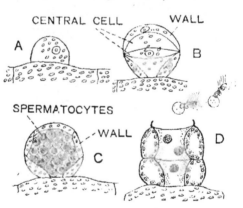

Fig. 172.—Development of Antheridium of Fern.

This is the female sexual cell, or gamete, called the **ovum**, *oosphere*, or egg-cell. Its nucleus is sometimes called the germinal vesicle, its nucleolus the germinal spot. The neck consists of four longitudinal rows of cells surrounding a central canal, which is at first closed at the apex, and which leads down into the venter. The neck is

* In some forms the antheridium has a short unicellular stalk.

not straight, but bends so as to face towards the antheridia. At the junction of venter and neck canal there is another smaller nucleated protoplast, the **ventral canal-cell**, and the canal of the neck itself is filled with a protoplasmic mass containing several nuclei, and called the **neck-canal-cell**.

§ 17. **Development.** The *antheridium* (fig. 172) is developed from a single cell of the prothallus. The cell grows out to form a papilla-like outgrowth, which is cut off by a cell-wall. It increases in size, and is divided by a series of divisions, which mark off a number of peripheral cells, forming the wall, from a central cell in which the spermatocytes are formed. The *archegonium* also is developed from a single cell (fig.173, A-F). This cell

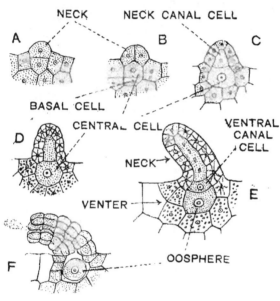

Fig. 173.—DEVELOPMENT OF ARCHEGONIUM OF FERN.

divides into three. The basal cell (B) forms a portion of the tissue surrounding the base of the archegonium. The outermost cell (B) is divided by two walls at right angles (only one can be shown in the figure) into four cells. These undergo further transverse divisions to form the four longitudinal rows of the neck (C-F). The protoplasm of the central cell (B) pushes its way between the neck-cells, and a small portion is cut off as the neck-canal-cell (C). The remainder of the protoplasm of the *central cell* undergoes division to form the oosphere and the ventral canal-cell (E). The cavity of the venter of the archegonium is partly lined by prothallus cells.

§ 18. **Fertilization** (see p. 40). When mature the antheridium absorbs the water present as a film on the under surface of the prothallus, and ruptures (fig. 172, D). The spermatozoids are thus set free. Each is seen to be a spirally coiled filament of protoplasm. It has a slender tail and stouter head. The tail bears laterally a number of extremely fine protoplasmic *vibratile* threads called **cilia**. The nucleus forms the greater part of the head, and it is developed from the nucleus of the spermatocyte. The protoplasm of the mother-cell forms the ciliated tail together with a little vesicle, containing starch grains, which remains for some time attached to the head of the spermatozoid. The spermatozoids move about in the water by means of their cilia. Sooner or later they come into the neighbourhood of the archegonia. When the archegonium is ripe the two canal-cells are disorganized and give rise to a mucilaginous substance which oozes out of the neck of the archegonium (fig. 173, F). This substance contains malic acid, which attracts the spermatozoids (chemiotaxis—see p. 198). They cluster round the neck of the archegonium, and, finally, one enters the canal and passes down to the venter. It penetrates the oosphere and its nucleus fuses with that of the ovum (cf. Angiosperm, p. 198). The fertilized oosphere forms a cell-wall and is then called the *oospore*. Although the prothallus bears as a rule both kinds of sexual organs, and is therefore hermaphrodite (p. 162), cross-fertilization generally takes place, the spermatozoids developed on one prothallus passing to the archegonia of another. This is necessary because antheridia and archegonia are not developed simultaneously on a prothallus (cf. dichogamy, p. 194). Sometimes, in badly nourished prothalli, only antheridia are developed.

§ 19. **Development of the Young Fern-Plant** (figs. 174, 175). The oospore begins to divide or segment, and this process of segmentation leads finally to the development of an embryo. The first division-wall is nearly parallel to the long axis of the archegonium. It is called the **basal wall**, and divides the oospore into anterior or **epibasal** and posterior or **hypobasal** halves. A second wall, the transverse or quadrant wall, at right angles to the basal wall, divides the oospore into upper (superior) and lower (inferior) halves. The oospore now consists of four cells (quadrants). Then a median or octant wall at right angles to the first two divisions divides the oospore into right and left halves. The oospore now consists of eight cells or **octants** (fig. 174). Of the two superior anterior octants one becomes the apical initial cell of the **primary stem**, the other takes no special part in the orientation of the members. The two inferior anterior octants give rise to the first leaf or **cotyledon**. Of the two inferior hypobasal octants, one becomes the apical cell of the **primary root**: it is diagonally opposite the cell which produces the primary stem. The two superior hypobasal octants give rise to an embryonic organ called the **foot**.

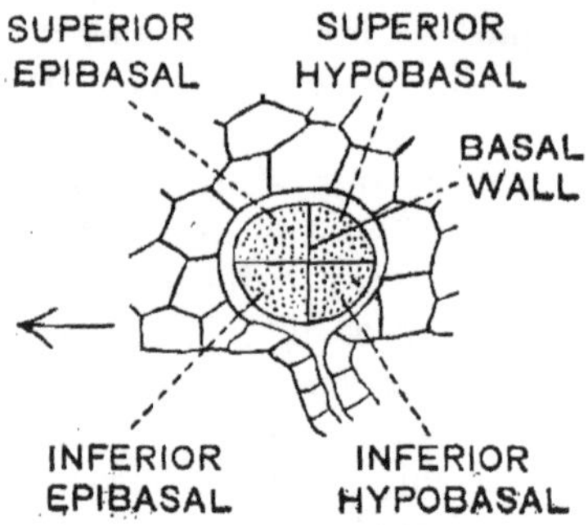

Fig. 174.—SEGMENTATION OF THE OOSPORE OF FERN.
The arrow points anteriorly. (Diagrammatic.)

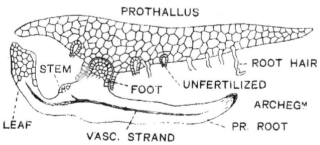

Fig. 175.—EMBRYO OF FERN ATTACHED TO PROTHALLUS.
(Longitudinal section.)

This is a massive structure which absorbs nourishment

from the prothallus for the developing embryo, till the latter can assimilate food material for itself. Further cell-division, of course, takes place in the octants marking out the plerome of root and stem (the young stem is monostelic, § 5). The primary stem and cotyledon break forth from the under surface of the prothallus, then bend upwards, make their way through the notch of the prothallus and come above ground, where they become green.* New leaves are developed, and gradually the stem becomes the rhizome of the fern plant. The primary root is not persistent. At a very early period its place is taken by adventitious roots developed from the stem or rhizome (cf. Monocotyledons). As a rule only one embryo is formed on a prothallus, which then dies. In some ferns, however, the prothallus has a longer life, and may reproduce itself vegetatively by means of branches, or little outgrowths of meristematic tissue called gemmae.

§ 20. **Sporophyte and Gametophyte—Alternation of Generations.** It will be noticed that in the life-history of the Fern there are really two plants to be considered. These are spoken of as the two *stages* or *generations* of the life-cycle. There is first the fern-plant, so called because it is by far the more conspicuous of the two. It is spoken of as the **sporophyte** or asexual generation, because it is the generation which bears the asexual reproductive organs, sporangia and spores. Then there is the prothallus, which is called the **gametophyte**† or sexual generation because it is the generation which bears the sexual reproductive organs and the sexual cells or *gametes* (the ovum and spermatozoid). Now it will be seen that a young sporophyte is *not derived directly* by a sexual process from the parent sporophyte, for a gametophyte generation is interposed between them. In the life-cycle there is an alternation of sporophyte and gametophyte. This is spoken of as the **alternation of generations.** The student must observe this phenomenon very carefully, as it is exhibited by all the

---

* The cotyledons of many ferns are able to turn green in darkness.
† It has also, but less correctly, been termed "oöphyte."

higher plants (Mosses, Vascular Cryptogams, and Phanerogams) in some form or other. We have explained it in connexion with the Fern because it is most clearly exhibited in the group of the Vascular Cryptogams, but it will be shown later (Chapter XVI.) that it is present also in a modified form in Flowering Plants.

§. 21. **The Reproductive Processes.** In the life-history of the Fern we have an illustration of the various kinds of reproduction mentioned in the introductory pages (p. 15), viz.—vegetative, asexual, and sexual reproduction. It will be noticed that both sporophyte and gametophyte begin their development from a single cell, the young sporophyte from an oospore formed as the result of a sexual process

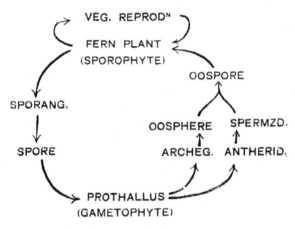

Fig. 176—LIFE-HISTORY OF FERN GRAPHICALLY REPRESENTED.

(fusion of gametes), the gametophyte from a spore formed asexually. This is characteristic of sporophyte and gametophyte wherever there is alternation of generations, the reproductive bodies formed in the one generation giving rise to the other. In the life-cycle of the Fern, then, there is an alternation of generations, and asexual spore-reproduction forms an integral part of it. Vegetative reproduction is very clearly distinguished. It has no share in the alternation of generations, and simply lengthens the life-cycle, either at the sporophyte stage or at the gametophyte stage, *i.e.* either generation may directly and indefinitely reproduce itself by vegetative methods, without the intervention of the other generation. These points may be illustrated and the general life-history shown graphically in the form of a diagram (fig. 176).

§ 22. **Apogamy and Apospory.** While the reproduction of sporophyte and gametophyte is usually effected as above described, there are in some ferns exceptional cases where either (*a*) the spore stage, or (*b*) the sexual process is, as it were, cut out of the life-cycle. The former condition is called *apospory*; the latter *apogamy*. Thus in a few ferns the prothallus may arise vegetatively from the tissue of the sporophyll, instead of from a spore.

Various grades of apospory exist: thus—(1) the formation of spores may be suppressed and the prothalli may develop directly from the young sporangia,—(2) prothalli may develop from the placentas in the place of sporangia,—(3) they may develop vegetatively from any portion of the frond, without any indications of the formation of sori, sporangia or spores being exhibited.

In apogamy the young sporophyte may arise directly as a bud from the tissue of the prothallus without the intervention of sexual organs. This is vegetative apogamy (one species of Pteris, and one variety of *Aspidium filix-mas*). Or it may be developed from the *unfertilized* oosphere. This phenomenon of the development of an embryo from an unfertilized egg-cell is called *parthenogenesis*, and we have here, therefore, a case of parthenogenetic apogamy. It will be noticed that these conditions tend to shorten the life-cycle, and to replace ordinary spore-reproduction or sexual reproduction by a kind of vegetative reproduction.

It has recently been found that the prothalli of a few cultivated ferns may actually produce normal sporangia containing fertile spores, and thus reproduce themselves by asexual means. The spores on germination produce the usual prothalli, which may ultimately give rise to a sexually produced fern plant. By such means as these the asexual generation may be suppressed for one or more life-cycles, which latter may subsequently resume their normal succession.

CHAPTER XV.

EQUISETUM AND SELAGINELLA.

§ 1. BEFORE proceeding to the consideration of the Gymnosperm it will be advisable to lay before the student some interesting points in connexion with the life-histories of two other types of the Vascular Cryptogams. These are Equisetum, the horse-tail, and Selaginella. These types, more especially Selaginella, are in many respects intermediate between the Ferns and Flowering Plants, and some acquaintance with their life histories is absolutely necessary if we wish to trace clearly the morphological and developmental resemblances, i.e. *homologies*, which exist between the Vascular Cryptogams and Flowering Plants.

A. EQUISETUM.

§ 2. **General Structure.** The plant which we call the horse-tail is the *sporophyte*. It consists of a branching horizontal rhizome giving off *aerial* shoots and numerous adventitious roots. The leaves borne on the aerial shoots are small and scaly. They are arranged in whorls, and in each whorl are fused together to form a sheath, which invests the base of the internode above. Whorls of axillary branches may be produced at the nodes. At the apex of the aerial shoot the leaf-organs have a totally different appearance. They are free and have the form of *stalked peltate discs* (fig. 177), which are closely packed together in whorls, so that a cone-like mass is formed at the apex of the shoot. These peltate leaves are **sporophylls**. This specialization and aggregation of sporophylls should be

carefully noticed. The reproductive region
of the shoot is quite distinct from the
vegetative portion (see p. 7). Each sporophyll has on its under surface a group of
sporangia containing **spores**. The sporangium is developed from *a group of cells*, not
from a single **cell** as in the Fern. After
the wall of the sporangium and the tapetal
layers have been formed a single large
archesporial cell remains, which divides to
form spores. The sporangiferous "spikes,"
as the cone-like structures of *Equisetum* are
called, are borne either at the apices of
ordinary vegetative shoots, or, in some
species, on special fertile or reproductive
shoots which resemble the others except
that they are unbranched and contain little
or no chlorophyll.

§ 3. **General Life-History.** The spores
are all of one kind, Equisetum, like the Fern,
being *homosporous*. When the spores fall
to the ground and germinate, however, they

Fig. 177,—EQUISE-
TUM.
A, Apex of fertile
shoot; B, A
Sporophyll.

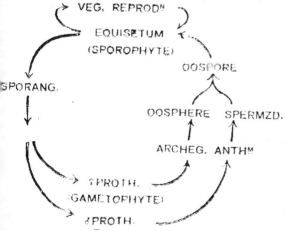

Fig. 178,—LIFE-HISTORY OF EQUISETUM GRAPHICALLY
REPRESENTED.

usually give
rise to **pro-
thalli** *of two
kinds*. Some
of the spores
produce pro-
thalli bearing
antheridia
only; others
produce pro-
thalli bearing
archegonia
only. The

student will remember we have an indication of the same thing occasionally in the Fern (p. 263). It has become the rule in Equisetum. The differentiation of sex has, as it were, been carried back from the sexual organs (antheridia and archegonia) to the structures bearing these sexual organs, so that we may now speak of male and female prothalli. The prothalli are unisexual, and the gametophyte is represented by two plants. *As a rule the male prothalli are much smaller than the female ones.* In other characters they resemble those of the Fern, as do also the sexual organs. Fertilization is effected and an embryo sporophyte developed in much the same way. The life-history may be graphically represented as in fig. 178.

B. SELAGINELLA.

§ 4. General Characters (fig. 179). The plant is the sporophyte. The external vegetative characters vary considerably in the different species, which number over three hundred, and of which only one is a native of Britain—namely, *S. spinosa* (or *selaginoides*). Many species are small, moss-like plants, with creeping stems and dorsiventral symmetry. Others are larger, and more or less erect and isobilateral. In the single British species, while the main stem is creeping, the *branches* are erect and have radial symmetry. Some foreign species are climbing plants, often growing to a great height.

The slender **stem** usually bears four rows of leaves—two rows of small dorsal leaves on the upper surface, two of larger ventral leaves at the sides of the stem. The arrangement of the leaves seems to be opposite and decussate, one large and one small leaf apparently arising at each node. The leaves, however, are somewhat twisted, and close examination shows that each leaf arises from it own node, the dorsal leaf being slightly above the opposite ventral one. In the British species all the leaves are of the same size and the spiral arrangement is immediately obvious. In all the species a small membranous **ligule** is developed on the upper surface of the leaf at its base.

The **branches** are developed from lateral buds which

become visible near to the apex of the stem, but since they develop almost as rapidly and as strongly as it does, the branching assumes a dichotomous appearance. The branching is *not axillary*, and the branches all lie in one plane.

The **roots** in some species are developed adventitiously on the stem; in other species they are borne on peculiar specialized branches, called **rhizophores**. These organs are intermediate in structure and development between roots and stems. Like stems, they have no root-cap, and are developed exogenously; they resemble the true roots in internal structure and in the fact that they bear no leaves or reproductive organs. We may infer, however, that they are specialized stem-branches from the fact that *occasionally* they are found to develop into ordinary shoots.

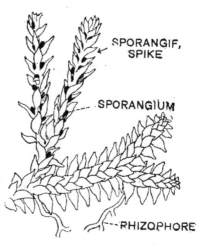

Fig. 179.—SELAGINELLA HELVETICA.

The rhizophores, when present, are given off from the lower surface of the stem, one from below each point where an ordinary branch arises. They grow down to the surface of the soil without branching, but on reaching the soil give rise, by endogenous development, to a number of true roots.

The **reproductive organs** (figs. 179, 181) are produced at certain periods towards the apices of fertile or reproductive shoots. These reproductive shoots are more or less *erect*, and, in nearly all the species, bear spirally arranged sporophylls, which do not differ very markedly from ordinary foliage leaves. The reproductive organs are **sporangia** and **spores**. *One* sporangium is developed *in the axil* of each leaf of the reproductive shoot. The sporangia are of two kinds, **megasporangia** and **microsporangia**, each of the former containing usually four large **megaspores**, each of the latter a large number of small **microspores**. Thus Selaginella is *heterosporous*. The same sporangiferous "spike" usually bears both kinds, the microsporangia in the axils of the upper leaves, the megasporangia in the axils of the lower, although this is not always the case, for the megasporangia may occur in the middle of the cone. There is no special vegetative propagation.

§ 5. **Stem** (fig. 180). In some species there is a single apical cell as in the Fern; in others, a group of two or three initial cells which by their divisions give rise to the tissue of the stem. This may be regarded as an intermediate condition between the single cell of the Fern, and the small-celled meristem, showing dermatogen, periblem and plerome, which is characteristic of Flowering Plants. In some species the stem is *monostelic*; in most species, *polystelic*, with two or three steles. Each **stele** is suspended in the middle of a large air space by a number of delicate trabeculæ, which represent the

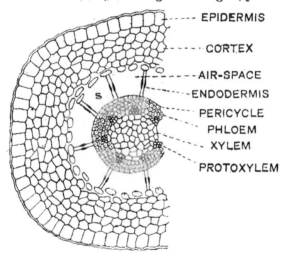

Fig. 180.—Stem of SELAGINELLA SPINOSA
(Tranverse section.)

stretched endodermis. These often have silica deposited upon them in the form of irregular plates, or annular deposits (fig. 180, s). The stele is concentric. The central wood, or **xylem**, consists of slender scalariform tracheides. According to the species one, two, or a number of **protoxylem** groups can be distinguished, consisting of small annular and spiral tracheides—that is, the stele may be mon-arch, di-arch, or poly-arch. A striking peculiarity with regard to the protoxylem is that it lies at the periphery of the xylem, so that the differentiation of xylem tissue is centripetal, *i.e.* moves towards the centre of the stele (cf. differentiation in roots, p. 99). The **phloem** consists of thin-walled elongated cells, representing the sieve-tubes of higher types. The sieve-plates are lateral. Outside the phloem is a **pericycle**, consisting of either one or two layers of cells. The steles run to the apex of the stem, and never exhibit any secondary growth. In the British species there is a single polyarch stele (fig. 180).

The **ground-tissue** of the stem consists of comparatively thin-walled, more or less prosenchymatous cells, without intercellular spaces (thin-walled prosenchyma, p. 42). The **epidermis** also consists of elongated pointed cells, and has no stomata.

§ 6. **The Leaf** is entire, and has a very simple structure. The epidermal cells contain chloroplasts. Stomata are usually confined to the lower surface. The ground-tissue (mesophyll) is not clearly differentiated into palisade and spongy layers. A single vascular strand, given off opposite a protoxylem group from the stele of the stem, runs through it. This strand is concentric. The central xylem is surrounded by a layer of phloem, outside which is an endodermis.

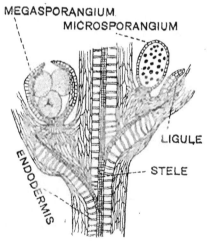

Fig. 181.—Part of "Sporangiferous Spike" of Selaginella. (Longitudinal section.)

§ 7. **Rhizophore and root** are both monostelic, and agree in internal structure. The stele is mon-arch. The root grows by a single apical cell; the rhizophore may have a single cell or a group of cells, according to the species, like the stem.

§ 8. **Sporangia and Spores** (fig. 181). The sporangium consists of a capsule borne on a short, stout stalk. The wall of the capsule consists of two layers of cells, and has no annulus. The megasporangium is somewhat larger than the microsporangium. The spores as usual have two coats—endosporium and exosporium, the latter being cuticularized. A large amount of food substance is stored up inside the megaspore, consisting chiefly of oil. Owing to the fact that the spores are developed by tetrahedral division (see below) they are pointed at one end.

§ 9. **Development of Sporangium** (fig. 182). The sporangium is developed from a group of meristematic cells. It is first seen as a little papilla-like outgrowth in the axil of a young leaf, close to the apex of the sporangiferous shoot. The outermost layer forms the wall of the sporangium. But at an early stage beneath this outermost layer *a row of large cells* can be recognized, and this is the **archesporium**. The archesporial cells by division form a **tapetal layer** towards the apex of the papilla. In the lower part of the developing sporangium the tapetum is formed from the cells surrounding the archesporium. Then the archesporium, by repeated division, gives rise to spore mother-cells, as in the Fern. So far the development of both mega- and microsporangia is the same, but from this point differences are observed. In the microsporangium the mother-cells separate from each other, and float freely in a nutritive fluid formed by the disorganization of the tapetal cells. Then

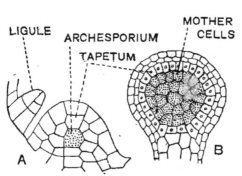

Fig. 182.—SELAGINELLA. DEVELOPMENT OF SPORANGIUM.

In A only one archesporial cell is shown.

in each mother-cell four *special mother-cells* are formed, and each of these gives rise to a microspore in the same way as in the Fern. The micropores are tetrahedrally arranged; they are not formed in the same plane. In the megasporangium one of the mother-cells increases in size, and produces four megaspores in the same way as a mother-cell produces microspores. The other mother-cells are disorganized and serve as nourishment to the developing megaspores.

At an early stage in development the outermost layer of the papilla divides into two, so that the wall of the sporangium is double. The stalk of the sporangium is formed by division and growth of cells at the base of the original papilla. One layer of tapetal cells persists, so that the wall seems to consist of three layers.

§ 10. **Germination of the Spores—Prothalli and Sexual Organs.** The germination of the megaspore begins before it is set free from the sporangium. The nucleus of the spore divides into two. One daughter-nucleus passes to the apex or pointed region of the spore, the other to the basal region. A process of *free cell-formation* then begins. It is most active in the apical region, and there results in the production of a tiny mass of small-celled tissue. In the lower region the process is much less active, and actual cell-formation may not take place there till after the

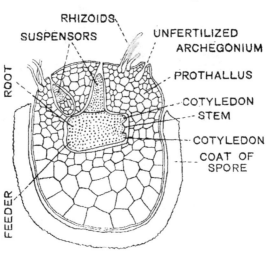

Fig. 183.—FEMALE PROTHALLUS OF SELAGINELLA. (Longitudinal section.) An old stage, showing two developing embryos.

spore has fallen to the ground. The cells formed in this region are larger and filled with food-material. The megasporangium ruptures near the apex by a transverse slit, and the spores fall to the ground. The megaspore ruptures at the apex by a triradiate fissure which exposes the small-celled tissue immediately underneath. On this an **archegonium** is developed, and others are formed later if fertilization is not effected. It is evident that the tissue formed as described in the megaspore is the **female prothallus** (fig. 183).* It protrudes slightly, turns green in the presence of light, and may even develop one or two root-hairs, but it isnot set free from the spore as an independent plant, like the prothallus of the Fern or of Equisetum. It is nourished by the food-material stored up in the spore. This reduction of the female prothallus to a minute and practically dependent structure should be carefully noticed.

The structure and development of the archegonium (fig. 184) are practically the same as in the Fern. The only difference is that the neck is shorter, consisting of only eight cells; each of the four longitudinal rows of the neck consists of only two cells.

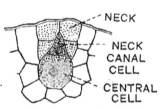

Fig. 184.—Young Archegonium of Selaginella.

The microsporangium is ruptured in the same way as the megasporangium, and the microspores fall to the ground and germinate. The microspore increases in size, and a small cell is cut off at the pointed end (fig. 185). Then the rest of the spore divides into ten or twelve cells, eight peripheral cells surrounding either two or four central cells (according to the species). The *central* cells undergo further division, and the small cells thus produced are the mother-cells of spermatozoids. In each a *biciliate* spermatozoid is formed in exactly the

* Some speak of the small-celled apical portion only as the female prothallus, and call the later-formed lower portion, which consists of larger cells distinctly marked off from the small-celled portion, the secondary prothallus or endosperm.

same way as in the Fern. The small cell first cut off represents an extremely rudimentary **male prothallus**, and may be called the *prothallus-cell*. The eight peripheral cells represent the wall of an **antheridium**, inside which the spermatozoids are produced. The extreme reduction of the male prothallus is of great interest, and is correlated with the small size of the microspore. The megaspores remain large because of the necessity of providing nourishment for the young sporophyte, and owing to this fact they are able to produce a number of recognizable archegonia. During the course of the above development the exosporium is ruptured. Later the peripheral cells undergo disorganization and nourish the spermatozoid mother cells. Eventually the spermatozoids are set free.

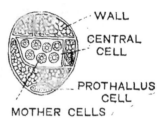

Fig. 185.—Germination of Microspore of Selaginella.

§ 11. **Fertilization and Development of Embryo** (figs. 186, 183, 187). The process of fertilization is essentially the same as in the Fern and Equisetum. A spermatozoid enters the ovum and fuses with it, the male and female nuclei amalgamating into one. The **oospore** which is thus formed segments and develops into an embryo sporophyte. The first division is at right angles to the axis of the archegonium It divides the oospore into an upper or *hypobasal* half, and a lower or *epibasal*. The hypobasal half either remains unicellular or undergoes only a few divisions, the resulting structure being called the **suspensor**. The epibasal half is divided into quadrants by two walls at right angles to each other, and from its apical tier (fig. 186) are developed the stem and two cotyledons. The base of the stem (hypocotyl)

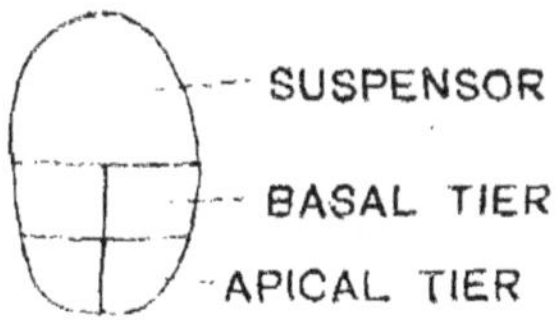

Fig. 186.—Segmentation of Oospore of Selaginella. (Diagrammatic.)

The epibasal half is divided transversely into basal and apical tiers of cells; the basal tier forms the hypocotyl—the apical, the stem and cotyledons.

becomes enlarged and forms a massive organ serving as a "feeder," and formerly regarded as equivalent to the *foot* of the Fern (p. 264). The true homologue of the foot, however, is the suspensor developed like the foot of the Fern from the hypobasal half of the oospore. There is *no primary root* in Selaginella. The *first* root is developed from the base of the stem, and is therefore adventitious. The developing embryo grows down into the lower part of the prothallus; the "feeder" absorbs the food-material. Eventually the stem and cotyledons escape from the spore

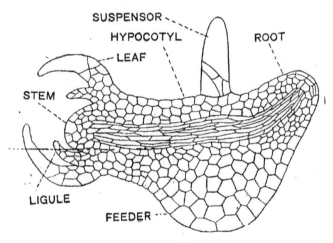

Fig. 187.—EMBRYO OF SELAGINELLA.
(Longitudinal section.)

and grow above ground, while the first and other adventitious roots pass down into the soil.

The life-history may be graphically represented as in fig. 188.

§ 12. The following important points should be noticed. In the first place the differentiation of sex has been carried back another stage. We not only have two kinds of prothalli, as in Equisetum, but these prothalli are developed from spores of quite different appearance. The student will now have some perception of the origin of the heterosporous condition in plants. In the second place, the *reduction* of the male and female prothalli, *i.e.* the gametophyte, has to be noticed. In Selaginella the gametophyte, so far as nourishment is concerned, is not independent of the sporophyte.

The life-histories of Fern, Equisetum, and Selaginella should be very carefully compared. While there are many differences, by reason of which these three types are placed in different classes of the Vascular Cryptogams, the student will recognize that the *general* course of the life-history is very similar. In all three there is a more or less distinct alternation of generations and equivalent or *homologous* structures occur at the same points in the life-history.

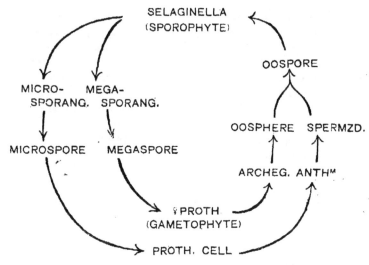

Fig. 188.—Life-History of Selaginella graphically represented.

The graphical life-histories which have been given will enable the student to grasp more readily the more important homologies discernible. The degeneration of the gametophyte does not cease at this point, but is continued so far, that in Flowering Plants the entire sexual stage is included in the spore, and the microspores (pollen-grains) and megaspores (embryo-sacs) become practically sexual bodies, from a physiological point of view (cf. Chap. XVII.).

CHAPTER XVI.

THE GYMNOSPERMS—STRUCTURE AND LIFE-HISTORY OF PINUS.

§ 1. **General.** The Gymnospermous Flowering Plants are not so highly differentiated as the Angiosperms, and in many respects they resemble the Vascular Cryptogams, forming as it were an intermediate group. They are large plants, either shrubs or trees, and include the Cycads, which in many of their characters approach the Vascular Cryptogams, the Conifers, and a third group, the Gnetaceæ, which approach nearer to the Angiosperms. Like the Angiosperms, their reproductive organs are aggregated to form flowers, which differ markedly, however, from the flowers of the Angiosperms in the fact that when carpels are present the ovules are borne freely exposed on their upper surfaces. The carpels are not closed up to form ovary, style, and stigma. Hence the name *Gymnosperm* (Gr. γυμνος, naked, σπερμα, a seed).* The flowers are always unisexual; the plants usually monœcious, sometimes diœcious (yew, juniper).

The most important group of the Gymnosperms is that of the **Coniferæ** or cone-bearing trees, so called because their flowers usually have the characteristic form known as *cones.* Of these our woods and shrubberies contain many familiar examples, such as the Pines, Firs, Larches, Yews, Spruces, Cypresses, Junipers, Araucarias, etc. The true pines constitute the genus *Pinus,* many species of which are grown in Britain, but only one is native, viz.

* Angiosperm from Gr. αγγος, a vessel, σπερμα, a seed.

GYMNOSPERMS—STRUCTURE AND LIFE-HISTORY OF PINUS. 281

P. sylvestris, familiarly known as the Scots *Fir*. As this is one of our commonest trees, we have selected it as our special type of the cone bearing Gymnosperms. Reference, however, will be made to other types.

§ 2. **External Characters of Pinus** (fig. 189). The full-grown plant is a large tree. It has an elongated **tap-root** giving off stout branched lateral roots in regular succession. The main stem is cylindrical and covered with a rugged

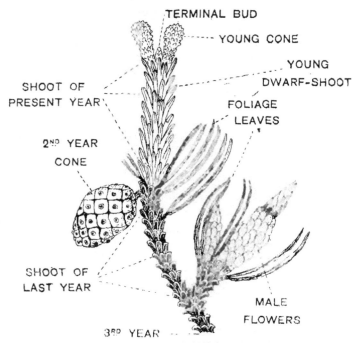

Fig. 189.—BRANCH OF *Pinus sylvestris* CUT IN MAY.
Most of the older dwarf-shoots and foliage leaves removed.

scaly bark (p. 89) Secondary growth takes place in the same way as in Dicotyledons, and hence the stem tapers towards the apex. The branches are formed in apparent whorls from lateral buds developed on each parent axis. These are formed in the axils of scale-leaves, at the end of each year's growth. This regular development of branches gives the tree a very symmetrical appearance, which is, however, frequently spoilt owing to the loss of many of the branches. In addition to the ordinary branches, which, as they grow indefinitely, are called *shoots of unlimited growth*, there are numerous **dwarf-shoots** (p. 66), or *shoots of limited growth*. These also arise in the axils of brown scale-leaves or cataphyllary leaves borne on the main branches. The leaves are of two kinds:—(*a*) the **cataphyllary** leaves just mentioned, which are the *only* ones borne on the shoots of unlimited growth, and which are also present on the dwarf-shoots; (*b*) green **acicular foliage-leaves**, popularly called "*needles*," which occur solely upon the dwarf-shoots; they are not borne directly on the shoots of unlimited growth. The dwarf shoots, with their clusters of green leaves, are called "*spurs*." The number of green leaves in each cluster varies according to the species. In *P. sylvestris* there are two, and the dwarf shoots together with their leaves are termed "bi-foliar spurs." These persist for a number of years, so that the tree is an evergreen. When they fall off, as they eventually do, it is the dwarf shoots which are shed, and the leaves fall with them. This phenomenon is called *cladoptosis*, *i.e.* the shedding of branches. Pinus has no power of vegetative reproduction.

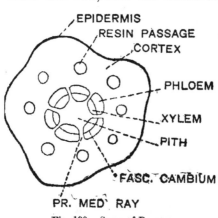

Fig. 190. Stem of Pinus.
(Transverse section: Diagrammatic.)

The presence of a tap-root is characteristic of Gymnosperms. Many, *e.g. Picea*, the Spruce, have leaves and shoots of one kind only. The branching is axillary, but buds are not formed in the axils of all the leaves. No buds are formed in the axils of the foliage leaves of Pinus.

§ 3. **Structure of the Stem.** In the general *arrangement* of tissues the stem of the Conifer closely resembles that of the Dicotyledon. It is **monostelic**. The **apical meristem** shows dermatogen, periblem, and plerome layers, though not so distinctly as the dicotyledonous stem. The **bundles** in the primary condition (fig. 190) are common, conjoint, collateral, and open, and form a ring in transverse section. The primary bundles in Pinus lie close together, so that the medullary rays between them are extremely narrow. The **ground-tissue** differentiates into pith, cortex, and medullary rays. The pericycle is parenchymatous, and therefore cannot readily be distinguished from the cortical tissue; the same is the case with regard to the endodermis, which is the innermost layer of cortical cells. The primary bundles have therefore no pericylic sclerenchyma or hard bast. Large **resin-passages** are present in the cortex, each surrounded by a layer of glandular secreting epithelial cells. As the transverse section of the young stem cuts through the bases of the dwarf-shoots its outline is irregular. A somewhat lignified hypodermal layer may be recognized in the outer cortex. **Secondary growth** is effected in exactly the same way as in Dicotyledons (pp. 83-89), the cambium ring giving rise to secondary wood and phloem, and the phellogen to cork and bark. The **phellogen** or cork-cambium originates in the cortical tissue near the surface, though not in the outermost layer, and, later, there is a repeated formation of tangential lines of phellogen cutting off strips of scaly bark (p. 89).

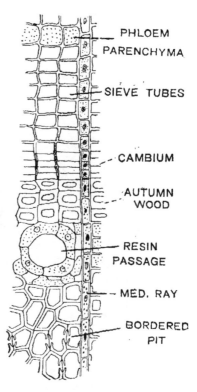

Fig. 191.—STEM OF PINUS. Portion of a transverse section after secondary growth.

§ 4. **The Tissues of the Stem** (figs. 191—193). The close resemblance to Dicotyledons will be recognized. The differences, however, are considerable, the vascular tissues especially being of simpler character. The wood or **xylem** contains *no true vessels*, but consists of tracheides (cf. V. Cryptogams) with very typical bordered pits. The **protoxylem**, however, is composed largely of annular and spiral tracheides. Small resin-passages are present in the primary and secondary wood, each with its lining epithelial layer. The **phloem** consists of sieve-tubes and phloem parenchyma;

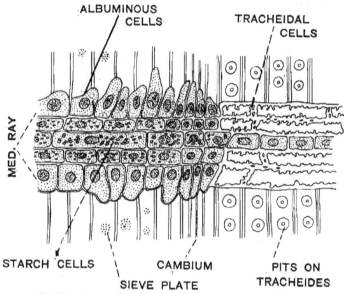

Fig. 192.—RADIAL LONGITUDINAL SECTION OF STEM OF PINUS.
The section is taken at the junction of secondary wood and phloem.

there are no true *companion cells* (cf. V. Cryptogams). The sieve-tubes consist of elongated, more or less pointed (prosenchymatous) cells, with lateral sieve-plates on the radial walls. The structure of the **medullary rays**, however, is more complex than that of Dicotyledons (figs. 192, 193). The rays in the secondary wood consist partly of cells containing starch, partly of *tracheides* running radially. These tracheides allow for the radial diffusion of watery fluids through the wood, and thus make up for the want of wood-parenchyma. In the secondary phloem the rays consist partly of starch-containing cells, partly of cells with albuminous contents. The medullary rays vary much in size; the smallest are only two cells high and one cell wide.

Figs. 191—193 show transverse, and radial and tangential longitudinal sections of the wood. A radial, longitudinal section is one which passes through the middle of the stem; a tangential longitudinal section, one taken peripherally. Thus a radial section runs parallel to the medullary rays in the region in which it is taken, while a tangential section cuts *across* them. This will explain the difference in the appearance of the medullary rays in the two sections. The differences in the appearance of the bordered pits are due to the fact that the tracheides are four-sided, two of the sides being approximately radial and two tangential, and that the bordered pits are confined to the radial walls. Thus, in radial section, the radial walls are not cut through and the pits are seen in surface view; while in tangential section, the radial walls are cut through and the pits are seen in section (fig. 193).

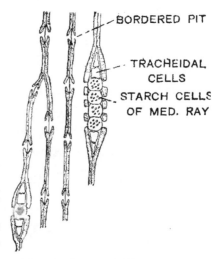

Fig. 193.—SECONDARY WOOD OF PINUS. Portion of a tangential longitudinal section.

§ 5. **The Root.** The growth and general arrangement resembles that of Dicotyledons. The **apical meristem** shows dermatogen, periblem, and plerome layers, and is covered by a root-cap, the innermost layer of which is continued back as the epiblema. In Pinus there are three to six **Y**-shaped **xylem** bundles, and an equal number of **phloem** bundles alternating with them; there is a resin-passage between the arms of each **Y**, and a pith is present, This arrangement, however, is not characteristic of Conifers as a class. In most cases the stele is di-arch (or tri-arch), and a pith is absent. In Gymnosperm roots the **pericycle** *consists of several layers*, but there is a single-layered *endodermis.* **Secondary growth** takes place as in the roots of Dicotyledons. The *phellogen* originates in the outermost layer of the pericycle. The wood and phloem have the same structure as in the stem. **Lateral roots** are developed from the second layer of the pericycle; the outermost layer covering them helps in the formation of the digestive sac which enables them to burrow outwards through the cortex. Root-hairs are scantily developed.

§ 6. **The Leaf.** Fig. 194 shows a transverse section of the foliage-leaf. The **epidermis** consists of extremely thick-walled cells with a strong cuticle. Owing to the erect position of the leaf stomata are developed all over

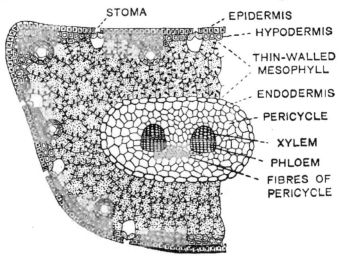

Fig. 194.—Transverse section of Leaf of Pinus.

its surface. The guard-cells are sunk beneath the level of the epidermis, so that there is an outer cavity leading down to the stoma. Beneath the epidermis there is a fibrous sclerenchymatous **hypodermis** interrupted beneath the stomata. The parenchymatous **mesophyll** consists of thin-walled cells, whose walls show numerous peg-like ingrowths of cellulose projecting into their cavities. The presence of these is probably connected with the feeble development of air spaces in the leaf, for they increase the internal surface of the cell-wall, and hence also the excreting and absorbing surface of the protoplasm. The cells contain numerous chloroplasts and starch-grains. The mesophyll is not differentiated into palisade and spongy layers, so that the leaf approximates to the centric type (p. 9). In the mesophyll, immediately under the hypodermis, are a number of *resin-passages*, each with a thin-walled epithelial layer and an investing strengthening layer of sclerenchyma. In the middle of the leaf there is a conspicuous **endodermis** surrounding a many-layered **pericycle**, in which two **vascular bundles** are imbedded. The bundles and pericycle constitute a *meristele*. The bundles are collateral, the xylem facing towards the flat upper surface; there is a rudimentary cambium. The pericycle contains cells of two kinds :—(*a*) parenchymatous cells containing protoplasm, proteid, and starch; (*b*) similar cells with bordered pits and no contents, resembling tracheides and called tracheidal cells. This peculiar tissue is called **transfusion-tissue**. It is characteristic of the leaves of Gymnosperms. It helps in the transference of nutritive solutions, and thus makes up for the poor development of vascular tissue. The tracheidal cells serve for the passage of inorganic solutions from the vascular bundles to the mesophyll; the other cells for the diffusion of elaborated compounds from the mesophyll to the phloem. In addition to the transfusion-tissue, a number of fibres are developed in the pericycle near the bundles.

§ 7. **The Male Flowers** (fig. 189) appear early in the year —about the beginning or middle of May. They are produced in the axils of scale-leaves at the bases of the developing shoots of the same year, but not on all the shoots. They form a spike at the base of the shoot, and the latter, as it continues to grow, develops ordinary dwarf-shoots in the axils of the upper scales. In other words, the male flowers are produced at the base of the shoot instead of dwarf-shoots, and are homologous with them. Each male flower (fig. 195) consists of a somewhat elongated axis which corresponds to the thalamus, and which bears a number of spirally arranged scaly leaves. On the under side of each scale two pollen-sacs are developed and these are filled with **pollen-grains.** The scales of the male flower, therefore, are **stamens.** The pollen-grains are *at first* unicellular bodies with exosporium or exine, and endosporium or intine (p. 174). On each side of the pollen-grain the exosporium is inflated with air, forming two balloon-like expansions (fig. 195, c). The male flowers differ from the flowers of Angiosperms in that (*a*) the axis which corresponds to the thalamus is elongated, (*b*) the stamens are less highly differentiated, showing no distinction into filament and anther, (*c*) there are two instead of four pollen-sacs. It should also be noticed that the essential organs only are present. There is no perianth, but in some cases a few sterile scales may occur at the base of the cone.

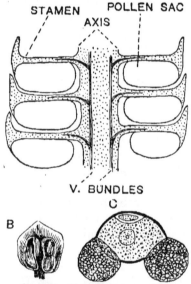

Fig. 195.—MALE FLOWER OF PINUS.
A, Part of a median longitudinal section (diagrammatic); B, Stamen (under surface); C, Pollen-grain (two-celled stage, highly magnified).

Most Conifers have male flowers similar to those of Pinus. The highest development is found in the yew, where the stamens are peltate, and may be regarded as showing a differentiation into filament and anther —the stalk, corresponding to the filament, the peltate head, bearing eight to ten pollen-sacs, to the anther. The number of pollen-sacs varies in different types.

§ 8. **The Female Flowers,** or cones (fig. 189), are found at the same time as the male flowers. They are developed laterally, one or two together, in the axils of scale-leaves at the apices of the young elongated shoots. They are usually borne on shoots which do not bear male flowers, and they take the place of shoots of unlimited growth.

The female flower (figs. 196, 197), if examined at this early period, is found to be a small, light-coloured structure **consisting of a stout central axis, bearing scales of two**

GYMNOSPERMS—STRUCTURE AND LIFE-HISTORY OF PINUS. 289

kinds:—(*a*) small scales arranged spirally and developed directly on the axis; these are called the **bract-** or **cover-scales**; (*b*) rather stouter scales developed one on the upper surface of each bract-scale; these are called **ovuliferous scales**, because each bears two ovules on its upper surface. As the ovules afterwards give rise to seeds, the ovuliferous scales are also called "seminiferous scales." Each bears at its apex a little protuberance called the *apophysis*.

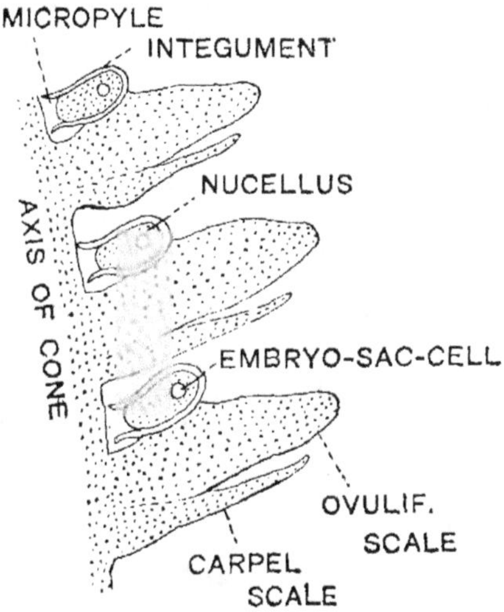

To compare this with the Angiosperm, the so-called bract-scale is regarded as equivalent to (*i.e.* the homologue of) a carpel, and should, therefore, be called a *carpellary* scale; while the ovuliferous scale is a large **placenta**. Here again the elongated axis should be noticed. The most important point of difference, however, is that the carpels do not assume the form of a closed ovary, with a style and stigma.

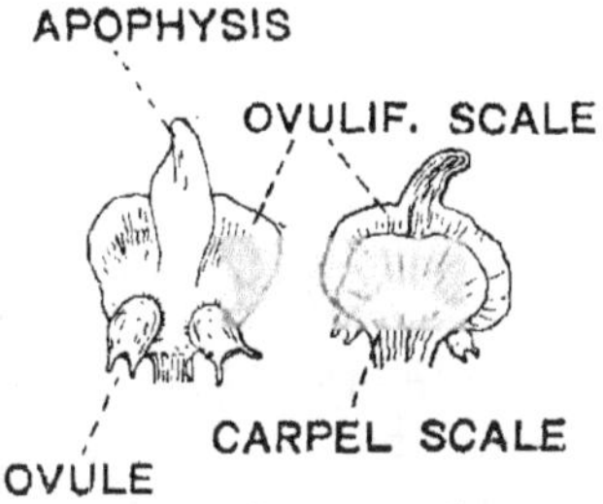

Fig. 196.—YOUNG FEMALE CONE OF PINUS. (Part of a longitudinal section: diagrammatic.)

Fig. 197.—SCALES OF FEMALE CONE. From above and below (Carpel-scale = cover-scale).

The female cones of the spruce and larch closely resemble those of Pinus. In the larch the bract-scales are considerably larger. In the cypress and the "tree of life" (*Arbor vitæ*) the scales (carpels) of the female cone do not bear ovuliferous scales: a number of ovules are borne directly on the surface of the scales, which are arranged, not spirally, but in opposite decussate pairs. In the yew there is no female cone; the female flower consists of a single ovule borne terminally on a short axillary shoot.

There is another view entertained by many with regard to the homologies of the female cone of Pinus. The bract-scale is regarded as really a bract ; and the ovuliferous scale with the two ovules, as a female flower of a very rudimentary type. According to this view the female cone would be, not a single flower, but an inflorescence. This view, however, involves us in several difficulties with regard to the homologies of the parts of the male cone.

§ 9. **Structure of the Ovule** (fig. 196). In the young female cone just described, the ovule consists of a small-celled mass of tissue, the **nucellus**, surrounded by a single integument. Towards the base of the nucellus, one large cell is developed called the **embryo-sac-cell**, homologous with the embryo-sac of the Angiosperm. There is a wide gaping *micropyle* directed towards the axis of the cone. The ovule is orthotropous.

§ 10. **Comparison with Vascular Cryptogams.** Having explained in the preceding sections the homologies between Pinus and the Angiosperm, we may now make the connecting link in the other direction by giving the homologies which can be recognized between Pinus and the Vascular Cryptogams.

They may be stated briefly thus :—
 (*a*) The **plant (Pinus)** is the *sporophyte*.
 (*b*) **Pollen-grain** = *microspore* ;
   Pollen-sac = *microsporangium* ;
   Stamen = *microsporophyll*.
 (*c*) **Embryo-sac-cell** = *megaspore* ;
   Ovule = *megasporangium* ;
   Carpel (bract-scale) = *megasporophyll*.

These homologies should be very carefully noticed. In the description of the flowers of Pinus, either set of terms may be used.

The terms pollen-grain, pollen-sac, etc., were given to these structures before their homologies were revealed by a study of their development and position in the life-history, and by a careful comparison with such types as Selaginella.

The student must clearly comprehend that the recognition of homologies is based on a comparative study of development. To emphasize this we shall now describe the development of pollen-sac and ovule in Pinus ; it will be found to be essentially similar to the development of micro- and mega-sporangia of Selaginella.

GYMNOSPERMS—STRUCTURE AND LIFE-HISTORY OF PINUS.

Further evidence will be forthcoming when we pass to the life-history of Pinus.

§ 11. **Development of Pollen-Sac and Ovule.** The pollen-sac is developed *from a group of cells* (cf. Equisetum and Selaginella) on the under surface of the young stamen. The wall of the pollen-sac is formed from the superficial or epidermal layer; it remains single. Several hypodermal cells, forming the **archesporium**, begin to divide rapidly. Tapetal cells are segmented from them towards the wall of the pollen-sac, and the remaining central portion forms a mass of **pollen-** or **spore-mother-cells**. Each mother-cell gives rise, exactly as in the Fern or Selaginella to four "**special mother-cells**" in each of which a pollen-grain or microspore is produced. The pollen-grain has the structure of a spore; its outer coat corresponds to the exosporium, its inner one to the endosporium. The tapetal cells are disorganized during the development of the pollen-grains. Here the homology is perfect.

The **ovule** arises as a small cellular protuberance on the upper surface of the young placenta. It increases in size and becomes the nucellus. The single integument arises from the base of the nucellus and gradually invests in. At the apex of the young nucellus a single hypodermal cell, the **archesporium**, can be recognized. This begins to divide and forms **tapetal cells** towards the apex of the nucellus. The archesporium proper remains unicellular, and owing to continued division of the overlying tapetal and nucellus cells it comes eventually to lie near the base of the nucellus. It undergoes no division and itself becomes the **embryo-sac-cell** or *megaspore*. Here the homology with the megasporangium of Selaginella is not so perfect. The nucellus is probably equivalent to the proper wall of the megasporangium. The integument is by many regarded as of the nature of an *indusium* (cf. the Fern), in this case surrounding a single sporangium; it is not represented in Selaginella. The greatest difficulty, however, is met with in the fact that there is only one mother-cell, and it does not show the usual division into four, but becomes directly the embryo-sac-cell or megaspore. Other

Gymnosperms, however, show intermediate conditions, and the evidence will be strengthened when we come to consider the fate of the embryo-sac-cell in the life-history.

§ 12. **Pollination.** As in Angiosperms, the pollen-grains or microspores have to be transferred to the neighbourhood of the embryo-sac or megaspore. In Pinus this is effected by means of the wind, the transference being facilitated by the balloon-like expansions of the exosporium already described. Pinus, therefore, is *anemophilous*. Pollination takes place towards the end of May or beginning of June. At this period the scales of the female cone open out, and separate from each other. Much of the pollen is wasted, but some of the grains are blown between the scales of the female cone and fall near the ovules. A mucilaginous secretion is given out from the micropyle. In this the pollen-grains are entangled, and, as the mucilage dries up, they are drawn down the micropyle and finally come to rest on the apex of the nucellus. In the Gymnosperms pollination consists in the transference of the pollen-grains, not to a stigma as in Angiosperms, but directly to the surface of the nucellus. The scales of the female cone close up after pollination.

§ 13. **Male Gametophyte.** If the pollen-grain is really a microspore we should find it giving rise on germination to something equivalent to a male prothallus. At first the pollen-grain is unicellular. Even before it leaves the pollen-sac division begins, and it is completed on the surface of the nucellus. Two very minute rudimentary cells, the **prothallus-cells**, are cut off on one side. The remainder then segments into a small **antheridial cell** and a large **vegetative cell** (fig. 198, A). Comparing this with the germination of the microspore in Selaginella, there is an obvious suggestion that the small prothallus-cells represent a very rudimentary or reduced **male prothallus**. The antheridial cell is so called because, as will presently be explained, it produces the male sexual cell, and is, therefore, the equivalent of the central cell of the antheridium of the Vascular Cryptogam, the male sexual cell being the equivalent or homologue of the spermatozoid.

GYMNOSPERMS—STRUCTURE AND LIFE-HISTORY OF PINUS.

The **male cell**, however, is not motile, *i.e.* it is not a spermatozoid, and hence the term antheridial cell is preferable to that of "*antheridium*." The large *undivided* vegetative cell probably corresponds to the peripheral cells which form the wall of the antheridium in Selaginella. In Gymnosperms there are definite cellulose walls formed between all these cells in the pollen-grain.

In the further germination of the pollen-grain on the apex of the nucellus, the exosporium bursts open and the large vegetative cell protrudes and elongates to form a slender pollen-tube (fig. 198, B). This is not represented

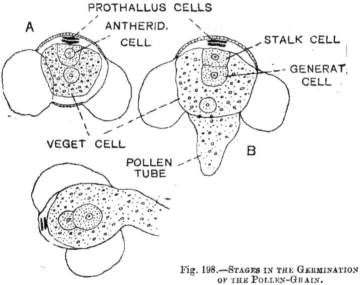

Fig. 198.—STAGES IN THE GERMINATION OF THE POLLEN-GRAIN.

A, B, early stages in *Picea*, the Spruce, where the antheridial cell divides shortly after pollination; C, Late stage in Pinus.

in Selaginella; its development in Phanerogams, as will afterwards be explained (p. 309), is an adaptation to entirely different conditions. This pollen-tube grows down into the tissue of the nucellus. Its destiny will be traced presently.

§ 14. **Growth of Female Cone—The Female Gametophyte.**
At first, as we have seen, the female cone is comparatively small; while the ovule consists simply of the integument,

the nucellus, and the embryo-sac-cell or megaspore. Although pollination is effected at this stage, fertilization does not take place in Pinus till about a year later—some time in June of the second year. This, however, is not characteristic of Gymnosperms; in most of them fertilization is effected in the same year as pollination. In Pinus, during this protracted interval between pollination and fertilization, many important changes go on in the ovule and in the cone as a whole. The cone increases in size and becomes green. During the winter these green cones are found at the apices of the shoots, just below the terminal winter bud. This increase in size is due to the enormous growth of the axis and of the ovuliferous scales. The carpellary scales remain small, and are completely concealed. The growth is continued rapidly in the second year (fig. 189). At the time of fertilization the cones are large green structures, the rhomboidal areas on their outer surface being the outlines of the apices of the ovuliferous scales.

Inside the ovule the embryo-sac-cell becomes much larger, and by *free cell-formation* there is formed *inside it* a mass of thin-walled parenchymatous tissue. This tissue was formerly called the "*primary endosperm*" because it becomes laden with food-material and constitutes the endosperm of the seed; but, if the *germination* of the embryo-sac-cell be compared with that of the megaspore of Selaginella, there will be no difficulty in recognizing that

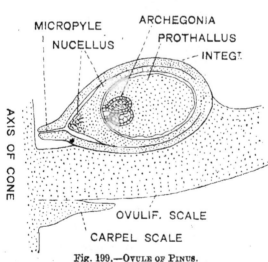

Fig. 199.—Ovule of Pinus.
(Longitudinal section—about the time of fertilization.)

this tissue is the **female prothallus** (fig. 199), a name which it now receives. Practically, the only important difference is that the megaspore in Pinus is not set free from the megasporangium, as it is in Selaginella. But even in Selaginella, the student will remember, the germination of the megaspore begins inside the sporangium. The female prothallus in Pinus is enclosed in the nucellus. It has neither chlorophyll nor root-hairs. At its micropylar end are developed two, or occasionally three, **archegonia**—formerly called *corpuscula*. This completes the evidence of homology. The archegonium consists of a *venter* and a short *neck*. **Oosphere** and *ventral canal-cell* are present, but there is no neck-canal-cell. The protoplasm of the oosphere presents a frothy appearance, owing to extensive vacuolation; it has no cell-wall. Except that no neck-canal-cell is formed, the development is essentially the same as in the Fern or Selaginella.

§ 15. **Fertilization.** In the first year the growth of the pollen-tube is arrested after it has penetrated a short distance into the nucellus. In the second year it again begins to grow. Some time in April the antheridial cell (§ 13) divides into two (fig. 198, B), a barren cell called the **stalk-cell**, and a **generative cell** which produces the male sexual element. The protoplasmic contents and nuclei of the large vegetative cell, the stalk-cell, and the generative cell, all pass down to the apex of the pollen-tube. Before this can take place, of course, the intervening cell-walls must be absorbed. In the pollen-tube the naked generative cell divides into two. The pollen-tube ultimately reaches and enters an archegonium. One of the generative cells only is concerned in the act of fertilization. As the **male sexual element** (or gamete), it passes from the pollen-tube into the ovum, and its nucleus together with a small amount of cytoplasm fuses with the nucleus and cytoplasm of the ovum. The fertilized ovum forms a cell-wall and becomes the **oospore**.

§ 16. **Development of Embryo** (fig. 200). The nucleus passes to the lower end of the oospore, and there, by

repeated karyokinetic division, gives rise to four nuclei. These again divide, forming eight nuclei. Cell-walls are laid down between the four basal nuclei, and by this free cell-formation four small cells are formed at the lower end of the oospore. Its upper part, with the other four (free) nuclei, which are afterwards disorganized, takes no share in the development; it contains food material. The development, therefore, is said to be *meroblastic* (or partial), and this is characteristic of Gymnosperms. In most plants the development is *holoblastic*, *i.e.* the whole of the oospore undergoes division to form the embryo. The four small cells are divided into four rows of cells by two transverse walls, each row consisting of three cells. These four rows begin to elongate and push their way downwards into the tissue of the female prothallus (endosperm, § 14). The elongation is due to the formation of a long unicellular **suspensor** by the growth of the middle cell of each row (2 in fig. 200). The four suspensors *separate from each other*. The cell borne at the end of each (3 in fig. 200) is the **embryonal cell**; it continues to divide and forms a potential embryo. It will be noticed that four potential embryos are formed from each oospore.*
This phenomenon is known as *polyembryony*; it is very

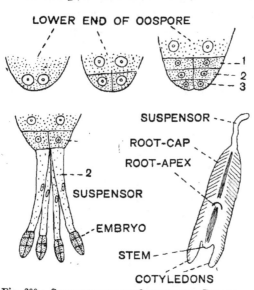

Fig. 200.—Segmentation of Oospore and Development of Embryo of Pinus.
Only half the number of nuclei, cells, and rows of cells, is, of course, shown in the early stages.

* In a few Coniferæ, *e.g.* the spruce, only one suspensor and one embryo are produced.

characteristic of Coniferæ, and not unknown in Angiosperms. As more than one oosphere may be fertilized, many potential embryos may be present in one ovule. Only one, however, develops; the others die off. The embryo which is developed consists of a **primary root**, a tiny **plumule**, and *a number of* **cotyledons**. The whole of it is derived from the embryonal mass, the suspensor taking no part in the development.

It should be carefully noticed that the **endosperm** is simply the tissue of the female prothallus laden with food-material which diffuses into it from the placenta.

The nucellar tissue is almost entirely crushed and disorganized owing to the expansion of the endosperm and embryo. A thin layer of it persists, and contains food material, forming a small amount of *perisperm* (p. 202).

§ 17. **Seed and Fruit.** Thus, as in Angiosperms, a seed is formed (fig. 201). The integument of the ovule becomes the **testa**. The seed contains, not only endosperm, but also a small amount of perisperm. The **embryo** is straight and lies in the middle of the endosperm. The suspensors disappear. The seed has a thin membranous *wing*, which assists in its dissemination. The wing is derived from the surface of the placenta, not from the testa.

The female cone, when it reaches maturity *in the third year*, is dry, brown, and woody. The scales (placentas) gape apart and allow the seeds to escape. The cone, as a fruit, is necessarily quite different from Angiospermous fruits, seeing that there is no ovary.

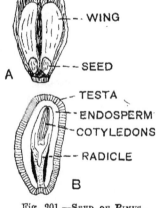

Fig. 201.—Seed of Pinus.
A, Surface view; B. Longitudinal section.

The fruits of most Conifers are dry, woody cones. In some, however, the carpels become fleshy and form a berry-like fruit, *e.g.*

298 BOTANY.

juniper. The seed of the yew is invested by a bright red fleshy aril developed as a second integument after fertilization.

§ 18. **Germination of Seed.** The cotyledons of Pinus may become green while still enclosed in the seed-coat. They gradually absorb the endosperm, and are epigeal, the seed-coat being carried by them above ground. The primary root passes downward and forms the tap-root

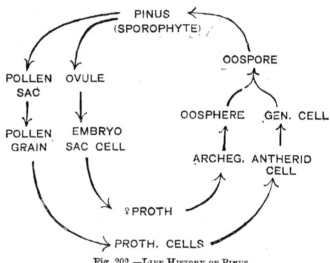

Fig. 202.—LIFE HISTORY OF PINUS.
Graphical representation. (Cf. figs. 176, 178, 188.)

system. The first year's shoot, formed by the elongation of the plumule, has no scale-leaves or dwarf-shoots; it has acicular green leaves, spirally arranged, thus indicating what was probably the primitive leaf arrangement.

The life-history of Pinus may be graphically represented as in fig. 202.

§ 19. **Physiology.** Most Conifers are xerophytic (p. 157), and many (including Pinus) have mycorhizæ (p. 151). The narrow acicular form of the leaves, their thick cuticle, the sunken stomata, the presence of a strong hypodermis, the simple vascular system, are marked xerophytic characters, all tending to reduce transpiration to a minimum.

CHAPTER XVII.

HOMOLOGIES IN ANGIOSPERMS.

§ 1. **The Sporophyte in the Angiosperm.** We are now in a position to correlate the main facts in the life-history of the Angiosperm with those of the Vascular Cryptogam and Gymnosperm. From what has been said in the preceding chapter the following homologies will be evident:—

(a) The **Angiospermous plant** is the *sporophyte*.
(b) **Stamen** = '*microsporophyll*';
   Pollen-sac = *microsporangium*;
   Pollen-grain = *microspore*.
(c) **Carpel** = '*megasporophyll*';
   Ovule = *megasporangium*;
   Embryo-sac = *megaspore*.

The sporophyte in the Angiosperm is still more highly differentiated than that of the Gymnosperm. As in the Gymnosperm, the sporophylls are aggregated to form **flowers**. In Angiosperms the flowers have undergone extreme specialization. In addition to the sporophylls or essential organs they usually have accessory structures—the floral envelopes or **perianth**—which play an important part in the production of seed.

To complete the evidence of homology, we may briefly describe the development of pollen-sac and ovule of the Angiosperm. It is essentially similar to the development in Gymnosperms and Vascular Cryptogams.

§ 2. **Development of Pollen-Sac** (fig. 203). The stamen in the Angiosperm arises as a protuberance on the thalamus. It consists of meristematic tissue, and soon shows a distinction into filament and anther. The two anther-lobes can be recognized at an early stage, and a procambial strand makes its appearance in the region of the connective. The dermatogen layer of the anther-lobes remains undivided, but, in *each* anther-lobe, two little groups of periblem cells lying immediately beneath the dermatogen begin to divide. They form usually three layers of cells underneath the dermatogen layer. The outermost of the three

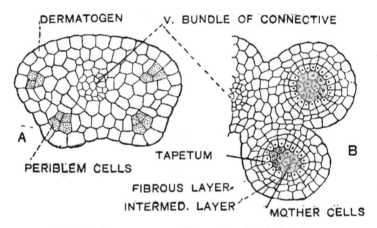

Fig. 203.—Development of Pollen-Sacs in Angiosperm.
Transverse sections of young anthers.

layers becomes the **fibrous layer** (p. 174) of the pollen-sac. The innermost layer consists of large granular cells and forms a **tapetum**; together with the intermediate layer it is disorganized during the development of the pollen-grains. The remaining central cells of each meristematic group constitute the **archesporium**. Thus in each anther-lobe there are two archesporia. The tapetal layer completely surrounds each archesporium. The archesporial cells divide in the usual way to form spore- (or pollen-) mother-cells. In Dicotyledons, the **special mother-cells** are formed in much the same way as in the case of the Fern (p. 259); but in Monocotyledons they are formed by ordinary cell-division, *i.e.* the mother-cell divides into two and then these two into four. The microspores or pollen-grains in both are formed in the usual manner.

§ 3. **Development of Ovule** (fig. 204). The **nucellus** arises as a tiny cellular protuberance on the placenta and gradually increases in size. The **integuments** arise, one after the other, as outgrowths from the base (chalaza) of the nucellus. This basal region also elongates to form the *funicle*. At an early stage the **archesporium** is recognized as a single *hypodermal* cell at the apex of the young nucellus. It *usually* divides into two — an upper *tapetal cell* and a lower cell; the *archesporium proper*. The tapetal cell *may* divide again, but sometimes no tapetal cell is formed. The archesporial cell sometimes remains undivided and directly becomes the embryo-sac; frequently, however, it divides into a row of four potential mother-cells, the lowest of which becomes the embryo-sac or megaspore, the other three, called "*cap-cells*," being aborted. As in Gymnosperms, no division of the mother-cell into four occurs and only one megaspore is formed.

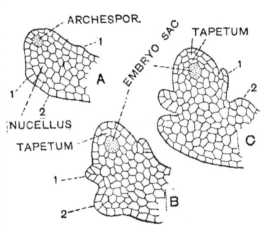

Fig. 204.—Development of an Anatropous Ovule.
1, 2, = 1st and 2nd integuments.

The embryo-sac or megaspore is at first a typical cell with a single nucleus; but before fertilization a process of free cell-formation takes place. The nucleus divides karyokinetically into two. One daughter-nucleus passes to the micropylar end, the other to the chalazal end of the embryo-sac. Each by further division gives rise to four nuclei. Three of the nuclei at the micropylar end become surrounded by protoplasm, and form the **oosphere and synergidæ** or egg apparatus: three at the chalazal end are surrounded by protoplasm and cell-walls, and form the antipodal cells. A nucleus remains at each end. These are called *polar nuclei;* they pass to the centre of the embryo-sac and fuse to form the **secondary nucleus.**

§ 4. **The Flower.** We must now bring distinctly before the student the fact that the flower is simply a specialized reproductive shoot (see p. 7) bearing an aggregation of sporophylls. *Morphologically,* the flower is not a structure peculiar to the Phanerogams. It has its morphological equivalent, *i.e.* its homologue, amongst the Vascular Cryptogams, *e.g.* the sporangiferous heads of Equisetum and Selaginella. Many, indeed, have extended the application of the term *flower* to these and similar structures amongst the Vascular Cryptogams. On this view, the term *Flowering Plant* applied to the Phanerogam would be a misnomer. Others, however, would restrict the term flower to the Phanerogams. This could be done by giving the following definition:—The flower is a specialized reproductive shoot bearing sporophylls and sporangia which are concerned in the production of seed. The point is of small importance practically; the important thing is the recognition of the morphological equivalent of the flower.

§ 5. **The Gametophyte in the Angiosperm.** The male gametophyte is completely reduced. The vegetative and generative cells of the pollen grain (p. 197) represent all that there is of male prothallus and antheridium. The vegetative cell is probably equivalent to the peripheral cells forming the wall of the antheridium in Selaginella. Otherwise the antheridium of the Vascular Cryptogam is represented only by a generative cell which, as in Gymnosperms (p. 295), divides to form two gametes, corresponding to the spermatozoids.

Remembering the process of free cell-formation which takes place in the megaspore of Gymnosperms and of Selaginella, we must recognize that the formation of antipodal cells and egg-apparatus in Angiosperms is, as it were, an attempt at the formation of a *female prothallus.* The process, however, comes abruptly to an end. After fertilization it is continued (see p. 200), and leads to the formation of the endosperm tissue. The endosperm tissue is the female prothallus *formed after fertilization*. The secondary nucleus may be considered as a resting nucleus set apart before fertilization to continue the formation after fertilization. Some regard the antipodal cells as a rudimentary formation of prothallus tissue; others as equivalent to an abortive egg-apparatus. The egg-apparatus probably represents three *reduced* archegonia. The female *organs* or

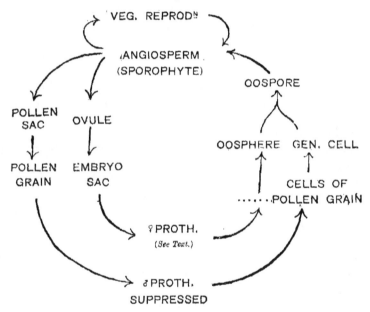

Fig. 205.—Life-History of Angiosperm Graphically Represented.

archegonia themselves are lost, but their essential cells (oospheres) have persisted. Two of these, the synergidæ, are functionless. In a few plants, however, one or other of the synergidæ may be fertilized. That these interpretations with regard to the female prothallus, etc., are approximately correct may be recognized from the fact that interesting intermediate conditions are found in the highest group of the Gymnosperms, the Gnetaceæ (p. 280).

The life-history of the Angiosperm may be represented as in fig. 205.

§ 6. **The Seed.** The student should notice carefully that the seed is a highly specialized reproductive body consisting of structures representing three generations:—(*a*) the parent sporophyte, viz. the integument of the ovule forming the seed-coat; (*b*) the female gametophyte—the endosperm tissue; and (*c*) the new sporophyte in embryo.

§ 7. **Comparative Summary.** In *Phanerogams*, as in Vascular Cryptogams, there is an **alternation of generations**, but it is very much less distinct. The male and female prothalli are even more reduced than in Selaginella. This extreme reduction of the gametophyte is characteristic of the Flowering Plants. More important differences, however, have to be noticed. The megaspore (embryo-sac) is not set free from the sporangium (ovule). The female prothallus is developed inside the nucellus. The other differences are correlated with this. The special process of pollination is necessary in order to bring the microspore into the neighbourhood of the megaspore. There is a new method of fertilization; the male element is no longer a spermatozoid, but a motionless gamete which is transferred to the ovum by a special organ, the pollen-tube. Finally, there is the most characteristic difference of all—the formation of a seed. This also is clearly due to the retention of the megaspore in the ovule. The non-formation of a seed in Vascular Cryptogams is correlated with the liberation and independent germination of the spores.

§ 8. **"Double Fertilization."** Recently it has been discovered that, in some plants at least, the second male gamete (see p. 199), whose fate was hitherto unknown, fuses with the secondary nucleus or with the micropylar polar nucleus before the formation of the secondary nucleus. The significance of this is not yet understood. It may be an act of fertilisation (hence the term double fertilization), in which case the endosperm would be an undifferentiated embryo devoted to supplying the actual embryo with food. At present, however, we may accept provisionally the other and more probable view that the fusion merely provides the necessary stimulus to the continued development of the female prothallus.

CHAPTER XVIII.

RELATIONSHIP BETWEEN VASCULAR CRYPTOGAM AND FLOWERING PLANT.

§ 1. **Homology based on Relationship.** While studying the types discussed in the preceding pages the following questions have probably occurred to the mind of the student——What is the meaning of these homologies? Why should there be such resemblances in the development and life-history of plants which in many other respects differ so widely from each other? The detailed consideration of such questions is beyond our present purpose. We shall merely try to suggest the answer.

We give the now generally accepted theory (**evolution theory**) when we say that these homologies are due to an actual **relationship** existing between plants—in other words, to the fact that these higher plants which we have been considering are *descended from common ancestors*. To make this clear, the preliminary consideration of a few important principles is necessary.

§ 2. **Struggle for Existence.** Let the student consider how numerous are the seeds produced by any one plant, and then reflect that the number of individuals in any one species remains approximately the same from one year to another. Evidently very few of the seeds develop into mature plants. Some fail to reach a suitable soil; others produce seedlings which, however, are crowded out and killed by other and hardier plants—and so on. There is, evidently, a very keen *struggle for existence*, in which those which are most favourably circumstanced will survive.

§ 3. **Heredity and Variation.** It is well known that the characters of parents are transmitted to their offspring, so that the offspring resemble their parents more or less closely. This is the principle of *heredity*.

Now if the offspring were exact copies of their parents and all *equally vigorous*, it would be a matter of *chance* which of them survived in the struggle for existence; those would survive which *happened* to be most favourably situated. But the offspring are not exact copies of the parents. As we know, they exhibit individual differences, some of which may appear for the first time in the history of the race. These characters in which offspring differ from their ancestors are called *variations*.

Variation, *i.e.* the fact that these variations do occur, introduces a new factor into the struggle for existence. Survival is not to the same extent a matter of chance Some of the variations exhibited by plants may be useful or advantageous, *i.e.* they may give the individuals possessing them an advantage over other individuals in the struggle for existence. Evidently those variations will be such as will enable the individual to cope better with the external conditions to which it is subjected. Thus in the struggle for existence the fittest will survive and their advantageous variations will tend to be transmitted, perpetuated, and even intensified in their offspring.

§ 4. **Natural Selection.** In each generation, the unconscious influence of the external conditions, by killing of the weaklings, *selects*, as it were, those which by reason o some useful variation are more fitted for survival, just a a gardener might select the plants showing the variation which he wishes to intensify. The sum-total of externa conditions constitutes what we call "*Nature*," and therefore to this unconscious operative cause the name "*natura selection*" has been appropriately given.

§ 5. **Origin of Species, Genera, etc.** If the student has followe what has been said above, he should have no difficulty in realizin that the forms of plants would become immensely altered or modifie during the lapse of long periods of time. Within the limits of on original species (see p. 219) several distinct variations might appear marking out certain groups which would be called "**varieties**" of the species. After a still longer period the characters which distinguish these varieties would be intensified and would eventually be constantly transmitted. The varieties would then be elevated to the dignity of species. In this way the original species would give rise to a group of species, *i.e.* a genus. Similarly the genus might give rise to a group of genera, forming a natural order ; and so on.

§ 6. **Adaptation.** Thus, in the evolution of plants, new characters would arise as small useful variations preserved by natural selection and intensified in the course of generations. These characters would be suited to the environment of the plant—in other words, they would be adaptive characters, otherwise they would not have been preserved by natural selection. The student will now have a clearer idea of the origin and significance of adaptation.

The structural characters of a plant are adaptive, because they are the characters which, being useful, have been *selected*, and have determined the existence of the plant.

Under changing conditions of environment, species which are incapable of appropriate adaptive variation tend to become extinct.

§ 7. **Homology and Analogy.** While in the course of evolution the forms of plant-members would undergo extensive adaptive modification, many of *the general developmental characters of original ancestral forms would be constantly transmitted. These would be recognized as homologies in their descendants.* The nearer the relationship the more numerous and complete would be the homologies. On the other hand, it is because adaptive modification may *similarly* affect *different* members that we so frequently find examples of members which are *analogous*, but not homologous (see p. 16). It is evident that analogy is no indication of relationship. Aquatic plants, for example, have many features in common, but nevertheless no general relationship exists between them.

§ 8. **Evolution of the Higher Plants.** The types which we have studied will serve as an illustration of the above principles. We may suppose that the common ancestors of the Vascular Cryptogams and Phanerogams were plants showing a distinct alternation of generations—the sporophyte differentiated into root, stem, and leaf, and producing asexual spores—the gametophyte a green thallus, bearing both antheridia and archegonia. Some of the ancestral characters have been inherited in common by the descendants. These constitute the homologies recognizable, *e.g.* the similarity in the *general* course of the life-history, the resemblances discernible in the development of members such as roots, stems, leaves, sporangia, etc. The differences or modifications have arisen as different adaptations to changing conditions of environment, *e.g.* the forms of stem, root and leaf. Let us see, however, if we can trace some of the more important of these.

We must first ask the student to notice that the sporophyte is evidently a plant adapted to aerial conditions, while the gametophyte, as we find it in the Fern or Equisetum, and, presumably, as it existed in the ancestral forms, is adapted to moist conditions. Further, that in the Flowering Plants we have plants *completely* adapted to aerial conditions.* On the view that cross-fertilization is of advantage to plants, we can understand the gradual evolution of unisexual prothalli (*e.g.* Equisetum). But why the reduction of prothalli and the heterosporous condition? This is probably traceable to the more complete adaptation of the plant to aerial conditions. Evidently, if there was any uncertainty as to the conditions being sufficiently moist for the development of well-formed prothalli, the storing up of food-material inside the spore must have been a useful variation. It was more necessary in the case of the female gametophyte because it had to nourish the embryo—hence the differentiation of a megaspore. From this stage onwards the spores and prothalli may be regarded simply as organs having for their function the production of a new sporophyte. The advantage of the retention of the megaspore in a secure and definite position in the

* Aquatic Angiosperms have *returned* to aquatic conditions. Their ancestors were aerial types—cf. the whale among air-breathing vertebrates.

sporangium will now be perceived. The microspores being small would be readily blown about by the wind; and we can imagine that there were special means for catching them (cf. secretion of mucilage from the micropyle of Pinus). Here we trace the beginning of pollination. At first, probably, the microspore germinated on the surface of the female prothallus, and the spermatozoid made its way to the archegonium in water present on the surface of the prothallus and probably in part excreted by it. We can understand the gradual enclosure of the megaspore on the view that the embryo would be better protected and have a better chance of surviving. The development of a pollen-tube would probably be stimulated by the presence of water on the surface of the partly covered megaspore or prothallus. The advantage of the thickening of the wall of the sporangium (nucellus), the development of an integument, and the formation of a seed as a highly efficient organ of reproduction, will be easily followed. In Gymnosperms we have clear evidence that a number of distinct megaspores and embryos (polyembryony, p. 296) were present in primitive seeds. The advantage of the reduction to one is evident, for one strong and well-nourished seedling is much more likely to survive than two weakly ones. After the complete enclosure of the megaspore, it lost its cuticularized coat, which is still found in Cycads, and the archegonium, being now a useless organ, gradually disappeared, leaving the ovum behind it. Finally, the late development of endosperm in Angiosperms is of distinct advantage, as it is not required if fertilization is not effected.

All this is, of course, hypothetical; but we have grounds for believing that the above account gives approximately the course of evolution of flower, seed, and fruit. Thus, for example, quite recently it has been found that in two Cycads, *Cycas* and *Zamia*, and in *Ginkgo*, the only survivor of an ancient group of Gymnosperms, two multiciliate spermatozoids are developed in the pollen-tube instead of two generative cells, and that these reach the necks of the archegonia by swimming through a drop of water secreted on the surface of the female prothallus.

§ 9. **Specialized Characters of Flowers and Fruits.** We cannot doubt that the primitive flowers were anemophilous, like those of Pinus. Originally, insects probably visited flowers to feed on the pollen. A slight secretion of sweet substance by the floral leaves, serving as a further attraction to insects, no doubt led to the gradual evolution of **nectar-glands**; the close association of bracts around the flower may have been the precursor of a calyx or perianth; the sterilization of the outermost series of male sporophylls (stamens) probably led to the evolution of a **corolla**, and slight variations in form, facilitating the movements of insects in pollination, to the evolution of **mechanical contrivances**. Similarly we can picture the evolution of **succulent fruits** through natural selection by herbivorous birds and mammals.

§ 10. The student must be careful to avoid the idea that the Angiosperm has directly descended from a Gymnosperm like Pinus, or Pinus from Selaginella, and so on. The various groups of Vascular Cryptogams and Gymnosperms are offshoots from the main stem of evolution.

Questions on the V. Cryptogams and Flowering Plants

(Many of these were set at London University Examinations.)

1. Give a general sketch of the life-history of a fern from the germination of the spore to the formation of the fertile frond.

2. Describe in detail the structure of the stem in Pinus after three years' growth in thickness. Explain how this growth in thickness takes place.

3. Describe the general structure of the frond of a fern and state in what respects it differs from the leaf of a Flowering Plant.

4. Describe the structure of the mature ovule of a Gymnosperm, and point out the differences between it and the mature ovule of an Angiosperm.

5. Explain *fully* the reasons on which the conclusion is based, that the pollen-sac of the Angiosperm is a microsporangium, and the pollen-grain a microspore.

V. CRYPTOGAM AND FLOWERING PLANT.

6. Give an account of the structure and function of the prothallium of a fern.

7. In what important respects does the life-history of a Vascular Cryptogam (*a*) resemble, (*b*) differ from, that of a Flowering Plant? What is the significance of these resemblances and differences?

8. Give an account of the life-history of Selaginella, and indicate important differences from that of the Fern.

9. What is meant by an "alternation of generations" in the life-history of a plant? Illustrate your answer by reference to the life-history of a fern.

10. What is a gametophyte? Give an account of the gametophyte of Pinus.

11. What are the resemblances and what the differences between the floral organs of an Angiosperm and those of Pinus?

12. Describe and compare the sexual reproductive organs of a fern and an Angiosperm. State how fertilization is effected in the two cases.

13. Describe (*a*) the general form of the gametophyte generation in a fern; (*b*) the parts by which it is nourished, and how they operate; (*c*) the sexual organs, and the process of fertilization; (*d*) compare a fern in respect of these features with a Gymnosperm.

14. Give a comparative account of the development of sporangia and spores in the Fern, Equisetum, and Selaginella. Mention important differences in the development of the megasporangium and megaspore of Flowering Plants.

15. Give an account of the structure and development of roots in ferns, and mention important differences as compared with the roots of Angiosperms.

16. What is polystely? Describe the structure of a stele in the Male Shield Fern or Bracken.

17. Give a comparative account of the "nursing" of the embryo (*i.e.* the way in which it is nourished) in Vascular Cryptogams and Flowering Plants.

PART IV.—THE LOWER CRYPTOGAMS.

CHAPTER XIX.

LIVERWORTS AND MOSSES.

§ 1. The Group **Bryophyta** or **Muscineæ** is divided into two Classes—the **Hepaticæ** or *liverworts* and the **Musci** or *mosses* (see p. 4). We shall consider *Marchantia* as a type of the former, *Funaria* as a type of the latter.

A. Marchantia Polymorpha.

§ 2. **External Characters and General Life History** (fig. 206). The plant, which is commonly found on damp ground or other moist surface, is a green dorsiventral *dichotomously branching* **thallus** (p. 6), giving off numerous unicellular, hair-like rhizoids, from its under (ventral) surface, which also bears violet-coloured flattened scales consisting of a single layer of cells. The thallus has a distinct midrib. Special **reproductive branches** spring from the upper (dorsal) surface of the thallus. These bear the **sexual organs**, antheridia and archegonia. It follows that *the plant is the gametophyte*, and is equivalent to, *i.e.* homologous with, the prothallus of the Fern. This is an important point which must be carefully borne in mind. We shall see that the gametophyte of the liverwort is much more highly differentiated than that of the Fern. The branches bearing antheridia are called **antheridiophores**;

those bearing archegonia, **archegoniophores**. They are borne on different plants, so that Marchantia is unisexual. Each consists of a stalk bearing a terminal disc or head, **the receptacle**. In both cases the stalk has two furrows which bear rhizoids. At its first appearance the furrows are distinctly ventral. This indicates that the reproductive branch is formed by the fusion of the two branches of a

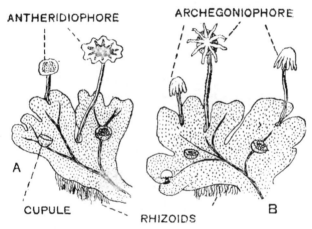

Fig. 206. —MARCHANTIA POLYMORPHA.
A, Male plant; B, Female.

bifurcation of the thallus. The *receptacle* of the male shoot is rounded and flattened and has a wavy margin; that of the female is star-shaped, bearing a number of spreading rays, resembling the ribs of an umbrella. The oospore produced by the fertilization of the oosphere of the archegonium develops into a structure called the **sporogonium**, which gives rise to **asexual spores**. Evidently the sporogonium is the homologue of the **sporophyte** of the higher plants; but in Muscineæ it is never differentiated into stem and leaves, and is never set free from the gametophyte. From the asexual spore a new gametophyte is produced. The gametophyte reproduces itself *vegetatively* by means of **gemmæ**, multicellular bodies developed in circular cup-shaped organs with membranous margins, called **cupules**, formed on the upper surface of the thallus (fig. 206).

§ 3. **The structure of the thallus** is shown in fig. 207.
The great bulk of the thallus is composed of thin walled
parenchyma. The cells towards the lower surface show
reticulate markings on their walls. Those nearer the
upper surface often contain numerous starch-grains. Here
and there single cells with mucilaginous or oily contents
occur. The upper superficial layer of cells (so-called
"epidermis") contains chloroplasts; there is no cuticle.
Immediately beneath it there are large diamond-shaped

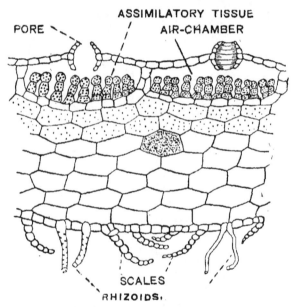

Fig. 207.—Thallus of Marchantia.
(Transverse section.)

air-cavities separated from each other by thin cellular
partitions which connect the superficial layer with the
underlying parenchyma. Short rows of small oval cells
packed with chloroplasts grow up from the floor of each
air-cavity; these form the chief **assimilating tissue** of the
plant. In the middle of the roof of each cavity there is a
large **pore** (figs. 207, 208), bounded by several tiers of
cells. Corresponding to the air-spaces the surface of the
thallus is marked out into a number of diamond-shaped

areas, in the middle of each of which the pore shows as a dot. The pores differ altogether from stomata in their development, although they are analogous organs having the same function. True stomata are not developed on any gametophyte. The absence of vascular tissue should especially be noticed (see p. 18).

The ventral scales and rhizoids are developed more especially on the midrib. The rhizoids in Marchantia are of two kinds:—(*a*) long slender hairs, with curious peg-like thickenings developed internally on their walls; (*b*) stouter hairs without thickenings. These rhizoids absorb inorganic solutions in the usual way and fix the plant to the soil.

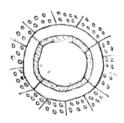

Fig. 208.—Pore seen from above.

The growth of the thallus is effected by a group of initial cells at each growing-point.

§ 4. **The Gemmæ** (fig. 209) are flattened isobilateral bodies with a notch on each side. Most of the cells contain chloroplasts, but here and there are larger clear cells, which are capable of forming rhizoids. The growing-points are situated in the notches and grow into new thalli.

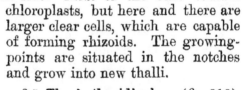

Fig. 209.—Gemma of Marchantia. (Surface view.)

§ 5. **The Antheridiophore** (fig. 210). The lower surface of the receptacle of the antheridiophore bears rhizoids and scales like the thallus. The upper surface is flat. The tissue of the receptacle is similar to that of the thallus. Air-cavities are present, opening by means of pores on the upper surface. But in addition to these there are large **flask-shaped cavities**, also opening by small apertures called **ostioles** on the upper surface. These cavities are arranged in rows radiating out from the centre, and an **antheridium** is developed on the floor of each. The antheridium (fig. 211) is oval in form, and is borne on a short multicellular stalk. Its wall consists of a single layer of cells containing chloroplasts. Inside there is a mass of **spermatocytes**, each of which gives rise to a bi-ciliate **spermatozoid**.

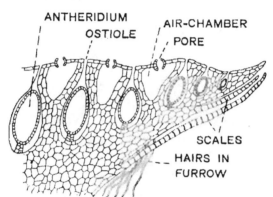

Fig. 210.—Part of Antheridiophore of Marchantia. (Vertical section of receptacle.)

In the development of the antheridium a cell grows out from the upper surface of the young receptacle. It is immediately divided into two. The upper cell forms the body, the lower the stalk of the antheridium. The antheridium is gradually enclosed in a flask-shaped cavity owing to the growth of the surrounding tissue.

§ 6. The Archegoniophore

(fig. 212). The tissue of the receptacle of the archegoniophore is similar to that of the thallus. There are large air-cavities opening by pores on the upper surface, and large mucilage-cells are present. The **archegonia** are developed on the under surface of the receptacle. They are arranged in radial rows which alternate with the rays of the receptacle. The youngest lie nearest to the stalk of the archegoniophore. The margins of the rays grow down and form curtain-like membranes, the **perichætia**, which envelop the archegonia.

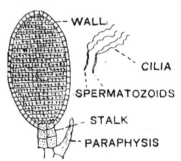

Fig. 211.—Antheridium of Marchantia.

The *archegonium* is borne on a short stout stalk and consists of a dilated **venter** and a very long **neck**. The wall of the venter consists of one layer of cells. It contains the **oosphere** and the **ventral canal-cell.** In the canal of the neck there is a row of **neck-canal-cells.** The neck consists of many longitudinal rows of cells surrounding the canal. The terminal cells, called **lid-cells,** are at first united, so that in the young archegonium the apex of the neck is closed. When the archegonium is mature the ventral canal-cell and the row of neck-canal-cells become disorganized, and are converted into mucilage which absorbs water, forces open the lid-cells and oozes out of the neck.

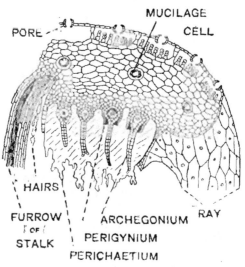

Fig. 212.—Archegoniophore of Marchantia. (Section between two rays.)

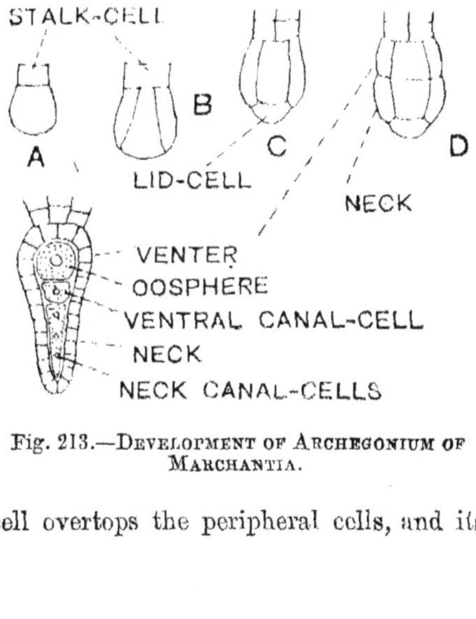

The archegonium is developed as a protuberance from a single cell (fig. 213). This grows out and is cut off by a wall. It is then divided transversely into two. The basal cell undergoes a few divisions and forms the stalk. The other cell is the mother-cell of the archegonium. It is divided by three longitudinal walls into three *peripheral* cells and one *central* cell. The central cell overtops the peripheral cells, and its apical portion is cut off as

Fig. 213.—Development of Archegonium of Marchantia.

the lid-cell, which afterwards by further division forms the lid-cells of the neck. The peripheral cells are further divided longitudinally into six, which are called *envelope-cells*. The six envelope-cells and the single central cell are then divided transversely into two storeys. The lower storey forms the venter; its envelope-cells further divide to form the wall; its central cell divides into oosphere and ventral canal-cell. The upper storey forms the neck; *its* central cell forms by division the row of 8—16 neck-canal-cells.

§ 7. Fertilization, and Development of Sporogonium

(fig. 214). Fertilization takes place when the plants are wet with rain or dew, and is effected in exactly the same way as in the Fern. The antheridium bursts at the apex, and the spermatozoids are set free. They are attracted to the archegonia by some organic substance present in the mucilage which oozes out of the neck. The effects of fertilization are not confined to the oospore. The whole archegoniophore elongates and becomes very much larger (cf. fruit-formation in Flowering Plants). The venter of the archegonium continues to grow; it is now called the **calyptra**, and forms an investment round the devoloping **embryo**. A loose cup-shaped structure, the **perigynium**, which makes its appearance just before fertilization (fig. 212) as an outgrowth of the stalk at the base of the archegonium, also grows rapidly and surrounds the calyptra and embryo.

The oospore is first divided transversely by a basal wall into an upper or **epibasal** and a lower or **hypobasal cell** (A). Then two other walls, at right angles to each other and to the basal wall, divide the oospore into **octants** (cf. the Fern). The hypobasal octants being further divided give rise to a short bulbous **foot** (B) which absorbs nourishment from the receptacle (cf. Fern), and a short **stalk** which is produced rather late in development by intercalary growth. The epibasal half forms the **capsule** of the sporogonium. By a wall formed parallel to the surface in each epibasal octant a central region, the **endothecium**, is marked off from a peripheral layer, the **amphithecium** (B). The wall of the capsule, consisting of a single layer of cells, is formed from the amphithecium. The whole of the endothecium constitutes the **archesporium**. Some of the cells produced by the division of the archesporial cells are sterile; they become elongated and spindle-shaped, and spiral markings are developed on their walls. These curiously modified cells are called **elaters** (E); they are hygroscopic, *i.e.* easily influenced and stimulated to movement by moisture; when the sporogonium opens they aid in scattering the spores.

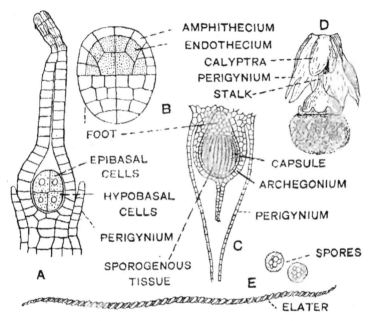

Fig. 214.—Segmentation of Oospore and Development of Sporogonium of Marchantia.

The other cells produced from the archesporium are **spore-mother-cells**. Each gives rise exactly as in the Fern to four **spores**.

Thus the **sporophyte generation** is represented by a sporogonium. It shows a rudimentary differentiation into root (foot) and shoot (stalk and capsule), but the shoot is not differentiated into stem and leaves. It is parasitic on the gametophyte. Eventually the capsule breaks through the archegonium (calyptra) owing to the elongation of the stalk. It splits longitudinally into a number of divisions or teeth, and the spores are set free (D).

320 BOTANY.

§ 8. **Germination of the Spore** (fig. 215). In germination the *exosporium* is ruptured, and the *endosporium* grows out to form a short tube (germ-tube). The further growth of this is at first filamentous, but eventually a small green cellular plate (germ-disc) is produced. The whole structure is called the **protonema**. The young Marchantia plant arises from this *as a lateral outgrowth*. The protonema then dies. It will be noticed that the development of the "plant" is **indirect** or **heteroblastic**. It is preceded by a **pro-embryo**, the protonema. A *trace* of this is probably to be recognized in the early filamentous stage of the fern-prothallus.

The life-history of Marchantia is represented graphically in fig. 216.

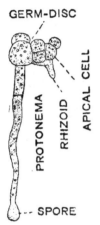

Fig. 215.—Protonema of Marchantia, showing young thallus developing from germ-disc.

B. FUNARIA HYGROMETRICA.

§ 9. **External Characters** (fig. 217). Funaria is a common moss which grows in dense tufts or patches on the surface of the ground, often on the tops of walls. The plants are small, being scarcely half an inch in height. They are differentiated into stem and leaf (leafy shoot), but there is *no true root*. The dark-coloured base of the shoot gives off numerous slender, brown, *multicellular* **rhizoids**, which pass down into the soil. The leaves are simple and more or less ovate. They show a distinct midrib and have a ⅜ spiral phyllotaxis. There is comparatively little branching: it is lateral, but not axillary; the branches are given off beneath the leaves.

§ 10. **General Life-History**. The plant, as in the Hepaticæ, is the **gametophyte**, but is much more highly differentiated. In the Mosses the gametophyte attains a high degree of development. The **antheridia** and **archegonia** are borne at the apices of the shoots, concealed amongst the leaves of the apical bud (fig. 219).

Funaria is unisexual. The apical buds containing antheridia can be more or less easily distinguished, as their leaves spread out and form rosette-like structures called **perigonia** or *perichætia*. The central leaves of the rosette are often reddish in colour. The plants bearing archegonia are

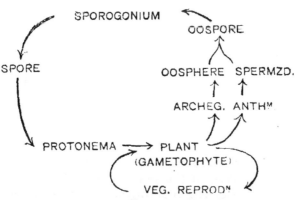

Fig. 216.—Life-History of Muscineæ.
In Marchantia and Funaria the gametophyte is represented by two plants.

longer than those bearing antheridia; their apical buds, however, are not specially distinguished.

It should be carefully noticed that the stem and leaves of the moss are not homologous with, but only analogous to the stem and leaves of the fern-plant; they belong to different generations.

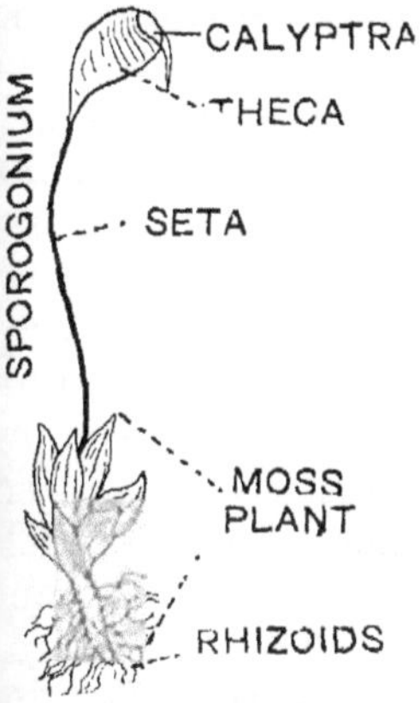

In Mosses, as in Hepaticæ, the sporophyte generation is represented by a **sporogonium** derived from the fertilized ovum. The sporogonium is rather more highly differentiated, and consists of a **capsule** or theca, a stalk called the **seta**, and a **foot**. The asexually produced spore gives rise to a **protonema** (fig. 224), which, however, is a much larger and longer-lived structure than that of Hepaticæ. It is a much-branched filament bearing an external resemblance to a green alga. The cells of most of the branches contain numerous chloroplasts, while others pass down into the soil, and are not to be

Fig. 27.—Funaria with Sporogonium.

distinguished from rhizoids. The rhizoids of many mosses may in fact form protonemata. The **moss-plant** is developed on the protonema as a lateral bud. The protonema continues to grow for some time, and produces numerous plants. Thus, as in Hepaticæ, the development of the gametophyte from the asexual spore is *indirect*.

Funaria has great powers of **vegetative reproduction.** Protonemata may be produced from any part—rhizoids, stem, leaf, and even from the sporogonium. In the last case we have an instance of **apospory** (p. 267). Some mosses form multicellular *gemmæ*, but this is not the case in Funaria.

§ 11. **Structure of the Stem** (fig. 218). The outermost layer of cells of the stem is marked off as an "*epidermis.*" Beneath this is a many-layered **cortical region,** surrounding a **central strand** of elongated thin-walled cells. The cells of the cortex contain chloroplasts, and in the outer region their walls are thickened. The central strand is a conducting tissue, and must be considered as representing a rudimentary *vascular cylinder, analogous* to the stele of the sporophyte in higher types. In some mosses, such as Polytrichum, but not in Funaria, the conducting strand shows a central region of more or less thick-walled cells surrounded by a region of thin-walled cells. This would be analogous to the differentiation into xylem and phloem.

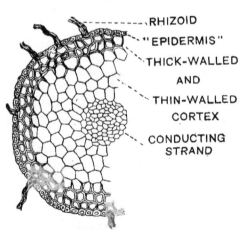

Fig. 218.—Stem of a Moss.
(Tranverse section.)

The growth of the stem in mosses is effected by a three-sided apical cell like that of the Fern. The segments cut off from it divide into inner and outer halves, of which the former give rise to the central conducting tissue. Each outer half is divided into upper and lower parts. The upper part protrudes as a two-sided apical cell, and developes into a leaf. The lower part forms the cortical tissue of an internode. If branching occurs the branch is formed from this lower part.

§ 12. **Structure of the Leaf** (fig. 219). Except at the midrib the leaf consists of a single layer of cells containing chloroplasts. This is the **assimilating tissue**. The leaf is thickened at the midrib, which contains a strand of thin-

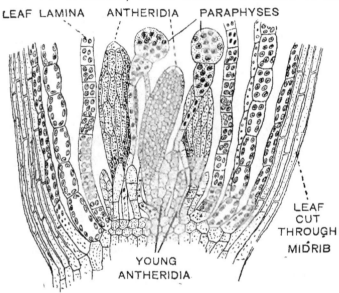

Fig. 219.—APEX OF MALE PLANT OF FUNARIA.
(Longitudinal section.)

walled conducting cells like those of the stem. In some mosses these strands join on to the central strand of the stem, but in Funaria this is not the case; here there are no leaf-traces.

§ 13. **The Antheridia** (figs. 219, 220A) are club-shaped bodies, borne on stout multicellular stalks. The wall consists of a single layer of cells within which are numerous **spermatocytes**. On the access of water the antheridium bursts at the apex and the spermatocytes are liberated. Their walls become mucilaginous and the **spermatozoids** (fig. 220, c) escape. They are biciliate like those of Hepaticæ.

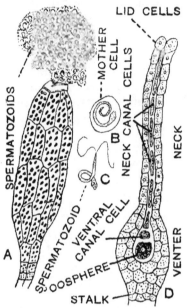

Fig. 220.—SEXUAL ORGANS OF FUNARIA. A, Antheridium; D, Archegonium.

The antheridia are developed from single cells at the apex of the shoot—including even the apical cell. The cell grows out and is divided into two. The lower cell forms the stalk. The upper grows like an apical cell and gives off two series of segments, which are divided into central cells, from which the spermatocytes are developed, and peripheral cells, forming the wall. This mode of apical growth is unusual.

§ 14. **The Archegonium** (fig. 220, D) is like that of Hepaticæ; but the stalk is more strongly developed, and the wall of the venter consists of two layers of cells. There is a long twisted neck, consisting of six longitudinal rows of cells, surrounding the central canal.

The archegonium (fig. 221) is developed from a single cell which may be the apical cell of the shoot. The cell grows out and the protuberance is divided into two. The *lower cell* forms the stalk. The *upper cell* functions as an apical cell. It shows continued growth, and gives off segments forming, after further division, tiers of cells, each tier consisting of a central cell and surrounding peripheral cells. The lowest central cell forms the oosphere and ventral canal-cell; the other central cells form the neck-canal-cells. The peripheral cells form the cells of the neck and venter. Here also the continued apical growth is highly remarkable.

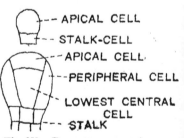

Fig. 221.—DEVELOPMENT OF ARCHEGONIUM OF MOSS.

§ 15. **Fertilization** is effected in the usual way. The spermatozoids make their way to the archegonia when the plants are wet with dew or rain. The attracting substance here seems to be some form of sugar. The oospore develops into the sporogonium.

§ 16. **Structure of Sporogonium** (fig. 217). The sporogonium representing the sporophyte generation is differentiated into root and shoot, but not into stem and leaves. It consists of foot, seta, and capsule. The **foot** (fig. 223, E) is a small conical structure which buries itself in the apex of the female plant, and serves for the absorption of nutriment. It is invested by a membranous sheath, the **vaginula**, representing the lower half of the archegonium, which is ruptured during the development of the sporogonium. The **seta** is a long slender structure of a reddish colour. It has an "epidermis," a thick-walled cortex, and a conducting strand like that of the moss-plant.

The **capsule** (fig. 222) is a pear-shaped structure. Its solid basal region is called the **apophysis**. The epidermis of the apophysis has *true stomata*. At each end of the pore the wall between the two guard-cells breaks down, so that the pore seems to be surrounded by a single ring-shaped cell. The parenchymatous cells beneath the epidermis contain chloroplasts. The conducting strand of the seta is continued into the apophysis. It has been proved that the sporogonium can assimilate all the carbon that it requires, so that only inorganic solutions are absorbed by the foot. It may be regarded as only semi-parasitic. The **wall** of the capsule consists of several layers of cells; the inner layers contain chloroplasts. Internal to this is a large **air-space** traversed by delicate strands of cells. Next comes the **spore-sac**, surrounding a sterile central column, the **columella**. The outer wall of the spore-sac (sometimes called the *outer spore-sac*) consists of two or three layers of cells. The inner wall (sometimes called the *inner spore-sac*) lies next the columella. At the apex of the capsule is a sort of lid, the **operculum**, which separates off when the capsule dehisces. The dehiscence is effected by the rupture of a ring of cuticularized epidermal cells, the **annulus**,

326   BOTANY.

round the base of the operculum, immediately above the upper end of the spore-sac. When the operculum comes away a number of yellow thickened tooth-like structures, constituting the **peristome**, project. These are hygroscopic, and allow the spores to escape only when the air is dry.

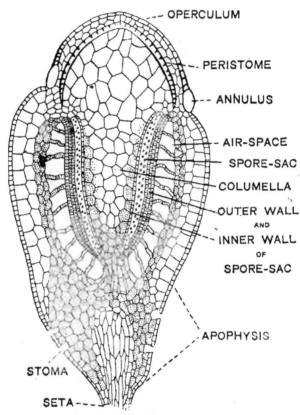

Fig. 222.—CAPSULE OF FUNARIA.
(Longitudinal section.)

In Funaria there are two rows of peristome teeth (outer and inner). They represent the outer and inner thickened and cuticularized regions of the walls of a plate of cells which have otherwise broken down. The apex of the capsule is covered by a membranous cap, the **calyptra** (fig. 223, E), representing the upper half of the ruptured archegonium.

§ 17. **Development of Sporogonium** (fig. 223). The oospore is first divided by a basal wall into **hypo-** and **epi-basal cells**. By further division a two-sided apical cell is formed at each end. The two rows of segments cut off from the hypobasal end form the foot (A). The epibasal half also forms two rows of segments (A, B). The segments are divided into inner and outer halves (A, C). In the region

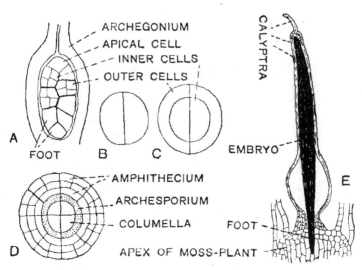

Fig. 223.—DEVELOPMENT OF SPOROGONIUM OF FUNARIA.
B, C, D, transverse sections; D in region of Capsule.

of the seta the inner halves form the central conducting tissue; the outer halves the cortical tissue. In the region of the capsule, which is not distinctly marked off from the seta till the embryo has elongated considerably, the outer halves constitute the **amphithecium**; the inner halves the **endothecium** (D). The **archesporium** is the *outermost layer* of the endothecium, the rest of the endothecium forming the columella. Everything outside the sporogenous tissue, including the outer wall of the spore-sac, is derived from the amphithecium. The operculum slowly differentiates, and the innermost layer of the amphithecium over the region covered by the operculum gives rise to the peristome. The archegonium is ruptured during the elongation of the sporogonium. The **spores** are developed from the mother-cells in the usual way. There are no elaters.

§ 18. **Germination of the Spore** (fig. 224). When the spore germinates, the exosporium is ruptured and the endosporium grows out at each end into a germ-tube. At one end the germ-tube forms a **rhizoid**. The other develops into the **protonema**. The growth of each branch of the protonema is effected by means of its apical cell.

§ 19. **The Young Moss-Plant** (fig. 224) arises as a little bud from a cell of the protonema close to a septum. In this protuberance, oblique divisions appear and these separate the apical cell of the young plant from those which give rise to the first leaf and the first rhizoid.

The life-history of the moss is represented graphically in fig. 216.

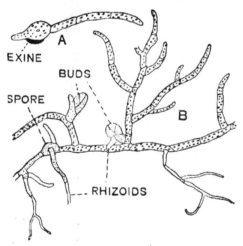

Fig. 224.—A, Germinating Spore: B, Protonema of Funaria.

§ 20. **Summary and Conclusions.** Thus, in the Bryophyta there is a distinct alternation of generations. The gametophyte is *the plant*; the sporophyte generation is represented by a sporogonium parasitic or semi-parasitic on the female gametophyte. The relative importance of the two generations has been reversed as compared with what we find in Vascular Cryptogams and

Flowering Plants. The sporophyte is practically only a sporogenous capsule; there are no distinct sporangia. A seta is developed simply to elevate the capsule, and a foot to absorb nourishment.

There is a very wide gap between the Bryophyta and the Pteridophyta — wider even than that between the Pteridophyta and Phanerogams.

Nevertheless, the homologies discernible between the two groups—*e.g.* the alternation of generations, the similarity in the *general* course of the life-history, the development of the spores—leave no doubt that they are genetically connected, in other words, that the two groups are descended from common ancestors. There is now reason to believe that the Hepaticæ are the living forms which most nearly approach these primitive ancestral forms. On this view the Pteridophyta would represent the main line of evolution in which the sporophyte became more highly differentiated; while the Musci would represent a divergent line in which the gametophyte attained a high development. We can readily understand that the higher evolution of the sporophyte, along the Pteridophyte line, was initiated by the sporogonium borne on some *thalloid form* sending down absorbing organs (roots) into the soil. During the early imperfect stages of evolution the foot would continue to carry on its absorbing functions; but gradually its period of functional activity would be restricted to embryonic stages, as in the Fern.

CHAPTER XX.

THE ALGÆ.

§ 1. THE Algæ constitute one of the two important Classes into which the Thallophyta are subdivided. They are aquatic plants. Many are fresh-water forms, but the great majority live in the sea, and constitute the assemblage of organisms called the marine algæ or *seaweeds*. The vegetative body is usually a **thallus**: but many show a more or less well-marked differentiation into root and shoot and some also into stem and leaf. Structurally, the thallus consists entirely of living cells, although, in some of the larger forms, distinct conducting and assimilating tissues are developed. In essential points the processes of nutrition resemble those of the ordinary green plant, while differing widely in detail (see p. 14). All contain chlorophyll, but, in many, the green colour is masked by the presence of other colouring matters. These are chiefly a brown colouring matter (phycophæin), and a red colouring matter (erythrin). They are present in the plastids along with the chlorophyll. The differences in colour are correlated with important differences in development and life-history—hence the convenient division of the Algæ into green (Chlorophyceæ), brown (Phæophyceæ), and red (Rhodophyceæ). The brown and red Algæ are mostly marine.

The plant is the **gametophyte**. The *sporophyte* is either unrepresented or represented only in a rudimentary form. Hence, there is at most only an indication of an alteration of generations.

Sexual reproduction is general. The **gametes** and the **gametangia**, *i.e.* the organs producing the gametes, may or may not be differentiated into male and female. If the sexual process consists in the conjugation of similar gametes it is said to be **isogamous**, and the **zygote** formed is termed a **zygospore**. If it consists in the fertilization of an

THE ALGÆ.

oosphere by a male element it is **heterogamous** (as in higher types) and the zygote is an **oospore**. This applies to the green and brown Algæ; in the red Algæ the sexual processes are very peculiar and highly specialized. The gametophyte can also in most cases reproduce itself asexually. The asexual organs of reproduction will be described in connexion with our special types.

SPIROGYRA.

§ 2. **General Characters.** Spirogyra is one of the green fresh-water Algæ. It grows in dark-green slimy masses in ponds, springs, or slow-running streams. Each Spirogyra plant has an extremely simple structure. Its vegetative body is an *unbranched* filamentous **thallus** (fig. 225), consisting of short cylindrical cells placed end on end and showing no distinction of base and apex. The filament increases in length by ordinary cell-division and growth of the cells. All the cells have the same structure and all are capable of division. Here we have an example of a multicellular plant which shows little or no division of labour. Indeed, we might, physiologically, regard each cell as an individual plant, and the whole filament as a colony of such individuals, for each cell carries on exactly the same vital functions as are necessary for the maintenance of the whole. The filament is, in most species, invested by a delicate mucilaginous sheath secreted by the cells. It is this which makes a mass of Spirogyra filaments feel slimy to the touch.

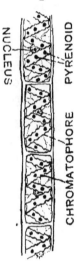

Fig. 225.—PART OF A FILAMENT OF SPIROGYRA.

§ 3. **Structure of the Cell** (fig. 225). Each cell is cylindrical in form with transverse end-walls, and has the structure characteristic of typical parenchymatous cells. The wall is a cellulose membrane. Inside the cell there is a "**primordial utricle**," from which delicate **protoplasmic strands** run across a central **vacuole** to the centre of the cell. The **nucleus**, containing a distinct nucleolus, is usually embedded in the small central mass of protoplasm. The most conspicuous structures in the cell are certain green spirally coiled bodies called *chlorophyll bands* or **chromatophores**. There may be one or several of these in a cell. Each is a specialized protoplasmic body containing chlorophyll diffused through its substance. It lies in the primordial utricle, coiled as it were round the central vacuole. In its substance, at intervals, are distinct rounded granules, also protoplasmic in nature. These are the **pyrenoids**. They are specially active in the formation of starch; each becomes surrounded by starch-grains when the cell contains surplus carbohydrates. Pyrenoids are frequently found in the Algæ.

Numerous species of Spirogyra are distinguished according to the size and form of the cells, the number of chromatophores, etc.

§ 4. **Reproduction.** There is no special method of asexual reproduction; but if a filament happens to be broken into a number of pieces, consisting of one or several cells, these

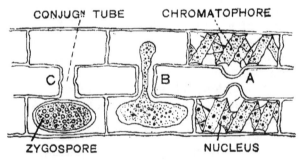

Fig. 226.—SPIROGYRA IN CONJUGATION.

by ordinary cell-division may form new filaments (vegetative reproduction). Voluntary fragmentation of this kind often occurs under abnormal nutritive conditions.

Sexual reproduction (fig. 226) seems to occur whenever the vigour of the filaments becomes impaired by age, by prolonged division, or by unfavourable external conditions.

THE ALGÆ. 333

It is **isogamous**. In this process two filaments, lying side by side, send out little outgrowths from corresponding cells. These increase in size, meet, and finally fuse between the two filaments, so that the corresponding cells are joined by short **conjugation-tubes**. In the meanwhile the contents of each cell have contracted, losing water and forming a **gamete**, in which the characteristic form and appearance of the chromatophores can no longer be distinguished. Then the gametes of one filament pass over, by means of the conjugation tubes, into the cells of the other filament, and fuse with the gametes in these cells. It should be specially noticed that the gametes bear no cilia (**aplanogametes**), and that any cell of a filament may function as a **gametangium**; also that **conjugation** is effected by means of conjugation-tubes. The more active gametes, which may be regarded as being male, are formed earlier than the more passive ones, which may be considered as female. Generally all the gametes of one filament pass over into the other filament, so that the filaments are unisexual. There are, however, exceptions to this; and conjugation tubes are occasionally formed between cells of the same filament. In a family closely allied to Spirogyra conjugation is effected in the middle of the tube, and the gametes are in all respects alike.

§ 5. **The Zygospore.** The result of conjugation is in each case the formation of a zygospore. This is a large oval body, at first green, but later becoming dark brown. After a more or less lengthy period of rest, the zygospores germinate. Each contains food-material in the form of oil and other substances. In germination the two outer coats of the spore

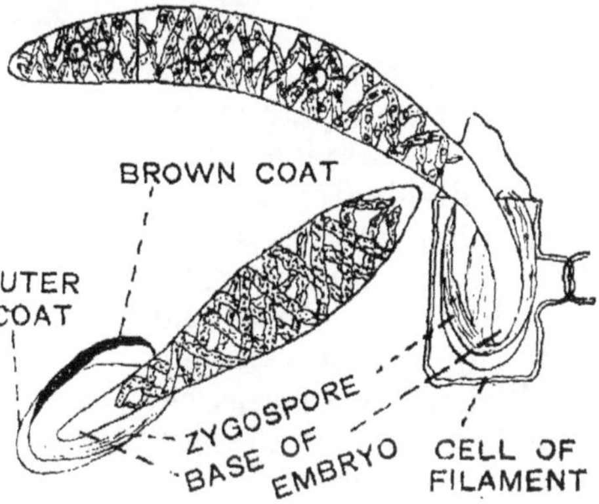

Fig. 227.—SPIROGYRA; GERMINATION OF ZYGOSPORE.

(fig. 227) are ruptured, and the contents enclosed in the innermost coat grow out into a tube which is divided into two cells by a transverse septum. The cell which protrudes from the spore is green, and grows out to form the filament. The other cell is colourless. Thus, at first, there is a distinction of base and apex, but this is soon lost.

Sometimes, when conjugation does not take place, a gamete may develop directly into a spore called an azygospore. This may be compared with parthenogenesis (p. 267).

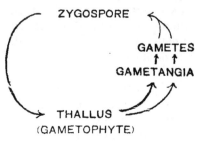

Fig. 228.—Graphical Life-History of Spirogyra.

§ 6. It will be readily recognized that the Spirogyra plant represents the gametophyte. There is no sporophyte, and therefore no alternation of generations. The zygospore, instead of giving rise to a purely asexual plant, produces a new gametophyte. The life-history is graphically represented in fig. 228.

Ulothrix Zonata.

§ 7. **Structural Characters.** This is a bright green Alga, masses of which are commonly found, in spring and early summer, in running streams, fountains, cattle-troughs, etc. The individual plant consists of a simple, *unbranched* **filamentous thallus**, attached to stones or other substratum by a clear, colourless **root-cell** (fig. 229, A). It thus shows a distinction into base and apex.

The filament is extremely slender, and consists of numerous small short cells placed end on end. The growth of the filament is effected by the elongation and division of all its cells.

Each cell has a **primordial utricle**, containing the **nucleus** and a ring-shaped **chromatophore** with numerous **pyrenoids**, and enclosing a **vacuole** filled with cell-sap. The chromatophore is placed transversely about the middle of the cell, forming as it were a zone or girdle to the vacuole—hence the name *U. zonata*.

§ 8. **Reproduction.** The plant is the gametophyte.* The
sexual process is **isogamous.** The **zygospore** does not directly
reproduce the gametophyte. It gives rise to a very minute
plant, a **rudimentary
sporophyte,** in which are
developed a number of
motile spores called
zoospores, from which
the new gametophytes
are produced. Thus
there is an indication
of an **alternation of
generations.** The game-
tophyte also reproduces
itself **asexually**—chiefly
by means of special
motile protoplasts called
zoogonidia. We must
now consider these pro-
cesses in detail.

§ 9. **Asexual Repro-
duction.** The zoogonidia
are also called zoospores
(or megazoospores). Be-
fore proceeding further
it will be well to under-
stand what they really
are. In many Algæ (red
seaweeds and others) re-

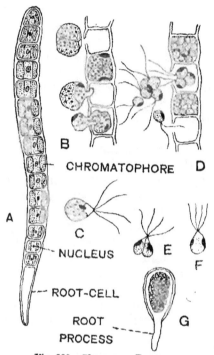

Fig. 229.—ULOTHRIX ZONATA.
A, Young plant; B, C, Zoogonidia; D, Formation and escape of gametes; E, F, Conjugation; G, Rudimentary sporophyte.

productive cells of a similar nature are produced, but being
non-motile they are called **gonidia** not **zoogonidia.** These
gonidia and zoogonidia, being specialized reproductive cells
capable of giving rise to a new organism, are *spores.* They are
produced, however, on the gametophyte, and directly reproduce it. In these respects they differ from the spores of the

* It is by no means certain that what is here termed a "gametophyte"
corresponds phylogenetically to the gametophyte of the Vascular
Cryptogams.

higher plants (p. 266), which are produced on the sporophyte, and give rise on germination to a gametophyte. This is why they have received the special name "gonidia."[†] Similarly, the organ in which such spores are formed is called a **gonidangium**, just as the organ in which the spores of higher plants are developed is called a sporangium.

In Ulothrix any cell of the filament may function as a gonidangium (fig. 229, B). It may produce one, two, four, or occasionally eight zoogonidia. If one, the greater part of the contents are simply rounded off (rejuvenescence, p. 39) to form it; if two, the contents are first divided transversely; if four or eight, there are further *longitudinal* divisions.

The **zoogonidia** (fig. 229, B, C) are set free by the rupture of the wall of the gonidangium. Each is a pear-shaped protoplasmic body with a single nucleus. The clear, pointed anterior end bears four cilia, and has a red **pigment-spot** (called the "eye-spot"), and a **contractile vacuole**, *i.e.* a vacuole which alternately expands and collapses. The rounded posterior end contains the chromatophore. The zoogonidium swims about actively for an hour or two by means of its cilia. It is sensitive to light, moving in the direction of light of moderate intensity, away from too intense light. It ultimately settles down, the cilia are withdrawn, a cell wall is formed, and a new filament is gradually developed.

Sometimes in unfavourable conditions of growth, chiefly periods of drought or similar conditions, the filaments break up into groups of rounded cells which can multiply by division. These were formerly mistaken for a distinct Alga, and called Palmella. It may be spoken of as the **palmella-stage** of Ulothrix. On the return of favourable conditions, the cells either germinate directly or give rise to zoogonidia. It may be considered as another method of asexual reproduction.

[†] We have seen however, that certain fern prothalli can reproduce themselves by means of spores produced in sporangia which are identical with those of the sporophyte (p. 267). Hence gametophytes also may reproduce themselves by means of spores; and since we may regard the Thallophyta as plants in which no definite alternation of generations has as yet been developed, and in which therefore the gametophyte and sporophyte generations lie dormant in each individual plant, it is not absolutely necessary to give these spores any special name.

§ 10. **The Sexual Cycle.** Any cell may function as a **gametangium.** The gametes (fig. 229, D) are ciliated motile protoplasts (**planogametes**—*i.e.* wandering gametes). They are also termed *microzoospores.* Except that they are smaller, and bear only two cilia, they resemble the zoogonidia or megazoospores. The gametangium may produce eight, sixteen, or thirty-two of them. They are developed, set free, and move about in the same way as the zoogonidia. If two gametes *from different gametangia* meet, **conjugation** takes place (fig. 229, E, F). The gametes become attached by their cilia; the clear pointed ends fuse first, and then the gametes come to lie side by side and fuse laterally. The two nuclei fuse together. The **zygospore**, when first formed, has four cilia, but these are soon withdrawn, and a cell-wall is formed. *If the gametes fail to conjugate, they can germinate directly like zoogonidia.*

The zygospore germinates and a clear process resembling a root-hair is given out at one end (fig. 229, G). This becomes attached to some object. The spore increases in size, but does not divide, and rests during the summer. In the autumn a varying number of zoospores (2-14) are produced by a process of *free cell-formation.*

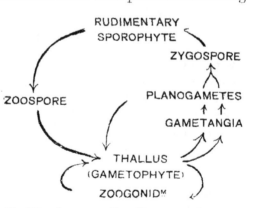

Fig. 230.—Graphical Life-History of Ulothrix.

They resemble the zoogonidia, and, when set free, come to rest and give rise to new filaments. The tiny plant produced from the zygospore is considered to represent, in a rudimentary form, the **sporophyte** of higher types. The ciliated protoplasts developed in it would be, according to this view, zoospores in the strict sense, and would need to be clearly distinguished from the zoogonidia and gametes, which, as already explained, are also called zoospores. The graphical representation of the life-history given in fig. 230 will make this clear.

§ 11. **Origin of Sexuality.** A careful consideration of the fact that the gametes may germinate like the zoogonidia or megazoospores inevitably leads to the conclusion that they have been derived from the latter—that they, in fact, *are* zoogonidia in which the tendency to, and capacity for, conjugation has been evolved. On the view that conjugation means increased vigour and vitality, we can understand how this tendency would be evolved. In Ulothrix, however (and some other Algæ), it is scarcely a fixed hereditary character; the gametes may still germinate without conjugation, although they usually form weaker plants. *Parthenogenesis* probably represents a trace of this in higher types.

Similar reasoning with regard to the zoospores produced from the zygote might also lead to striking conclusions. It *may be* that, in its first beginnings, the sporophyte was simply a reduced gametophyte—a gametophyte reduced to the condition of an organ producing asexual spores and nothing else. But this is too abstruse a subject to be discussed here.

VAUCHERIA.

§ 12. **Structure.** Most of the species of this green Alga grow in fresh water, or on the damp surface of the soil. A few species are marine. *V. sessilis* and *V. terrestris* are commonly found, mixed with other Algæ and the protonemata of mosses, in the form of a green tangled felt on the damp soil of neglected flower-pots.

The thallus (fig. 231, c) consists of long, rather coarse, tubular threads, branched at considerable intervals, and fixed by means of a branched colourless **root-process**. The tubes are **non-septate**, *i.e.* they are not divided by cell-walls into distinct cells. Septa, however, are formed in connexion with the development of reproductive organs. The cellulose wall of each tube has a continuous **protoplasmic lining**. A **vacuole** filled with cell-sap runs up the middle of the tube. In the outer region of the protoplasm there are numerous oval **chloroplasts**, and large numbers of small **nuclei** are found in the layer internal to this. Small refractive **oil-globules** are associated with the plastids. It is interesting to notice that usually no starch is present; here the storage product of metabolism is oil. Vaucheria was formerly described as a *unicellular* Alga. The branched tubes, however, are not cells, but **cœnocytes**; here we have a good example of cœnocytic structure (p. 39). The branches of the cœnocyte exhibit apical growth.

§ 13. **Asexual reproduction** is commonly effected by means of zoogonidia. In the formation of a zoogonidium the apex of a branch swells up and becomes club-shaped (fig. 231, A), owing to the aggregation of protoplasmic substance in it.

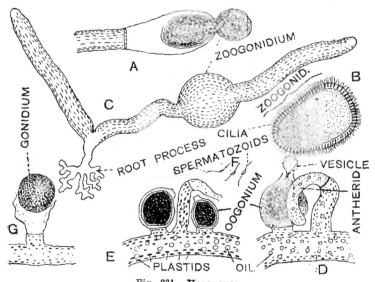

Fig. 231.—VAUCHERIA.
A, B, The zoogonidium; C, Young plant formed from zoogonidum;
D, E, F, Sexual organs.

This club-shaped body, which is the **gonidangium**, is separated from the rest of the tube by a distinct septum. It ruptures at the apex, and the protoplasmic contents escape as a zoogonidium. The opening is very narrow, and, as the protoplasmic body makes its way out, it is frequently constricted and divided into two zoogonidia. The **zoogonidium** (fig. 231, B) is a large oval body which can be seen by the naked eye. It shows a central region, *the endoplasm*, containing numerous chloroplasts, and a clear outer region, *the ectoplasm*, with numerous small nuclei. Evidently it also is cœnocytic and not a single protoplast. It might be called a *zoocœnocyte*. It is covered with **cilia**, a pair being developed opposite to each nucleus. (In some plants allied to Vaucheria numerous zoogonidia are produced, each having one nucleus and two cilia.) The zoogonidium, after moving about for a short time, develops a cellulose wall and comes to rest. The cilia are withdrawn, and germination takes place. Two tubes grow out. One branches and produces the colourless root-process, the other develops into the green tubular thread.

Some species of Vaucheria (not *V. sessilis*) produce **gonidia** or non-motile spores. The apex of a tube or short lateral branch swells up and becomes more or less globular. This body, the gonidium, is cut off by a septum, and germinates directly without producing a zoogonidium. Sometimes the gonidium is formed in this outgrowth by rejuvenescence and the formation of a new cell-wall (fig. 231, G).

In unfavourable conditions, such as conditions of drought, the tubes may become septate. They are divided up into short cells which develop thick walls. When first seen, this was mistaken for a distinct Alga and called *Gongrosira*. It is now called the **gongrosira condition** of Vaucheria. It is a resting, protective condition. On the return of favourable conditions the cells of the gongrosira germinate and produce new plants. This may be considered as a purely vegetative method of reproduction.

§ 14. **Sexual reproduction** is heterogamous (fig. 231, D—F). The male organ is an **antheridium**; the female organ is called the **oogonium**. They arise as outgrowths, either of the tube itself (*V. sessilis*) or of a special short branch, and are both borne on the same plant. The number of oogonia and antheridia associated together varies in the different species. In *V. sessilis* there is frequently one antheridium between two oogonia.

The outgrowth which becomes the antheridium contains numerous chloroplasts and small nuclei. As development proceeds the nuclei aggregate in the central region of the protoplasm and give rise to a large number of very minute biciliate spermatozoids. The chloroplasts pass to the base of the outgrowth, and are cut off by a septum from the portion containing the spermatozoids, which is the antheridium proper. When fully formed the antheridium is a colourless tubular structure curved like a horn (D). It ruptures at the apex, and the spermatozoids are set free (E, F). The outgrowth which forms the oogonium at first contains numerous nuclei in addition to chloroplasts. One nucleus passes to the centre of the protoplasmic mass, and becomes the nucleus of the **oosphere**; the others pass back into the tube. The oogonium is then separated off by a septum. A little protuberance or beak appears near the apex. It bursts open and a small portion of the protoplasmic contents is given off (D). The rest of the contents form the oosphere, which contains numerous chloroplasts. It shows a clear spot—the **receptive spot**—near the beak where the protoplasmic vesicle was given off. The fully formed oogonium is sessile, and more or less ovate in form. It has a simple cellulose wall and contains one oosphere. It is a much less complex female organ than the archegonium.

Fertilization is effected by a spermatozoid entering the ovum at the receptive spot, and fusing with it. The **oospore** develops a thick wall and enters **on a period of rest.** When it germinates it produces a new plant directly.

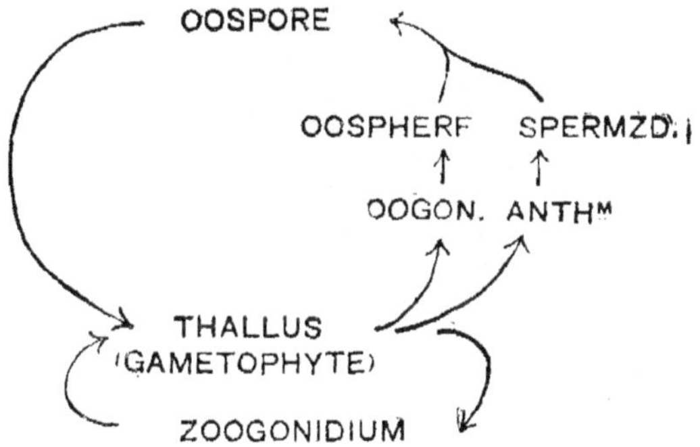

§ 15. Thus, in Vaucheria, the plant is the gametophyte. There is no alternation of generations. Perhaps the most striking feature about Vaucheria is the association of highly differentiated sexual organs with a very simple vegetative body. The life-history is graphically represented in fig. 232.

Fig. 232.—GRAPHICAL LIFE-HISTORY OF VAUCHERIA.

FUCUS.

§ 16. **General Appearance and Habit.** Fucus is one of the *brown* seaweeds, and it includes several common species differing in certain minor characters. The bladder-wrack (*Ascophyllum nodosum*) so common on our coasts is a closely allied plant. Fucus is one of the larger forms of Algæ. When full-grown, the plant consists of a basal, branching, *root-like organ* of attachment, a straight, cylindrical *stalk-like portion*, and a dichotomously branched *membranous expansion* (fig. 233). The **vegetative body** is essentially thalloid, although in the organ of attachment there is a distinct indication of the differentiation of root from shoot. It should be noticed, however, that the root-process is simply a fixing organ; it has no absorbing function. Running up the middle of each flattened lobe of the thallus is a sort of midrib, due to the thickening of the tissue in that region. In the older parts of the thallus the marginal portion dies away and leaves only the midrib. This is the mode of origin of the cylindrical stalk which represents the persistent thickened midrib of the older part of the thallus. A disstinct stalk is not recognizable in the young plant.

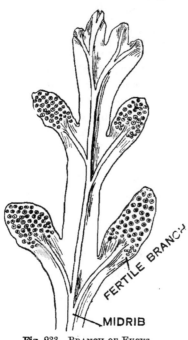

Fig. 233. Branch of Fucus Platycarpus.

Fucus inhabits the zone between low and high tide-marks. It is exposed at low tide, covered at high tide. It is interesting to notice how well the structure of the plant is adapted to this habitat, where it is exposed to the full force of the waves. The strong fixing organ attaches the plant to rocks, etc., and prevents it from being washed away. It is preserved from injury by the pliant nature of its stalk, and the yielding, flattened character of its branches. Some species (*F. vesiculosus*) are rendered still

THE ALGÆ. 343

more buoyant by the development of **air-bladders**—hollow dilatations of the thallus filled with air. The plant is saved from excessive desiccation, when exposed between tides, by the mucilaginous character of its tissues.

If we examine the apices of the branches at certain periods it will be found that they are covered with small papillæ, and are more or less swollen (fig. 233). These papillæ mark the position of internal flask-shaped cavities, called **conceptacles**, in which the sexual organs are developed, and hence protected by being enclosed in the tissue of the thallus.

§ 17. **Structure and Growth of the Thallus.** A section, passing through both surfaces and taken at any part of the thallus, will show that a central or **medullary region** of tissue can be distinguished from an outer or **cortical region** on each side (fig. 234). The outermost layer of the cortex (**outer limiting layer**) resembles an epidermis,

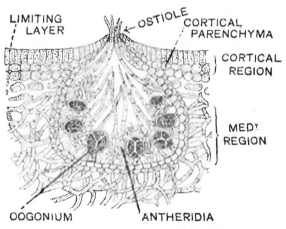

Fig. 234.—*FUCUS PLATYCARPUS.*
(Transverse section, through a conceptacle.)

but it is quite different from a true epidermis. Its cells are *meristematic* and by division give rise to new cortical cells. This is the assimilating region of the thallus. The subjacent cells are parenchymatous, with distinctly pitted walls; they constitute the **cortical parenchyma**. It is the storage region of the thallus. No starch is formed in Fucus. The product of metabolism is a carbohydrate somewhat resembling starch, but which is not stained blue by iodine. The medulla consists of an interlacing network of filaments or hyphæ embedded in a clear mucilaginous matrix. The filaments are divided at intervals by septa. It is a tissue in which the cell-walls, to a large extent, have become mucilaginous. The rows of cells still enclosed in the inner layers of their cell-walls constitute the filaments. The medulla is essentially the conducting tissue of the plant. In many of their characters the hyphæ resemble sieve-tubes, and in Laminaria, a large Alga closely allied to Fucus, distinct sieve-tubes are developed.

In the older parts of the thallus there is a similar structure, but the outer limiting layer is lost, and the cortical parenchyma by meristematic activity gives rise to secondary increase in thickness. This can be observed in the region of the stalk. The hyphæ of the medullary region also increase in number.

At each growing point of the thallus there is a row or group of **apical cells**, each of which assumes the form of a four-sided truncated pyramid. The segments from the bases of these cells form medullary hyphæ, those from the sides form mainly cortex. At each branching the apical cells separate into two similar groups (true dichotomy).

From a study of Fucus alone, the student will recognize that in Algæ the gametophyte may be highly differentiated. But some of the allies of Fucus exhibit much greater complexity. Thus the well-known Sargassum, which gives the name to the Sea of Sargasso, where it occurs in great abundance, exhibits a considerable amount of differentiation, and the occurrence of sieve-tubes in Laminaria has already been mentioned.

§ 18. **Reproduction.** There is no special method of asexual reproduction in Fucus, but sometimes adventitious branches, formed at the base of the thallus by the activity of the cortical meristem, may be separated off and form new plants. There is abundant **sexual reproduction**. The plant is the gametophyte. The sexual organs are contained in the conceptacles mentioned above. Each **conceptacle** (fig. 234) opens on the surface of the thallus by a minute aperture called the **ostiole**. Numerous multicellular **hairs** are developed from the cells lining the conceptacles. Many of these form *paraphyses*; others bear the sexual organs.

In the development of a conceptacle one or several superficial cells of the thallus cease to grow and become disorganized. The surrounding tissue grows vigorously, and soon encloses a flask-shaped cavity.

§ 19. **The sexual organs** (figs. 234, 235) are antheridia and

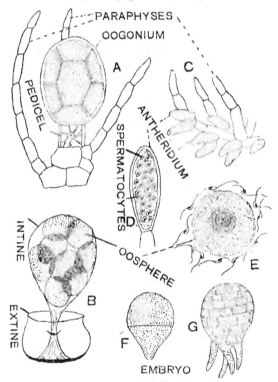

Fig. 235.—*FUCUS VESICULOSUS.*
A–D, Sexual organs; **E**, Fertilization; **F, G**, Embryo.

oogonia. The **antheridia** are borne in clusters, and represent the finer branches of much-branched hairs. Each antheridium is developed from a single cell. When fully formed it is a small, oval, orange-coloured sac, the wall of which consists of two thin membranes called the *intine* and *exine*. The contents have undergone division to form a large number (64) of small spermatocytes from which the biciliate **spermatozoids** are developed. The cilia are developed laterally, and each spermatozoid has an orange-coloured **chromatophore** (or two). The **oogonium** is a much larger, dark-coloured structure. It also is developed from a single cell. It is borne on a short unicellular pedicel representing an abbreviated hair. Its wall also consists of *intine* and *exine* and its contents divide to form eight **oospheres**.

The plants are usually *diœcious*, e.g. *F. vesiculosus*, recognized by its bladders, and *F. serratus*, recognized by its serrate margin; but in *F. platycarpus* (figs. 233, 234) the sexual organs are borne on the same plant and in the same conceptacle.

§ 20. **Fertilization** (fig. 235, E). When the spermatozoids and oospheres are ripe, the antheridia and oogonia become detached. The exine is ruptured (B), and the contents, enclosed in the intine, move towards the ostiole. This generally takes place when the plants are exposed at low tide. The hairs of the conceptacle secrete a mucilaginous substance. This oozes out of the ostiole, carrying with it the sacs of spermatozoids and oospheres. When the tide rises again the intine is ruptured, and the spermatozoids *and oospheres* are set free into the water. Spermatozoids cluster round each oosphere, which is thereby caused to rotate. Finally, one spermatozoid enters and its nucleus and cyptoplasm fuse with those of the oosphere. The resulting zygote is the **oospore**.

§ 21. **Germination of the Oospore** (fig, 235, F, G). Without any resting stage, germination takes place. The oospore becomes pear-shaped, and is divided by a wall into a pointed basal cell and a rounded apical cell. The basal cell sends out a number of *rhizoids* which fix the young plant and become wefted together to form the root-disc; the upper cell, by further division, gradually develops into the thallus. There is no alternation of generations.

§ 22. **Differentiation of Sex.** In connexion with the life-history of Ulothrix we recognized the probable origin of gametes and the evolution of sexuality. In Ulothrix, however, the gametes are similar, and, further, sexuality is not *completely* established. In another (brown) Alga, Ectocarpus, the gametes are also similar and sexuality is imperfect, but some of the gametes are less active and come to rest sooner than others. In some species of Ectocarpus, and in Cutleria (another brown Alga), the gametes are of two sizes. The larger ones are less active and come to rest sooner than the more active smaller ones which fuse with them. Here we have a series of Algal types which enable us to trace the probable evolution of sexual differentiation. The larger gametes are undoubtedly female. Their larger size and more passive character probably originated in a more abundant storage of food-material, making better provision for the embryo plant. We have the next stage in Fucus. Here the gametes are fully differentiated and the ovum is not ciliated. But *both gametes are set free into the water*. This is a striking feature in the life-history of the Fucaceæ. The next step in the evolution is to the stage where the female gamete is retained in the female organ.

§ 23. **Relationship of the Algæ to the Higher Plants.** The Algæ are an interesting group, supplying many facts which help towards the elucidation of important problems. Amongst them there are many divergent lines of evolution, four of which are illustrated by our types. Along most of these lines we recognize a gradual transition from isogamous to heterogamous reproduction. There can be no doubt that the primitive ancestors of the Bryophyta and Pteridophyta were evolved from primitive Algal forms; but here the gap is too wide for us to bridge over. All that we can say is that the evolution of the sporophyte followed the line of adaptation to aerial conditions, and that those forms of Algæ which have a rudimentary sporophyte show, amongst living forms, closer affinities than others to higher types.

CHAPTER XXI.

THE FUNGI.

§ 1. The Fungi constitute the second Class into which the Thallophyta are subdivided. They are readily distinguished from the Algæ by the want of chlorophyll; chromatophores and starch also are entirely wanting. This *by itself* would not be a sufficient reason for separating the two classes. If this were all we might, with equal reason, separate those few Flowering Plants which have no chlorophyll from the rest. The Fungi, however, as a whole, are further distinguished by special characters in their structure, development, and life-history.

§ 2. **The Mycelium.** The typical vegetative structure of the Fungi is a filamentous and much-branched thallus called a *mycelium*. The filaments or threads, of which the mycelium consists, are called **hyphæ**. Sometimes we find more massive structures, more especially in the parts bearing the reproductive organs. When we examine these, however, we find that they consist of hyphæ woven together so as to form **a false tissue**, *i.e.* a tissue formed, not by the division of cells all originally connected with each other (true tissue), but by the interweaving of separate hyphæ. Occasionally, in very degenerate forms like yeast, the plant consists of separate cells. The mycelium (or its hyphæ) may be septate or non-septate; in most cases it is cœnocytic. Even where it is septate the different segments often contain several nuclei, and are hence cœnocytic cells. Their walls do not consist of ordinary cellulose, but of a substance called **fungus-cellulose**, which is not stained blue by iodine and sulphuric acid.

§ 3. **Nutrition.** The nutrition of fungi is quite different from that of ordinary green plants. Having no chlorophyll they can make no use of the carbon dioxide of the atmosphere. They derive their carbonaceous food-material from complex organic compounds which they obtain from external sources. This also, to a large extent, applies to the absorption of nitrogen. Fungi, however, can assimilate comparatively simple compounds of nitrogen; but they prefer ammonium compounds, *e.g.* ammonium tartrate, to nitrates. Altogether their food-materials are less complex than those of animals.

§ 4. **Mode of Life.** Fungi may live either as parasites or as saprophytes. Some are not confined to one mode of life, but may live as parasites or as saprophytes, according to circumstances. In a parasitic fungus the hyphæ may penetrate the living cells of the host, or simply ramify between the cells. The power of penetrating and disorganizing cell-walls which such hyphæ frequently possess is due to the secretion of a ferment which acts on cellulose. The hyphæ of a saprophyte ramify through decaying organic substance or grow immersed in organic solutions. The whole of the mycelium may be absorptive, but some parasites, whose hyphæ ramify between the cells of the host, develop special absorbing organs, called haustoria, which penetrate the cell-walls and enter the cells.

§ 5. **Reproduction and Life-History.** There can be no doubt that the Fungi were originally derived from Algal forms, and that their evolution has, from this original starting-point, taken place along a special line. We may suppose that the primitive ancestral types were simply Algæ which assumed a dependent mode of life, and lost their chlorophyll. As supporting this view, we find that one large group of Fungi bear a very close resemblance to green Algæ like Spirogyra or Vaucheria. Their reproductive organs are very similar, and frequently also their general structure presents a close resemblance. It is only in this particular group that *undoubted* sexual organs are met with amongst the Fungi. The other groups of Fungi diverge widely from the Algæ, and present very special characters, while only in one of these groups do we find anything resembling a sexual process. Sexuality has been lost in the more specialized groups of Fungi; their reproduction is entirely asexual. This, together with their low organization, is a mark of degeneracy, and this degeneracy is to be ascribed to their mode of life.

Amongst the Fungi we recognize that there has been a gradual adaptation to aerial conditions, becoming very perfect in the higher groups. Evidence of this is found in the fact that, even in the group in which sexual organs are developed, the gametes are not ciliated; and also in the fact that in the other groups the "spores" or gonidia, which are called conidia when they are cut off from a hypha, are eminently adapted for transport through the air.

The plant may be regarded as representing the gametophyte, although in the higher groups the sexual organs have been lost (but see footnote, p. 335). There is no distinct alternation of generations. At most there is but an indication of a sporophyte in some of the sexual forms.

The types whose structure and life-history we now proceed to describe, will serve to illustrate most of the above points.

Pythium.

§ 6. Structure and Mode of Life. If large numbers of cress-seedlings (*Lepidium sativum*) are grown together under very moist conditions, it is found that they become "diseased." They fall over, turn pale, and then brown, and, finally, begin to rot. The disease begins at certain points, and spreads in circles. It is spoken of as the "damping off" of seedlings, and is caused by the attack of a fungus called *Pythium de Baryanum*, one of the commonest species of Pythium. The fungus attacks the seedling near the base, and begins to eat into its tissues, penetrating upwards into the stem and downwards into the root. The disease can be stopped at an early stage by admitting abundant light and air, for Pythium can live only under moist conditions. If the moist conditions are maintained, the hyphæ of the fungus extend from one seedling to another until they are all reduced to a blackened rotting mass, covered by dense white felts of mycelia like spider's web. The fungus, beginning as a parasite, continues to live as a saprophyte. Other species of Pythium attack seedlings in the same way; some live normally as saprophytes.

The mycelium of Pythium (fig. 236) is a much-branched,

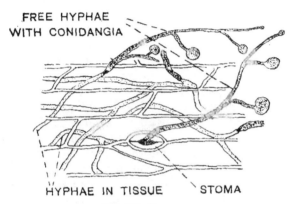

Fig. 236.—PYTHIUM.
Epidermis of a plant attacked by the fungus.

non-septate cœnocyte (cf. Vaucheria). Its protoplasm contains numerous small nuclei and oil globules. The hyphæ may eat right through the cells of the host or ramify between them.

§ 7. **Asexual Reproduction** (fig. 236). If a diseased seedling is placed in a watch-glass in water and kept under observation, it will be found that the ends of some of the hyphæ, which grow out from the surface of the plant, or of small branches of these, swell up and become globular. These globular swellings, due to aggregation of protoplasm in them, are cut off by distinct septa. They are asexual reproductive organs. They produce new mycelia in two ways, according to the conditions under which they are developed. Sometimes, if there is scarcity of water, they function as "spores" or **conidia**, and each without being set free sends out a germ-tube which grows into a mycelium directly. At other times, when there is abundance of water, they function as **gonidangia** (also called zoosporangia). Each (fig. 237, a-e) develops a little protuberance which expands to form a round thin-walled vesicle into which the protoplasmic contents pass. These divide into a number (9 or 10) of **zoogonidia** (also called zoospores, see p. 335), which are set free by the rupture of the wall of the vesicle. The zoogonidium is a very minute, colourless body with two cilia. After moving about for about a quarter of an hour it comes to rest, withdraws its cilia, rounds itself off, and forms an investing wall. A hypha is then sent out, which makes its way into another seedling, either by boring through the wall of an epidermal cell or by means of a stoma.

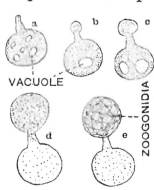

Fig. 237.—GONIDANGIUM OF PYTHIUM. Stages in the production of zoogonidia.

The production of zoogonidia is a distinct resemblance to what we find in the Algæ, but there is an *indication* of adaptation to aerial conditions in the fact that the reproductive bodies may germinate directly.

§ 8. **Sexual Reproduction** (fig. 238). The female organ is an **oogonium**. It may be formed on the end of a hypha (terminal), or on the course of a hypha (intercalary), either outside the seedling or in the tissues of the seedling. It arises as a globular swelling of the hypha, and bears a close resemblance to the asexual reproductive organ. It is cut off by a septum from the rest of the hypha. At first the protoplasmic contents have many nuclei, but soon a central region containing one nucleus, constituting the **oosphere**, is marked off from a peripheral region, called the **periplasm**, containing the other nuclei. The male organ, called the **pollinodium**, is developed on a branch which arises either on the same hypha as the oogonium, or on another hypha. It is separated off by a distinct septum, and its protoplasmic contents are differentiated into a central portion, the **male gamete**, and a peripheral *periplasm*. The male gamete has no cilia, and for this reason the male organ is usually called a pollinodium and not an antheridium. By some, however, it is called the antheridium.

The pollinodium is more or less club-shaped. It applies itself closely to the oogonium and develops a tubular process, the **fertilization-tube**, which pierces the wall of the oogonium and carries the male gamete to the ovum. This process can be readily observed in Pythium. The fertilized ovum forms a thick wall and becomes the **oospore**. The outer layer of the wall is formed from the periplasm.

The sexual organs and oospores are produced after the asexual organs, when the con-

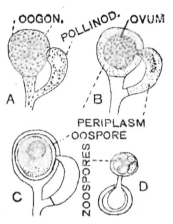

Fig. 238.—SEXUAL REPRODUCTION IN PYTHIUM.
B, Fertilization; D, Germination of oospore.

ditions for rapid growth and reproduction by asexual methods are becoming unfavourable. The oospores are essentially resting spores. They remain inactive during the winter and germinate in the following spring. This is why seedlings grown on the same ground as seedlings attacked the previous year are so liable to the disease. If the conditions are favourable the oospore sends out a hypha which directly develops into a mycelium; if unfavourable, the oospore either forms a number of small **zoospores** directly from its protoplasm or sends out a little tube or hypha, called a **promycelium**, in which zoospores are developed (fig. 238, D). The zoospores precisely resemble the zoogonidia and germinate in the same way. The promycelium has been considered to represent a very rudimentary sporophyte, but this is highly problematical. Where it is not developed the oospore in itself, or together with the zoospores it produces, may be supposed to represent the sporophyte (fig. 239).

Bot. 23

§ 9. The close resemblance between the structure of the mycelium and the reproductive organs of Pythium, and

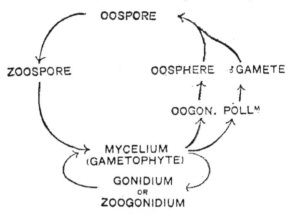

Fig. 239.—GRAPHICAL LIFE-HISTORY OF PYTHIUM.

the thallus and reproductive organs of Vaucheria, should be carefully noticed. Pythium belongs to the group of Fungi which approaches nearest to the Algæ, and in which the adaptation to aerial conditions is very imperfect.

EUROTIUM.

§ 10. **Habit and Structure.** Eurotium is a saprophyte. It lives on decomposing organic substance. It is frequently found on the surface of mouldy bread, damp fruit and vegetables, preserved fruit, etc., and belongs to the familiar group of Fungi known as moulds, which form filamentous growths on the substances on which they live. If a piece of dry, stale bread be kept under a bell-jar, one of the commonest species, *Eurotium Aspergillus-glaucus*, a green mould, will be found. At first, before the reproductive bodies are developed, the mould is white; but when it enters on the reproductive stage it assumes a greenish colour.

The **mycelium** (fig. 240) consists of a tangled mass of hyphæ ramifying in and on the surface of the nutrient substratum. It is much branched, and the hyphæ are divided at intervals by transverse septa. Each segment of a hypha contains a mass of granular vacuolated protoplasm, with several nuclei and oil globules. The structure is coenocytic. Starch and plastids are wanting. The hyphæ which ramify beneath the surface absorb nutrient organic substance.

§ 11. **Asexual Reproduction** (Fig. 240). From the mycelium numerous straight and, as a rule, non-septate branches pass up into the air. Each is given off from a hypha, usually at a point immediately behind a septum. These branches bear the asexual "spores," or **conidia** (gonidia), and are therefore called **conidiophores** (gonidiophores). The head of each conidiophore swells up and becomes spherical. On this spherical head arises a large number of peg-like outgrowths called **sterigmata**.

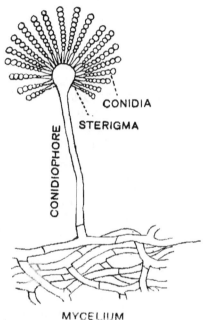

Fig. 240.—EUROTIUM. MYCELIUM AND CONIDIOPHORE.

From the apex of each sterigma as it elongates conidia are separated by abstriction one after the other. In this way rows or chains of conidia are formed on the sterigmata; the oldest conidia are at the apex of each row. The conidia are small oval bodies with a greenish colour and more or less spiny surface. Their protoplasm contains a nucleus and oil-globules, and they have two coats, *exosporium* and *endosporium*. These conidia are produced in enormous quantities, and are very readily carried through the air. It is because they are always present in the atmosphere that decaying substances so readily become infected with the fungus. On reaching a suitable substratum they germinate in the usual way, and produce new mycelia directly. They will germinate very readily in a *weak* decoction of plum juice. By means of these conidia rapid and extensive multiplication is possible.

§ 12. **Sexual Reproduction** (fig. 241). In Eurotium the same mycelium which has produced conidia eventually bears sexual organs. The female organ is called an **archicarp** or **ascogonium**. It differs from the female organs we have already become familiar with in that its protoplasm is not rounded off to form an ovum. The male organ is a **pollinodium**. In the development of the archicarp the end of a hypha becomes coiled, at first loosely, but later very closely. This tightly packed spiral organ, consisting of four or five coils, is the archicarp. In Eurotium it is at first non-septate. Several slender branches arise from the hypha beneath the archicarp. One of these develops faster than the others, and arches over to the apex of the archicarp. It is the pollinodium; like the archicarp, it is non-septate. It has been supposed that fertilization is effected by the contents of the pollinodium passing into the archicarp, but the actual process of transference has not been seen, in spite of the most careful examination. The other branches arising below the archicarp are sterile. At first few in number, they are afterwards numerously developed. They wrap round the archicarp, and, by interweaving and septation, give rise to a false tissue (pseudo-parenchyma), completely enclosing it. This false tissue sends in hyphal branches between the coils of the archicarp, and thus gives rise to a "filling tissue" which separates the coils. The fertilized (?) archicarp becomes septate and develops small outgrowths, the *ascogenous branches*, which penetrate into the false tissue. The ends of these outgrowths are cut off by septa and form unicellular sporangia called **asci**. Each ascus is oval in form, and, by a process of free cell-formation, gives rise to eight spores called **ascospores**. The peripheral protoplasm of the ascus (epiplasm) is not used up in the formation of ascospores. It contains a large quantity of a carbohydrate called glycogen, serving for the nourishment of the spores. During the development of the asci the whole of the central "filling tissue" is disorganized. Thus, from the archicarp and the investing sterile hyphæ, a **sporocarp** has been formed. The wall of

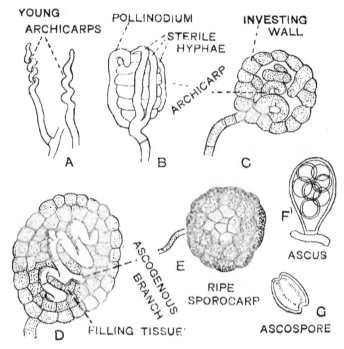

Fig. 241.—EUROTIUM. DEVELOPMENT OF SPOROCARP AND ASCOSPORES.

the sporocarp consists of small-celled pseudo-parenchyma, and encloses a number of eight-spored oval asci. It is also spoken of as an *ascocarp*. In the ripening of the sporocarp only the outermost layer of the wall persists; its cells become dry and firm and covered with an oily secretion, which gives the sporocarp a yellow colour. The asci are disorganized, and the ascospores are eventually liberated by the bursting of the wall of the sporocarp. The ascospore, which was oval when young, is biconvex when fully developed. In germination the exosporium is ruptured; the endosporium grows out and directly produces a mycelium.

§ 13. **Eurotium** is a type of a large group of Fungi called **Ascomycetes**, characterized by the production of asci and ascospores. It is a disputed point whether the archicarp and pollinodium are really sexual organs. Some would regard the archicarp simply as an organ producing another kind of asexual "spore" or conidium, and the supposed pollinodium as one of the sterile hyphæ. On this view there would be no indication of an alternation of generations, and Eurotium would exhibit a form of *polymorphism* quite different from that exhibited in the alternation of gametophyte and sporophyte—namely, the production at different stages of two kinds of asexual organs. If the archicarp and pollinodium are considered to be degenerate sexual organs, which they seem actually to be, there is an indication of an alternation of generations. The sporocarp would represent the sporophyte; or, rather, the asci and ascospores would, seeing that the wall of the sporocarp is derived from mycelial hyphæ. In support of this view there is the fact that in Fungi closely related to these moulds—namely, the mildews—there are similar organs, and the actual process of fertilization has now been observed. Considering the sexual degeneration which is characteristic of the higher groups of Fungi, we may conclude that in Eurotium the sexual organs are present in a degenerate form, but that the actual sexual process has been lost (cf. parthenogenesis).

§ 14. **Penicillium glaucum** is a blue mould very similar in habit and structure to Eurotium. The apex of each conidiophore, instead of being globular, divides into a number of finger-like processes bearing the sterigmata. In Penicillium sporocarps are developed only in special circumstances—namely, when free access of light and air is prevented. Here we have a more advanced stage of sexual degeneration. The conidia of this fungus, if grown in a saccharine solution, under certain conditions, give rise, not to a filamentous mycelium, but to isolated cells resembling yeast. This is called the **torula condition**. Like yeast, it excites alcoholic fermentation.

CLAVICEPS PURPUREA (ERGOT OF RYE, ETC.).

§ 15. **General Life-History.** The life-history of this fungus shows three well-marked stages or phases, and affords an illustration of the polymorphism which is s common amongst Fungi. (1) The **Sphacelia** or "**Honey dew**" **stage** is found on various cereals and grasses. It has been most carefully studied in the case of Rye (*Secale cereale*), but it also occurs on Barley, Wheat, etc., and on some of the grasses commonly found on roadsides, in meadows and in waste places (e.g. *Lolium perenne*). It is the active parasitic stage in which the fungus attacks the developing pistil, and forms a mycelium which grows vigorously and reproduces itself by means of conidia (fig. 242). (2) The **Sclerotium stage** is the resting or winter stage. In autumn, the *Sphacelia* forms a hard, slightly curved body, of a dark purple or black colour, which protrudes from between the paleae of the flower, and may attain the length of an inch or even more (fig. 244, A). It falls to the ground and undergoes no change till the following spring. It was to this body that the name ergot was first given in reference to its shape (from O. Fr. *argot*, a cock's spur). Hard resting bodies of this kind are formed by many fungi and are called **sclerotia**. (3) The **ascospore stage**. The sclerotium or ergot eventually produces a number of club-headed structures, called **stromata** (fig. 244, B), in which asci and ascospores are developed. From these the Sphacelia form again arises.

At first the connection between these three stages was not recognised. They were regarded as distinct fungi and received the generic names *Sphacelia*, *Sclerotium*, and *Claviceps** respectively. The last is now taken as the generic name of the fungus in all its forms.

§ 16. The **Sphacelia stage** (figs. 242, 243). The ascospores, which are peculiar in being very slender and threadlike (fig. 247, B), are liberated just when the flowers of the host plants are appearing. They are carried to the flowers by wind and there germinate. In the process of germination little swellings appear on the ascospore, and at these points the germ-tubes are developed. They pierce through the epidermis and make their way into the tissue at the base of the young ovary. The **mycelium** which is rapidly

---

* Latin *clava*, a club.

developed consists of septate hyphæ. It not only ramifies in the tissue of the ovary, but also, after a time, spreads over its surface and envelops the greater part of it with a dense white covering of interwoven hyphæ, showing numerous folds and convolutions. This investment is the **conidiophore**. The free ends of the hyphæ become slightly swollen, and form **sterigmata** from which small oval **conidia** are successively abstricted (fig. 242). The conidia

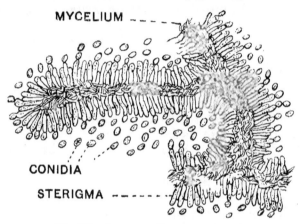

Fig. 242.—CLAVICEPS (SPHACELIA STAGE).
Section of Conidiophore.

are developed in great numbers, and are embedded in a sweet and somewhat viscid fluid which is formed by the disorganisation of the outer layers of the walls of both hyphæ and conidia. This fluid is called "**honey-dew**" and its appearance on rye and other crops is a source of anxiety to the farmer. Flies and other insects are fond of it. They are attracted to it by its odour, and by their agency the infection is spread to other plants. A conidium carried in this way to another flower sends out a germ-tube which pierces the base of the ovary and produces another mycelium.

§ 17. **The Sclerotium stage** (figs. 243-245). When the growth of the mycelium has continued for a considerable time, the mass of interwoven hyphæ at the base of the

THE FUNGI.

shrivelled pistil becomes densely compacted, and forms a false tissue which is the beginning of the sclerotium. The outer layers of this tissue become dark-coloured and growth now goes on actively in this region. The sclerotium rapidly elongates and assumes its curved horn-like shape. As it does so it bears at its apex the remains of the *Sphacelia* and the pistil (fig. 243), and finally these are thrown off. The sclerotium or ergot is fully formed just when the grain ripens It then protrudes from between the paleae, but is easily detached. Unless the crop is harvested early, it falls to the ground and there remains in a dormant condition till the following spring (fig. 244, A).

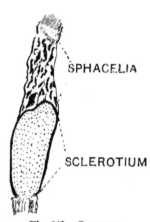

Fig. 243.—CLAVICEPS. Vertical section through developing sclerotium.

If a transverse section of the sclerotium be taken (fig. 245) it is found to consist of a dense pseudoparenchyma

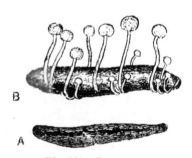

Fig. 244.—CLAVICEPS.

A, Sclerotium, resting stage; B, Sclerotium with stromata.

Fig. 245.—CLAVICEPS. Transverse section of germinating sclerotium showing a developing stroma.

formed of fine united hyphæ. The outline of the section is somewhat irregular and is interrupted in places by fissures or cracks. The outer layers are very dark-coloured. The cells of the central tissue are laden with oil, and also contain an alkaloid, called *ergotin*, and other poisonous substances, to which the peculiar properties of ergot are due.

§ 18. The **Ascospore stage.** In spring or early summer the sclerotia begin to develop their stromata. The process can be readily observed if at this time some of them (in good condition) be kept half buried in damp clean sand under a bell-jar. The first sign of development (it may not be till after the lapse of several weeks) is the appearance of a number of little swellings on the surface and sides of the sclerotium. Then the dark-coloured outer layer is gradually ruptured, and the light coloured heads of the stromata protrude (fig. 245). The development of each stroma is due to the outgrowth of a tuft of hyphae from the light coloured inner tissue of the sclerotium. Each stroma consists, when fully developed, of a light coloured purplish stalk, an inch or more in length, and a rounded head of a light brown or orange colour (fig. 244, B). As in the case of the sclerotium, the hyphae of the stroma are closely interwoven and united, and form a false tissue. The head is covered with minute papillae, on which are situated the openings or **ostioles** of numerous flask-shaped cavities, called **perithecia**, which lie in the peripheral tissue of the head. These perithecia can be readily seen in a vertical section of a stroma (fig. 246). From the cells in the floor of each perithecium are developed a number of elongated club-shaped **asci**, which project upwards towards the ostiole (fig. 247, A). The contents of each ascus divide to form from six to eight thread-like **ascospores** (fig. 247, B). When the asci are ripe they burst. The

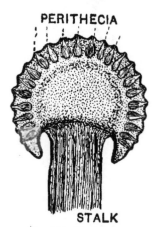

Fig. 246.—CLAVICEPS.
Vertical section through a stroma, showing the perithecia.

ascospores escape through the ostioles, and some of them, carried as already described to the young flowers of a grass, produce the *Sphacelia* again.

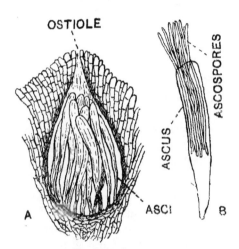

Fig. 247.—CLAVICEPS.
A, Section through a perithecium, showing the asci; B, A single ascus, ascospores escaping.

§ 19. **Notes on the Life-History.** Like *Eurotium* and *Penicillium*, *Claviceps* belongs to the Ascomycetes. It is, however, placed in a different division of that class. The classification is based on the character of the sporocarp or ascocarp (p. 357). In *Claviceps* it is a perithecium; in *Eurotium*, a *cleistocarp*, that is, a closed case.

The life-history of *Claviceps* is more complicated than that of *Eurotium*, owing to polymorphism. The mycelium of *Eurotium*, after a period of asexual reproduction by conidia, produces sexual organs, and is therefore the gametophyte. What corresponds to the gametophyte in *Claviceps* is represented by the three forms or stages described above. There is, however, in Claviceps no actual gametophyte. There are no sexual organs, and by comparison of this life-history with that of other **Ascomycetes**, the conclusion is reached that the ascocarps or perithecia are here produced apogamously. A graphical representation of the life-history is given in fig. 248.

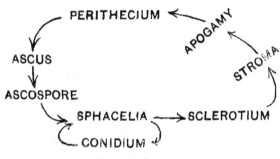

Fig. 248.—CLAVICEPS.
Graphical representation of life-history.

The life-history is also interesting from the fact that it exhibits the phenomenon known as **Lipoxeny**, that is the desertion or abandonment of host (Gr. λιπειν, to abandon, ξενος, host). The Sphacelia form, as we have seen, is parasitic on rye and other grasses, but the fungus then deserts its host, and the resting stage is passed on the ground.

The fungus in all its forms shows remarkable adaptation to the conditions of its existence. The abundan production of conidia and the method of their distribution (cf. insect-pollination in flowers) provide for the rapid spread of the active Sphacelia form. The sclerotium is developed and falls to the ground in time to prevent it being removed with the crop. The spores are produced just when the grass flowers appear, and by position and form are adapted for wind transport.

SACCHAROMYCES (= YEAST).

§ 20. **Habit and Structure.** This is the fungus which is the exciting cause of alcoholic fermentation in saccharin solutions. *S. cerevisiæ* is the beer-yeast (brewer's yeast) *S. ellipsoideus* is the species which sets up fermentation i grape-juice in the manufacture of wine. The yeast-plant is a saprophyte, and thrives best in saccharine solutions containing in addition small quantities of nitrogen- and sulphur-compounds.

The mycelium is not, as in most of the other Fungi, a branched filamentous structure, but consists of isolated cells or groups of cells (fig. 249). Each cell is more or less oval, sometimes almost spherical, and contains granular protoplasm with one or more small vacuoles filled with cell-sap. Nuclear substance is present, but can be demonstrated only by special methods of investigation.

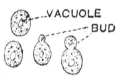

Fig. 249.—YEAST-CELLS, SHOWING PULLULATION.

§ 21. **Vegetative Reproduction.** The common method of reproduction is by vegetative budding. If yeast-cells in an active state of growth are examined, it is found that each gives rise to a tiny outgrowth which gradually increases in size, and is finally cut off as a separate yeast-cell. This process is known as **pullulation** or **gemmation** (fig. 249). It differs from ordinary cell-division (of which it may be regarded as a modification) only in that a cell is not directly divided into two daughter-cells, but that there is a gradual growth of a daughter-cell on the parent-cell. The daughter-cells, before being separated off, may repeat the process, and thus we have the formation of cell-groups.

§ 22. **Spore-Reproduction.** This is met with only under unfavourable conditions, more especially when there is an insufficiency of food-material, e.g. when yeast-cells are grown on the cut surface of a potato, or on moist plates of plaster-of-Paris, or left in a neglected condition in a jar. In these circumstances the process of pullulation ceases, and certain of the yeast-cells become larger and form sporangia. The nucleus of each sporangium usually divides into four nuclei. The protoplasm aggregates around each of these, and four spores are formed each with a firm, stout wall. In some cases eight spores, or less than four may be formed. They must be considered as resting spores specially developed to cope with unfavourable conditions of existence. When they germinate, on the return of favourable conditions, the outer coat of the spore is burst, and the process of pullulation begins.

The spores are developed in exactly the same way as the ascospores of the Ascomycetes. In both cases they are formed by free cell-formation, the peripheral layer of protoplasm of the sporangium not being utilized (§ 12). We have also seen that in some Ascomycetes, *e.g. Penicillium glaucum* (§ 14), the yeast or torula condition is met with in certain circumstances. On these facts is based the view that the spores of the yeast are *ascospores,* and the cell (sporangium) in which they are developed an *ascus*; and that, if we consider the ascospores of Ascomycetes to be true spores, the sporangia and spores of the yeast represent in a rudimentary form a sporophyte apogamously developed. According to this view Saccharomyces is a very degraded Ascomycete in which the torula condition has become fixed. Some, however, hold the idea that, after all, the species of yeast may represent only a particular form or stage in the life-history of filamentous Ascomycetes, otherwise normal—this particular form being fixed or permanent under certain peculiar conditions of the environment.

§ 23. **Alcoholic Fermentation.** In the process of alcoholic fermentation induced by yeast, grape-sugar is decomposed. The chief decomposition products are alcohol (C_2H_6O) and carbon dioxide, but there are also minute quantities of glycerine and succinic acid. The general equation might be represented thus:—

$$C_6H_{12}O_6 = 2C_2H_6O + 2CO_2.$$

It has been shown that while the yeast-cells thrive best in the presence of oxygen, *i.e.* while the growth and division of the yeast cells is most rapid in this condition, the weight of sugar transformed into alcohol and carbon dioxide *per unit weight of yeast* is much greater in the absence of oxygen; and that, when oxygen is abundantly supplied, *relatively* little alcohol is produced.

We have already had examples of similar decompositions in the digestion of stored food-material effected by the action of definite chemical substances called unorganized ferments, which can be extracted and act apart from living protoplasm. Fermentation processes, such as that effected by yeast, differ in that the decomposition seems to depend essentially on the presence of living organisms. The conclusion has been arrived at that the protoplasm of such organisms can itself act as a ferment, and the organisms are spoken of as **organized ferments**. Recently, however, there has been extracted from yeast a chemical substance which can induce alcoholic fermentation. The name *Zymase* has been given to it. This would tend to bring the processes of fermentation into line with those carried on by unorganized ferments.

§ 24. **The Meaning of Fermentation.** All plants must have a supply of energy for carrying on their metabolic processes, and for the performance of internal and external work. In most plants the necessary energy is liberated by oxidative decomposition (p. 144), to which the term of aerobic respiration is usually applied, and most plants die if deprived of oxygen. *Some* plants, however, can obtain a supply of energy in another way, which does not necessarily involve any consumption of oxygen—namely, by breaking down the complex substances of which their food is composed, which process is usually termed anaerobic respiration. In some cases it appears that the aid of ferments is invoked by the protoplasm (yeast plant, bacterium active in fermentation of urea) in order to induce these decompositions, although such aid is by no means always necessary. The term fermentation is a popular one, and is applied to cases of anaerobic respiration when the products of the latter are unusually conspicuous or abundant. The tendency to produce a ferment is so strong in the case of the yeast-plant that its production does not entirely cease even in the presence of oxygen, when the plant can obtain all the energy required by aerobic respiration. This is the explanation of the *relatively* greater activity of the process of fermentation in the absence of oxygen.

AGARICUS CAMPESTRIS—THE MUSHROOM.

§ 25. General. Agaricus is a very large genus, including many subgenera and species. Popularly most of the plants included in it are called "toadstools" and "mushrooms." Most of them are saprophytes, and are to be found in abundance in damp woods where the soil is rich in organic substance. A few, however, are parasitic, and very injurious to trees which they infest. Apart from this they all closely resemble each other in structure and in the general course of their life-history. *Agaricus campestris*, the common mushroom, is a very convenient form for description, and we therefore select it as our special type.

§ 26. Habit and Structure of the Mushroom (fig. 243). The mushroom is a saprophyte. It lives on decomposing organic substances (humus) in woods, fields, etc. The part of the fungus which is seen above ground—the part which is called the toadstool or mushroom—is only the reproductive structure, the fructification or **conidiophore**. This is developed on a delicate filamentous **mycelium**, the vegetative body or thallus, which ramifies through the soil and absorbs the organic compounds on which the fungus lives. The so-called "mushroom-spawn" so largely used in the cultivation of mushrooms, consists simply of compacted blocks of well-manured soil containing a tangled mass of mycelial hyphæ. If these be buried in a suitable

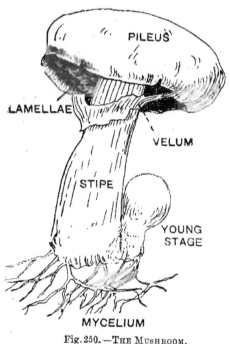
Fig. 250.—THE MUSHROOM.

locality (*i.e.* damp and rich in humus products), the mycelium grows and develops conidiophores. The much-branched filamentous **mycelium** is incompletely septate, *i.e.* the segments into which the septa divide the hyphæ are cœnocytic. The hyphæ are colourless, and contain vacuolated protoplasm with nuclei and oil-globules. Frequently the hyphæ may be found running in strands and anastomoses between them are not uncommon.

§ 27. **Reproduction.** The reproduction of Agaricus is asexual, being effected solely by means of **conidia**. Sexual reproduction has been completely lost in the group of Fungi to which Agaricus belongs.

§ 28. **The Conidiophore** (fig. 250), on which the conidia are produced, is a very massive organ. It seems altogether different in its structural characters from the filamentous mycelium on which it is developed. Examination shows, however, that it is really a false tissue (p. 348), consisting of compacted interwoven hyphæ resembling those of the mycelium. It consists of a massive circular umbrella-shaped head, called the **pileus**, borne on top of a stalk, the **stipe**. The upper surface of the pileus is more or less rounded and convex. In the different species of Agaricus it shows an immense variety of colour, due to the presence of colouring matters in the cell-walls. The under surface bears a large number of delicate vertical plates radiating from the stipe to the edge of the pileus. These bear an external resemblance to the gills of fishes, and are spoken of as the **gills** or **lamellæ**. They are flesh-coloured when young, but become a rich chocolate brown when fully developed, and covered by innumerable brown or black conidia. Encircling the stipe, close to the attachment of the pileus, are the remains of a membrane which originally extended from the stipe to the edge of the pileus and closed in the "gill-chamber." This torn membrane is called the **velum**.

§ 29. **Structure of the Conidiophore.** The hyphæ in the cortical region of the stipe are densely compacted, but the central or medullary region consists of loosely interwoven hyphæ with numerous spaces between them. If a vertical section be taken across a lamella it shows the following structure (fig. 251). There is a central core of interwoven hyphæ known as the **trama**. These hyphæ curve outwards towards the surface of the lamella, and end in small cells, which form what is known as the **subhymenial layer**. Outside these again, forming the superficial layer of the lamella, are larger, rather stout and slightly elongated cells. This is the **hymenium** or *hymenial*

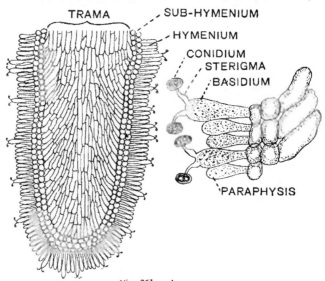

Fig. 251.—AGARICUS.
Section across one of the gills. The diagram to the right represents the hymenium and sub-hymenium more highly magnified.

layer. The cells of the hymenium are of two kinds: (*a*) barren cells called **paraphyses**; (*b*) cells called **basidia**. Each basidium bears at its apex two to four slender processes called **sterigmata**, and from each sterigma a small rounded conidium or basidio-spore is abstricted.

§ 30. **The conidia** are produced in great abundance. This can be demonstrated by laying a ripe pileus for some time on a sheet of paper. A sort of print of the under surface of the pileus is obtained owing to the thick deposit of conidia. The conidia when ripe fell off, and, if they

THE FUNGI.

reach a suitable soil, they germinate. Each sends out a hypha, which grows and branches and produces a new mycelium. The germination has only been observed after considerable difficulty. The growth of the mycelium from the conidium is slow, and conidiophores are not produced till after a lapse of seven or eight months.

§ 31. Development of the Condiophore.

Fig. 252 shows various stages in the development of the fructification. It makes its appearance on a strand of the mycelium as a tiny rounded or pear-shaped body, consisting of a tangle of hyphæ. At first there is no distinction between stipe and pileus, but, as growth goes on, the apex of the developing structure expands to form the pileus. In this, towards its lower surface and completely enclosed in the tissue, an annular cavity appears. In the roof of this cavity the lamellæ are differentiated, and its floor becomes thin and membranous, forming the velum, which is ruptured towards the close of development.

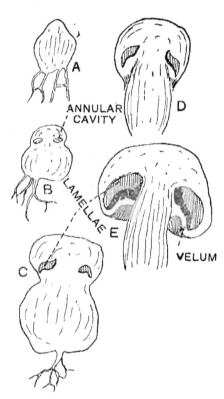

Fig. 252.—AGARICUS.
Stages in development of conidiophore.
(Diagrammatic longitudinal sections.)

BACTERIA (= SCHIZOMYCETES).

§ 32. General.

The Schizomycetes, commonly spoken of as bacteria, form a group of extremely minute organisms, destitute of chlorophyll, which play a part in organic nature altogether out of proportion to their size. By some they are placed among the Fungi but it is now usual to group them with a series of minute chlorophyllous forms, the *Cyanophyceæ*, in a special division called the *Schizophyta*. The organisms may be unicellular or multicellular. The multicellular forms may be filamentous, or form cell-plates or cell-masses; they are to be regarded, however, as essentially aggregations of unicellular forms. Bacteria are ubiquitous organisms, being found in the most unexpected media—river water, sulphur springs, etc.

The mode of life is parasitic or saprophytic. Many forms parasitic on animals are harmless, sometimes even beneficial; others (pathogenic forms) prejudicially affect the normal physiological processes and produce pathological or diseased conditions. Most of our infectious diseases have been shown to be due to the ravages of bacteria. The spores of these bacteria, popularly called *"germs"* or *"microbes,"* are present in the air or in various media, and in favourable circumstances reproduce themselves with great rapidity. A single germ may give rise to millions in the course of a week or two. This explains why many of these diseases are infectious and become epidemic.

The saprophytic forms thrive in various organic media and produce characteristic fermentative changes. Well-known examples are the souring of milk and the conversion of alcohol into acetic acid in the formation of vinegar. Bacteria, however, never induce alcoholic fermentation. The disastrous effects of infectious diseases are also in all probability due to the formation of poisonous waste products which accumulate in the blood.

In all these processes bacteria apparently act as **organized** ferments. From some bacteria, however, substances of the nature of unorganized ferments have been extracted, which are capable by themselves (*i.e.* apart from the living cells) of producing the characteristic fermentation. Most bacteria can live only in the presence of oxygen, and are said to be **aerobic**, but many cannot live in the presence of oxygen and are called **anaerobic**. Others again can live both in the presence and in the absence of oxygen. The significance of this capacity for doing without oxygen, and its relation to the process of fermentation, have been sufficiently dealt with in connection with the yeast-plant.

Putrefaction is a process of fermentation set up by certain species of bacteria in proteid substances, and is usually accompanied by the evolution of offensive gases. In this way these complex nitrogenous substances are gradually decomposed into ammonia and other compounds. In the soil, by the action of bacteria, the ammonia is acted on, and first nitrous and then nitric acid produced. This last process is called **nitrification**. It is effected by at least two different bacteria (nitrifying organisms) which obtain in this manner the energy necessary for the synthesis of organic food from carbon dioxide and water, in the absence of chlorophyll and without the aid of light being necessary. In this way dead organic substance is first decomposed, and then brought back into forms available for absorption by green plants. These are essentially processes of oxidation. The bacteria apparently act as carriers of oxygen.

According to some, the "bacteroids" present in the root-tubercles of leguminous plants are bacteria, and have been called *Bacillus radicicola*.

§ 33. **The Bacterium-Cell.** The cells are extremely microscopic, a diameter of less than $\frac{1}{10000}$ of an inch being not uncommon. They can be examined only under very high powers of the microscope, and even then only the coarser structural details can be made out. Each cell has a distinct cell-wall. In many cases this apparently does not consist of cellulose, but of a proteid substance. The cells contain granular protoplasm. A nucleus has been seen in some, and probably is present in all. There are no plastids. In a few forms, however, it is interesting to note that a granular substance has been observed, giving a blue or purple reaction with iodine, being probably therefore some kind of starch.

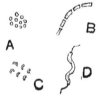

Fig. 253.—Forms of Bacterium-Cells. A, Micrococci; B, Bacilli; C, Commas; D, Spirillum with flagellæ.

There are many different forms of cells (fig. 253). Very minute spherical forms are called **cocci**, or *micrococci;* elongated rod-like forms, **bacilli**; spirally coiled forms, **spirilla**; comma-like forms, **commas**. These are the commonest. Sometimes filaments of bacteria are aggregated in enormous numbers and held together by mucilage, forming a sort of scum on decomposing liquids. This is known as the **zoogloea** condition (fig. 254, A). Sometimes the cells form cell-masses. These different forms are not to be considered as characteristic of different species. The same species probably passes through a number of forms at different stages. They are simply "**growth-forms**," *i.e.* forms assumed at different periods of growth. In other words, the bacteria are *polymorphic*.

Many bacteria have the power of independent movement. This is probably effected by means of very slender thread-like protoplasmic processes or outgrowths which are known as **flagellæ**. A cell may have one, two, or a number of these

§ 34. **Reproduction.** There are two methods of reproduction. Both are asexual. In the process known as "**fission**," the parent-cell undergoes division into two daughter-cells. It is simply a process of cell-division in a unicellular organism. A micrococcus cell simply becomes constricted and divides into two; a bacillus divides transversely. The other method of reproduction is that of "**spore**"-formation (fig. 254, A). It is commonly seen in bacilli. As a rule it takes place in the zoogloea stage. Studying this process, we find that the protoplasmic contents of the cells round themselves off. They withdraw from the cell-walls, and aggregate in the middle of the cells. A new cell-wall is then formed round the protoplasmic mass. When fully formed, this cell-wall is extremely thick and resistant. Thus the "spores" are produced, usually one in each cell, but sometimes more than one (endosporous formation). They may remain quiescent for a considerable time if the conditions are unfavourable, but are eventually set free by the decay of the walls of the parent cells. When they become active, fission begins, and ordinary bacterium cells are produced.

These spores can withstand great extremes of heat and cold without injury. Being extremely minute, they are easily blown about, and are always present in the air. For these reasons the "spores" are admirably adapted for the distribution of bacteria.

§ 35. **Bacillus subtilis** (fig. 254) will serve as an example. It is the hay-bacillus. If hay be chopped up and either soaked or boiled in water and kept for some little time, numerous bacillus cells can be recognized if the fluid be examined under a high power. Each cell is a tiny rod-shaped body having the structure above described. There is a flagella at each end. In this stage the cells multiply by fission, but after a time they pass to the surface and form a scum (zoogloea stage). If this be examined the cells will be found aggregated into long filaments embedded in a mucilaginous substance, formed by the disorganization of the outer layers of the cell-walls. It is in this stage that the spores are developed. They are extremely resistant, and can withstand boiling for a considerable time. When the spore germinates in a suitable solution, the outer membrane bursts, and the contents escape as a flagellate bacillus cell.

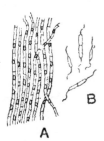

Fig. 254.—*BACILLUS SUBTILIS*.

A, Zoogloea stage with spore-formation; B, Motile stage.

Questions on the Muscineæ and Thallophyta.

1. What is a spermatozoid (antherozoid)? In what plants do spermatozoids appear? How are they developed and set free? What is their function, and how is it performed?

2. Give an account of the structure and life-history of Marchantia. State the more important differences presented in the structure and life-history of Funaria.

3. Is it possible to grow a mould and a Flowering Plant in artificially prepared aqueous solutions? State what should be the ingredients of such solutions, and explain in what important respect they must differ in the two cases.

4. Give an account of the structure and life-history of Vaucheria, and indicate the more important resemblances and differences presented in the structure and life-history of Pythium. How would you account for these resemblances and differences?

5. State clearly the grounds on which the conclusion is based that the moss-plant corresponds to the prothallus of the Fern, and the sporogonium to the fern-plant.

6. Compare and contrast Eurotium and Spirogyra as regards their nutrition and reproduction.

7. Give a full account of the differences which exist between an Alga and a Fungus with regard to the mode of their nutrition.

8. Describe and compare the sexual reproductive organs in Ulothrix, Fucus, and Vaucheria.

9. What are gonidia? In what plants are they found? In what respects do they differ from the asexual spores of fern or moss?

10. What is a parasite? Give an example. How is a parasite distinguished (*a*) from an epiphyte, (*b*) from a saprophyte? Give examples of epiphytes and saprophytes.

11. Describe the structure and life-history of the Yeast-Plant. Mention the points in which this plant resembles and differs from Eurotium or Penicillium.

12. Describe and contrast the development of asexual reproductive cells in Ulothrix, Vaucheria, Pythium, Eurotium, Agaricus, and Funaria.

13. Give a comparative account of sexual reproduction as found in Ulothrix, Spirogyra, Fucus, and an Angiosperm.

14. Give a short account of the structure and life-history of Agaricus.

15. What is fermentation? What is its probable significance in relation to the normal metabolic processes? Give an account of a typical example.

16. Name and illustrate by reference to particular types the various methods of reproduction found in plants.

17. What can you say with regard to (*a*) the evolution of sexuality in plants, (*b*) the differentiation of sex.

APPENDIX.

(Specially written for Students working alone.)

A. GENERAL ADVICE TO THE STUDENT.

§ 1. Reading. The necessity of *careful* reading cannot be too strenuously impressed on the mind of the biological student. He should be ever on his guard against the modern tendency to rapid reading and hasty assimilation. Before passing from one part of the subject to another he should make sure, *as far as he can*, that he has understood what he has read. A habit of skipping difficulties is easily acquired and not easily overcome. Of course there may be many cases where even the most diligent application will fail to clear up difficult points, and where persistence in wrestling with them only results in a waste of time. In such cases the student should for future reference make a note of the points he has failed to master. It may be that a wider knowledge of the subject will make them clear.

Special reference may here be made to the first two chapters of the present text-book. These chapters deal with general facts and principles; and it is not expected that, at the first reading, the student will acquire a perfect knowledge of their contents. They should be very carefully revised when the book has been once read through.

In Biological Science, however careful the reading be, it is perfectly useless unless accompanied by diligent practical work. This will be specially dealt with in the other sections of the Appendix. At present we may notice two other factors which cannot be neglected by the student who wishes to obtain a thorough grip of his subject. These are the drawing of diagrams and the answering of test-questions.

§ 2. **Drawing of Diagrams.** This is valuable in two ways. In the first place it helps the memory, and impresses on the student's mind not only the more important points, but also many details which might otherwise be overlooked. Secondly, it is an excellent way of letting the student see what he doesn't know about various parts of his subject. The student may begin by carefully copying the figures which are given in the text. He should next try to reproduce them from memory; and, finally, he should always devote some time to drawing the sections he cuts, or the dissections he makes in connexion with the practical work. In making these drawings the use of coloured pencils has many advantages. For this the Blaisdell pencils are recommended. These, however, can only be obtained in four colours.

§ 3. **Test-Questions.** Merely verbal knowledge should be avoided, *i.e.* knowledge which can be reproduced by the student only in words and sentences more or less nearly resembling those which he has read. He should be able to express his knowledge in words of his own. This faculty is one test of real knowledge, and will prove invaluable when the student has to deal with *general* questions,—that is to say, questions which are framed with a view to testing general knowledge, and which are not as a rule specially dealt with in text-books. Such, for example, are questions dealing with the comparison of certain types or structures. To answer these satisfactorily the student must be able to set down alongside of each other all the important resemblances or differences exhibited. It is in such general questions, as a rule, that the student who is a mere echo of a text-book comes to grief. The moral of all this is that the student should be constantly testing his knowledge by trying to write down what he knows on certain parts of the subject. For this purpose he will find the specimen questions, given on pp. 245, 310, 375, specially useful. Wherever it is possible he should make a point of illustrating his answers by outline drawings. It is here that the reproduction of figures from memory will prove of service.

§ 4. **Botanical Terms—Greek and Latin Roots.** Many botanical terms have departed so far from their original meanings, as implied in their etymology, that the student must get to know them in the same way as he would get to know the words in learning a new language. Frequently, however, a knowledge of the derivation of botanical terms is really helpful; for this reason we give here a table of Greek and Latin roots which may be of service to the student:—

GREEK.

a-, *without.*
acro-, *summit.*
actino-, *rayed.*
adelphos, *brother.*
amphi-, *both.*
ana-, *up.*
andr-, *of man or male.*
anemos, *wind.*
angos, *a vessel.*
anti-, *opposite.*
apo-, *away from.*
bio-, *life.*
blastos, *beginning.*
blema, *covering.*
carp, *fruit.*
chlamys, *a cloak.*
chloro-, *green.*
chromo-, *colour.*
cleisto-, *closed.*
cyto-, *cell.*
derma, *skin.*
di-, *twice.*
dich-, *apart.*
dynamis, *strength.*
e-, *without.*
ecto-, *outside.*
endo-, *within.*
epi-, *on.*
ergon, *work.*
gamos, *marriage*
gē, *earth.*
-gen, *producing.*
genesis, *growth, origin.*
gyn-, *of woman or female.*

helios, *sun.*
heteros, *different.*
histos, *web, tissue.*
homos, *same.*
hypo-, *under.*
kata-, *down.*
logos, *science.*
macro-, *large.*
mega-, *large.*
meros, *part.*
meso-, *middle.*
micro-, *little.*
mono-, *single.*
morphe, *form.*
-œcium (oikos), *house.*
-oid, *like.*
ōon, *egg.*
orthos, *straight.*
peri-, *around.*
phellos, *cork.*
-phile, *loving.*
-phore, *carrying.*
phyll, *leaf.*
phyte, *plant.*
plasma, *anything formed.*
pod, *foot.*
poly, *many.*
protos, *first.*
rhiza, *a root.*
sapros, *putrid.*
schizo-, *split.*
scleros, *hard.*
sperma, *seed.*
stichos, *a row.*

GREEK (continued).

syn-, *together with.*
taxis, *arrangement.*
tetra, *four.*
thēc, *a case.*

tropos, *direction.*
xero-, *dry.*
xylon, *wood.*

LATIN.

ad, *to.*
aster, *star.*
calyx, *cup.*
caulis, *stem.*
cid, *cut.*
ex, *without.*
-fid, *cleft.*
flos, *flower.*
fug, *flee.*
inter, *between.*
intra, *within.*
latex, *liquid.*
lignum, *wood.*
loculus, *little place.*
natus, *born.*

nodus, *a knot.*
nuto, *nod.*
par, *equal.*
paries, *a wall.*
partite, *divided.*
peto, *seek.*
radix, *root.*
scala, *ladder.*
-sect, *cut.*
suber, *cork.*
unguis, *a claw.*
utriculus, *a bladder.*
vas, *a vessel.*
verticillus, *whorl of a spindle.*

§ 5. **Supplementary Reading.** To such students as may be desirous of further extending their botanical studies the following books are recommended:—Vines' *Students' Textbook of Botany* (Swan Sonnenschein & Co., 15s.); Green's *Manual of Botany*, Vols. i. and ii. (J. & A. Churchill, 17s. 6d.); Willis' *Flowering Plants and Ferns*, Vols. i. and ii. (Cambridge Natural Science Manuals, 10s. 6d.); Scott's *Structural Botany*, Vols. i. and ii. (A. & C. Black, 3s. 6d. each); Strasburger's *Textbook of Botany* (Macmillan, 18s.).

B. MICROSCOPICAL WORK.

§ 6. **The Practical Text-Book recommended.** Here we cannot give full detailed directions serving to initiate the beginner in the mysteries of section-cutting, etc. Fortunately, however, it is unnecessary to do so. In Bower's *Practical Botany* (Macmillan, 10s. 6d.) the student will find most that he requires in this direction. To the student working alone this book is indispensable. It has one serious drawback: there are few illustrations. It is hoped, however, that the figures provided in the present work will to a large extent cover this deficiency.

All that remains for us to do is to indicate how the above practical text-book should be used, to map out the course the student should follow, and to offer a few suggestions on special points. Bower's text-book in many respects goes beyond the requirements of the elementary student. In the following paragraphs we shall endeavour to indicate the practical work which is indispensable, leaving it to the student to decide to what extent his time will allow of his following some of the more elaborate methods described by Bower.

§ 7. **Apparatus.** The following are necessary:—

(*a*) A good microscope with lenses $\frac{1}{2}$ and $\frac{1}{6}$ inch focal distance.

(*b*) Two *good* razors, slightly hollow ground.

(*c*) Glass slides, $3'' \times 1''$; cover-glasses, $\frac{7''}{8}$ diam. or sq.

(*d*) Small forceps, dissecting needles, and scalpels.

(*e*) A few deep watch-glasses; small brushes; a clean linen rag.

(*f*) Pickle jars, methylated spirit, formalin (see § 10).

(*g*) Small bottles (with dipping rods) containing iodine solution, aniline sulphate (or chloride), Schulze's solution, glycerine.

A really serviceable microscope for all ordinary work is Leitz's Students' Microscope, with eye-pieces 1 and 3, and objectives 3 and 7.

All information with regard to reagents will be found in Bower.

§ 8. **Cutting and Mounting Sections.** At first the student should content himself with sections simply stained with iodine or aniline sulphate and mounted in glycerine. Only if time permits, and after considerable experience in this method, should he attempt the more elaborate methods of staining and mounting described in Bower's text-book.

In taking a section, the tissue to be cut should be held between the thumb and fingers of the left hand; the razor in the right hand. The tips of the four right fingers should rest on the back of the razor, and the thumb in front, just behind the cutting edge. The cutting edge is therefore directed inwards, towards the operator. The arms should be brought close up to the body. Tissue and razor should both be wet with alcohol. The blade of the razor may rest gently on the forefinger of the left hand with the edge against the tissue. Then the razor should be drawn through the tissue with a sliding movement. *With practice*, extremely thin sections may be cut.

The sections should be removed from the razor by means of a brush, and placed in a watch-glass containing alcohol or water. Several may then be transferred to a slide and examined in water under the low power, so that the best may be selected. By means of the linen rag the excess of water may be removed, and iodine or other reagent added according to the special points which the student wishes to determine. The reagent should then be washed off with water, the excess of water removed, a drop of glycerine added, and finally the cover-glass put on. The section should always be mounted in the centre of the slide. The cover-glass should be rested on its edge and let down gradually by means of a needle. The section *must not be allowed to get dry* during the process, or air-bubbles will make their appearance. If these do appear, soaking the section for some time in alcohol will remove them. The cover-glass must be perfectly clean, and the upper surface dry.

Neatness and cleanliness are of great importance in practical work. At first the student will find that his sections are rather thick, and often obliquely cut. These are difficulties which can be got over only by care and practice. He should not attempt to draw a bad section.

Very slender or delicate tissues should be cut by embedding in pith or carrot. The more elaborate methods described in Bower may be passed over in early stages of practical work. If carrot be used, a piece $1'' \times \frac{1}{2}'' \times \frac{1}{4}''$ will be found convenient.

APPENDIX. 383

§ 9. **Reagents.** The effects produced on cell-walls and cell-contents by iodine, aniline sulphate, Schulze's solution, etc., should be studied. The tests for starch, proteid, cellulose, cork, mineral crystals, are important. All these points are fully dealt with in Bower's text-book.

§ 10. **Material.** Fresh material may and sometimes must be used. In many cases, however, it is better and more convenient to use "pickled" material. The pickling fluid used for ordinary work is ordinary methylated spirit. Stems, roots, leaves, etc., preserved in this way in glass jars are always ready for use. Delicate plants or parts of plants (flowers, etc.) may be preserved in four to six per cent. solution of formalin (formic aldehyde); formalin as sold is a forty per cent. solution. This method has the advantage of preserving colours. Types such as Vaucheria, Spirogyra, Ulothrix, Eurotium, Yeast, must be examined in the living state.

The student is advised to obtain material himself, as far as possible. With the Angiosperm, Pinus, Fern, and Agaricus there is no difficulty. Marchantia, Funaria, Ulothrix, Spirogyra, and Vaucheria are also easily found. Fern-prothalli can usually be obtained in abundance on the damp walls or on the soil of flower-pots in fern-houses. Directions for obtaining Pythium, Eurotium, etc., are given in Bower, and also in our descriptions of these types. Yeast may be obtained from a baker. If desired, however, all the necessary material can be obtained from James Backhouse and Son, Ltd., The Nurseries, York (see their catalogue).

§ 11. **Order of Practical Work.** The student should not begin practical work till he has a general acquaintance with the first two chapters of this book. Preliminary attempts at section-cutting on, say, a piece of pith or carrot, should precede the serious work. The student might then work through the seeds described in Chapter III., and after that take the other chapters in order. He should always make a point of finding out what he expects to see in his sections. Reading should precede practical work.

The practical work in connexion with the special types can be readily followed from Bower; but as the student may be in some doubt as to the practical work to be undertaken in connexion with the Angiosperm, the following list may be of service.

Stems:—Transverse and, in some cases, Longitudinal Sections of such Dicotyledon stems as sunflower, groundsel, white bryony, Clematis, mare's-tail, elder, elm, lime, and such Monocotyledon stems as Asparagus, butcher's broom, black bryony, maize. The list might be extended indefinitely. The beginning of secondary growth in Dicotyledons should be studied, *e.g.* in groundsel.

Buds:—Longitudinal sections, *e.g.* of the lilac. Try to make out the meristematic regions.

Roots:—Transverse sections of such roots as leek, maize, iris, primary roots of bean or castor-oil seedlings, sunflower, elm, willow, etc. The apical meristem can be studied in *median* longitudinal sections of the radicles in maize, almond, sunflower, and castor-oil seeds. The fibrous or tuberous roots of the lesser celandine (p. 223) are excellent for showing primary dicotyledonous structure, as there is no secondary growth.

Leaf:—Sections of petiole and lamina, *e.g.* beech, sunflower, horse-chestnut, lily; strip off part of epidermis to see stomata, etc.

Flower:—Sections of ovaries and anthers. The structure of the ovary, the placentation, the form and structure of the ovules, should be recognized. The form of the ovule can usually be readily made out in sections of ovaries, or, if the ovules are small, by examining entire in acetic acid.

Seeds:—In examining a seed the student should determine by means of sections and staining—(*a*) Dicotyledonous or Monocotyledonous, (*b*) albuminous or exalbuminous, (*c*) the nature of the food-material.

Contents of cells and cell-walls:—These should be very carefully studied in connexion with the above. To see such bodies as cystoliths and raphides the student should try to obtain the material mentioned in the text-book.

C. Descriptive Botany.

§ 12. Examination of Plants. The student should devote a considerable amount of time, whenever he has the opportunity, to the study of complete plants, and the examination of special parts. He should carefully notice the different kinds of roots and stems, and accustom himself to the use of the various terms explained in the text. Various bulbs, tubers, corms, rhizomes, suckers, etc., should be studied, and their special features recognized. The morphological value of spines or thorns, tendrils and other specialized structures, wherever specimens present themselves, should be clearly made out. The form, arrangement, and venation of leaves; branching; the position of buds, stipules, bracts, etc.; the forms, etc., of calyx, corolla, andrœcium, and gynæceum; the seed and fruit, etc.;—all these should be made the subject of careful scrutiny and observation. In the examination of minute or crowded parts, as, for example, frequently in the study of the flower, it will be found not only convenient, but necessary, to use a small **hand-lens**. A very convenient folding form, with three glasses, can be obtained from any dealer.

§ 13. **Description of Plants.** The art of describing plants consists simply in being able to make use of the proper terms in a neat, orderly way. An elaborate description is not expected of the elementary student. It will be sufficient if he can give a good general description. The following scheme simply indicates the order on which such a description should proceed. The student must not regard it as a form for plant-description, to be rigidly adhered to in all cases. It is merely intended as a guide:—

Root: tap or adventitious? branched or unbranched? the special form — tuberous, fleshy, fibrous, etc.? annual, biennial, or perennial?

Stem: kind of stem—*i.e.* is it erect, prostrate, or climbing, a rhizome, corm, or bulb, etc.? herbaceous or woody? cylindrical, angular, or compressed? hairy or glabrous? branched or unbranched (the branching may be described)? If herbaceous, is it solid, hollow (fistular), or jointed? If climbing, how does it climb? Does it bear cladodes, tubers, spines, etc.

Leaf: deciduous or evergreen? radical, cauline, or ramal? alternate, opposite (superposed or decussate), or verticillate? petiolate or sessile? stipulate or exstipulate (the stipules may be described)? sheathing, connate, perfoliate, ligulate, etc.? simple or compound?

If simple, outline of lamina (*i.e.* linear, oval, etc., or pinnatifid, palmatifid, etc.—if incised, the outline of the lobes, partitions, or segments may be indicated)? venation? margin? apex? surface (glaucous, hairy, etc.)?

If compound, pinnate or palmate? paripinnate or imparipinnate? number and arrangement of leaflets? *Leaflets* — sessile or stalked? outline? venation? margin? apex? surface?

Inflorescence: definite, indefinite, or mixed? kind of inflorescence?

Flowers: sessile or pedicellate? bracteate or ebracteate (if bracteate, the bracts may be described)? complete or incomplete? hermaphrodite or unisexual? actinomorphic, zygomorphic, or asymmetrical? cyclic, hemicyclic, or spiral? heterostylic? any other general character?

If there are two kinds of flowers, after giving common characters as above, describe separately.

Calyx: poly- or gamo-sepalous? green or petaloid? if polysepalous, the number, outline, and apex of the sepals? if gamosepalous, the special form or nature of the incision? inferior or superior? æstivation?

Corolla: regular or irregular? if irregular, zygomorphic or asymmetrical? poly- or gamo-petalous? if polypetalous, number and outline of petals, or any special terms? if gamopetalous, special form or incision? corona or other special features? hypogynous, perigynous, or epigynous? æstivation?

Perianth: described similarly, except that the terms poly- or gamo-phyllous must be used.

Androecium: number of stamens? or indefinite? polyandrous, syngenesious, or adelphous? epipetalous, epiphyllous, hypo-, peri-, or epi-gynous? special characters? filament? fixation of anther? dehiscence?

Gynæceum: mono- or poly-carpellary? if the latter apo- or syn-carpous? *Ovary*—unilocular or multilocular? superior or inferior? *Ovules*—number? or indefinite? form? *Placentation? Style? Stigma?*

Seed: albuminous or exalbuminous?

Fruit: kind of fruit?

Nectaries may be described in connexion with corolla, stamens, or pistil, as seems convenient.

§ 14. **Examples.** The following descriptions of well-known plants will serve as examples:—

(1) **Root**: a fibrous branched tap-root; perennial.
Stem: erect; branched; angular; herbaceous above, woody below; slightly hairy; perennial.
Leaves: herbaceous; cauline; shortly petiolate; exstipulate; simple; lanceolate; unicostate, reticulate; acute; entire; slightly hairy.
Inflorescence: indefinite—terminal and lateral, more or less corymbose racemes.
Flowers: pedicellate; ebracteate; isobilateral; hermaphrodite.
Calyx: polysepalous; four lanceolate, petaloid sepals in two series, the two inner (lateral) sepals slightly pouched (saccate); inferior.
Corolla: regular, polypetalous, cruciform, consisting of four unguiculate petals; *limb* obovate; hypogynous; imbricate.
Androecium: six stamens in two series; tetradynamous; two short lateral stamens; two pairs, anterior and posterior, of long stamens; hypogynous; *anthers* innate, introrse, with longitudinal dehiscence. *Nectaries*, having the form of green, rounded discs, are present at the base of the lateral stamens.
Gynæceum: bicarpellary, syncarpous; *ovary* bilocular, owing to a false septum developed between the placentas, superior; *ovules* ∞, campylotropous, on two parietal placentas; *style* short; *stigma* two-lobed.
Seed: exalbuminous. **Fruit**: an elongated, linear, slightly flattened siliqua.

(2)* **Roots**: adventitious, fibrous, and stout, yellowish.
Stem: woody, perennial, underground, covered with brown scales, and throwing out long slender runners which root at the nodes.
Leaves: radical, with long hairy petioles and membranous, lanceolate, petiolar stipules; compound, ternate; *leaflets* nearly sessile, roundish oblong, with unicostate reticulate venation and serrate margin.

* Adapted from Lindley's *Descriptive Botany*.

Inflorescence: panicled, more or less corymbose, cymes borne on erect slender scapes.

Flowers: pedicellate, with membranous bifid bracts; actinomorphic; complete; hermaphrodite; protogynous.

Calyx: gamosepalous, with five membranous, triangular, acuminate segments; green, persistent, inferior; an epicalyx is present consisting of five oblong segments alternating with those of the calyx proper.

Corolla: regular, polypetalous, rosaceous, consisting of five white roundish petals inserted perigynously.

Androecium: polyandrous; *stamens* ∞, persistent, perigynous; *filaments* short and stiff; *anthers* oval, more or less cordate, dehiscing at the edges.

Gynæceum: polycarpellary, apocarpous; *carpels* indefinite, and borne on a protuberance of the thalamus, with filiform styles and simple stigmas; *ovules* solitary, ascending.

Seed: exalbuminous, dicotyledonous. **Fruit:** a pseudocarp consisting of an etærio of achenes borne on a succulent thalamus.

D. ADDITIONAL NATURAL ORDERS.

(*For general table of classification see p.* 220.)

§ 15. **Evolution in the Angiosperms.** We have already in Chap. XVIII. (§§ 9-10) touched very briefly on this subject. A clearer statement with regard to it will help the student to a better understanding of the classification of Angiosperms and of the principles on which it is based.

There is no clear evidence to show at what point the Angiosperms took their origin, but it is now probable that they were derived from Pteridophyta by a route (or routes) distinct from that of Gymnosperms. Distinguished from the latter by the development of an ovary securing the better protection of the ovules and seeds, and by the late development of the female prothallus, they proved more progressive than the older type, and have become the dominant race at the present time. The relation between Monocotyledons and Dicotyledons as regards origin is also obscure. Some botanists believe that the Dicotyledons were derived from primitive Monocotyledons, but it is just as likely that the evolution of the two classes followed, from the first, distinct though parallel lines.

The early Angiospermous flowers no doubt simply consisted of more or less elongated axes bearing numerous sporophylls—the stamens and carpels. There is nothing to tell whether they were hermaphrodite or not, but it is commonly assumed that they were unisexual and wind-pollinated like those types of existing Dicotyledons such as the Birch and Hazel which, there is reason to believe, show primitive characters. On this view, if regard be had to the risks and lack of economy associated with unisexual flowers and wind-pollination, it is not difficult to understand the early transition to hermaphrodite flowers affording a possibility of self-pollination. In these hermaphrodite flowers the carpels would be above, the stamens below, the arrangement which, with modifications, still persists.

The evolution, first of a single series of perianth leaves for protective purposes, and then of a second for attraction, has been indicated on p. 310. The appearance of the corolla and the further evolution of the flower are intimately connected with insect visitation, and the evolution of the flower cannot be followed with any clearness except by reference to it, especially in the higher forms of Monocotyledons, and in the Gamopetalæ which show the highest degree of specialization in Dicotyledons.

The early flowers with double perianth were no doubt regular, the floral leaves all free, and the stamens and carpels numerous. To such flowers insects would have free access. Advance in organization has followed many lines and many important factors have influenced the course of evolution, which, it must be remembered, has always tended towards the more perfect adaptation of the flower for the functions which it has to discharge in relation to the conditions of environment. Some of the growth-processes which have led to modification of floral structure have been mentioned on p. 182. We have now to consider these modifications from the standpoint of evolution and to recognize their biological significance.

The enclosure of the ovary in the thalamus would secure the better protection of the ovules and seeds. Ready pollination would be facilitated by increase of the number of ovules in each ovary, or by their aggregation in a single compound ovary. Further, as pollination became more certain, there would be a reduction in the number of stamens and carpels. Hence the transition from hypogyny to epigyny, that from an apocarpous to a syncarpous condition, and reduction in the number of stamens and carpels, together with more efficient provisions for pollen protection and more elaborate adaptations for insect pollination (such as the concealment of honey, the development of zygomorphy, special mechanisms, etc.), which seem to have appeared at various points in the course of evolution, are the chief marks of an advance in organization, and characters of the greatest importance in classification. In this connection it must always be remembered that our existing orders cannot be arranged in a linear series, but are best regarded as the terminal twigs of branches given off along the main stem of Angiospermous development.

It is commonly believed that certain existing orders of Dicotyledons, although specialized in particular directions, have retained the primitive mono- or a-chlamydeous condition. Such orders are placed in the Incompletæ. Orders which are mono- or a-chlamydeous by reduction are excluded and must be classified according to their affinities as shown by their other characters.

Amongst the Polypetalæ there is a clearly marked advance from hypogyny to epigyny, and, amongst the hypogynous and perigynous forms, from the apocarpous to the syncarpous condition. Sometimes the transition is illustrated within the limits of a single order. It can be recognised also that the lower forms in both hypogynous and perigynous series have numerous stamens and carpels, while in the higher forms the number of essential organs is reduced. The classification of the Polypetalæ is in accordance with this. At various points specialization in adaptation to insect pollination has given rise to prominent orders, *e.g.* Ranunculaceæ, Caryophyllaceæ, Cruciferæ, Onagraceæ, Leguminosæ. In these it is interesting to observe how the various mechanisms or the arrangements for the concealment of honey, etc., have been produced.

The concealment of honey, for which, in the Polypetalæ, there are numerous contrivances, is usually secured in the Gamopetalæ by the gamopetalous corolla. This explains its evolution. The epipetalous condition of the stamens which is associated with it, except in some of the lower forms,* allows of the further narrowing of the tube. In the Gamopetalæ the evolution of hypogynous and epigynous series has followed two distinct lines with separate points of origin from the Polypetalæ. The highest degree of specialization in the evolution of hypogynous forms is reached in those bicarpellate orders which have showy zygomorphic flowers (Labiatæ). Evolution along the epigynous line reaches its highest stage in the Compositæ in which there are compact aggregations of small flowers showing marked division of labour. They represent the highest development in Flowering Plants.

Amongst the Monocotyledons a very similar progress can be recognised from a primitive to a petaloid and gamophyllous condition of the perianth, from hypogyny to epigyny, from an apocarpous to a syncarpous condition of the pistil. The position of many orders, however, is doubtful owing to the difficulty experienced here, as in Dicotyledons, in determining whether their characters are primitive or due to reduction. The evolution in the higher forms here also has been in close relation to insect pollination and reaches its highest point in the Orchidaceæ.

* Ericaceæ and Campanulaceæ (*q.v.*).

DICOTYLEDONS.

(a) INCOMPLETÆ.

§ 16. Cupuliferæ.

Distinguishing characters: Flowers mono- or a- chlamydeous, unisexual, monœcious, borne in catkins, singly or in groups of 2 or 3. Pistil bi- or tri-carpellary. Fruit dry, indehiscent and one-seeded, a nut or nutlet, frequently invested by a cupule formed by enlarged persistent bracteoles.

This is a primitive order of great interest, widely distributed in temperate regions. It consists of trees and shrubs with simple, alternate, stipulate leaves. The common British trees belonging to the order are the Birch (*Betula alba*), the Alder (*Alnus glutinosa*), the Hazel (*Corylus Avellana*), the Hornbeam (*Carpinus Betulus*), the Beech (*Fagus sylvatica*), the Oak (*Quercus Robur*), the Spanish or edible Chestnut (*Castanea vulgaris*—it is not indigenous).

The **inflorescences** are called catkins (p. 187). In its typical form the **catkin** consists of an elongated pendulous axis bearing numerous spirally arranged scales (bracts) in the axil of each of which are three flowers representing a sessile or reduced dichasium. The terminal (middle) flower has two lateral bracteoles, and in the axils of these (as bracts) arise the two lateral flowers which also may have bracteoles. Thus there are typically three flowers and six bracteoles in the axil of each bract (fig. 255). From this it is at once evident that the catkins typical of the order are not really simple pendulous spikes (see p. 187).

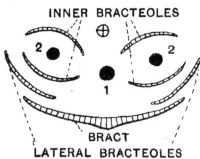

Fig. 255.—TYPICAL FLORAL DIAGRAM OF CUPULIFERÆ.

Showing arrangement of bracteoles and flowers: 1 = middle flower; 2, 2 = lateral flowers.

This typical form, however, is departed from, more or less, in the various genera. Sometimes only the middle flower or only the two lateral flowers are present, and some or even all of the bracteoles may be absent. The whole inflorescence is, in some cases, reduced to a cluster of flowers. The various modifications described below should be carefully studied. The female catkins persist till the fruits are ripe, or even longer as in the Alder.

The **flowers** are unisexual, monœcious, and borne (with rare exceptions—*e.g.* sometimes in Chestnut) in different catkins. They are anemophilous and, in correlation with this, they frequently come out before the leaves (Hazel and Alder), or just when the leaves are unfolding (Birch and Oak). A **perianth** is sometimes present, and occasionally well developed; when present it is epigynous. The **stamens** are two, four, or more; sometimes they are split or forked (fig. 258, c). The **gynæceum** is bicarpellary (Alder, Birch, Hazel, Hornbeam), or tricarpellary (Oak, Beech, Chestnut), syncarpous; the *ovary* at the time of fertilization bi- or tri-ocular, inferior; *ovules*, one (Alder, Birch, Hazel, Hornbeam) or two (the other genera) in each loculus, anatropous and usually pendulous.

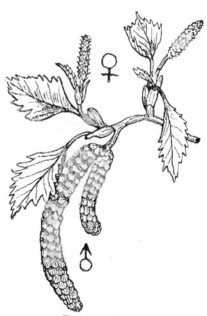

Fig. 256.—BIRCH.
Twig showing male and female catkins.

The **fruit** is dry, indehiscent, one-seeded—a nut or nutlet. In the Birch it has a membranous wing and is a samara (fig. 258, B). The fruits may be liberated from the coherent bract and bractcoles (Alder and Birch), or these may enclose one or more fruits as a cupule (the other genera). The seed is exalbuminous (fig. 263).

The following notes indicate the special characters of the various genera:—

Birch (figs. 256-258):—The male catkins appear in autumn at the ends of the shoots and are pendulous; the female catkins are borne on short

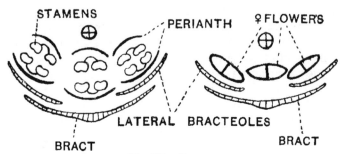

Fig. 257.—BIRCH.
Diagrams showing arrangement of bracteoles and flowers in male and female catkins.

lateral branches which are developed in spring and are erect. Flowering takes place in April or May. In both catkins each bract has three flowers. Only the two lateral bracteoles are present. Each male flower has a small perianth usually two-lobed, and two stamens the filaments of which are so deeply split that there appear to be four stamens. The female flower has no perianth. The pistil is bicarpellary and has two styles. The fruits are samaras. Bract and bracteoles become fused owing to continued basal growth. The three-lobed scale which they form falls off at fruiting, but does not invest the fruits.

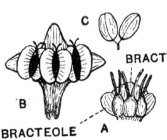

Fig. 258.—BIRCH.
A, Female flowers in axil of bract; B, Fruiting scale, with three samaras; C, Stamen from male flower.

Alder:—The male catkins are elongated; the female small and somewhat ovoid. Both appear in autumn, and are more or less erect. Flowering takes place in March or April. Each bract of the male catkin has three flowers, but in the female only the lateral flowers are developed. There are four bracteoles—the two lateral bracteoles and ne to each lateral flower on the side next the bract. The ♂ flower

APPENDIX. 395

has a four-lobed perianth and four stamens opposite the lobes. The ♀ flower resembles that of the Birch. The female catkins including the

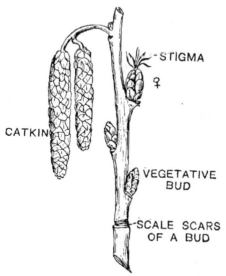

Fig. 259.—Male and Female Inflorescences of the Hazel.

hard five-lobed scales formed from the bracts and bracteoles remain on the tree after the fruits are set free. The fruits are not winged (nutlets).

Hazel (figs. 259-263):—The catkins appear in autumn. The pendulous

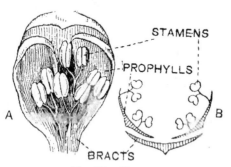

Fig. 260.—Hazel.
A, Male Flower; B, Diagram of same. (Prophylls = bracteoles).

male catkins are borne 1-3 together on a short axillary shoot. The female are solitary, axillary, and are not distinguishable from foliage buds till February or March when flowering takes place and the crimson styles protrude at the top. In the male only the median flower and the lateral bracteoles are developed in each scale. The

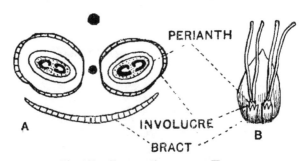

Fig. 261.—FEMALE FLOWERS OF HAZEL.
A, Diagram showing bract, bracteoles (involucre) and flowers; B, Bract and flowers.

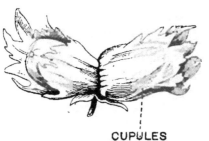

Fig. 262.—TWO HAZEL NUTS INVESTED BY CUPULES.

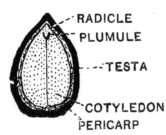

Fig. 263.—LONGITUDINAL SECTION OF HAZEL NUT.

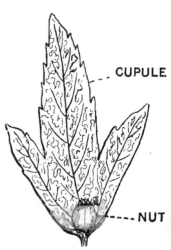

Fig. 264.—FRUIT OF HORNBEAM.

flower has four deeply split stamens and there is no perianth. In the female catkin the lower scales are sterile; in the upper fertile scales *all* the bracteoles, but only the lateral flowers, are present. Each female flower has a minute, toothed, greenish perianth on the top of the ovary; there are two styles. The two bracteoles of each flower with one of the lateral bracteoles fuse, owing to continued basal growth, to form an involucre which develops into the membranous cupule (husk—fig. 262).

Hornbeam:—The flowers are similar to those of the Hazel, but in the male catkins there are no bracteoles and the male flower has 4-10 split stamens. The cupule is large and trilobed (fig. 264).

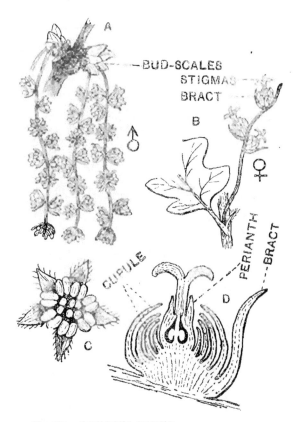

Fig. 265.—*QUERCUS ROBUR*, var. PEDUNCULATA.
A, Male; B, Female inflorescences; C, Male flower; D, Female flower in section.

Oak (fig. 265):—The catkins appear in the spring, the male in the axils of bud scales, the female in the axils of foliage leaves. Flowering occurs in April or May. The male catkin here is simply a long slender and pendulous spike, the flowers being borne singly in the axils of the bracts. They represent the median flowers and there are no bracteoles. Each (fig. 265, C) consists of a perianth of a varying number of bract-like segments (4-7) and as many or more stamens (frequently 10). A rudimentary ovary may be present. The female catkin contains only 2 or 3 flowers which may be in a cluster (*Quercus Robur*, var. *sessiliflora*) or may be separated by the elongation of the peduncle (*Q. Robur*, var. *pedunculata*). They are borne in the axils of bracts and represent the median flowers (in some allied species all three flowers are present). Each has an epigynous 3-8 toothed perianth and is surrounded by a number of imbricate scales forming an involucre which afterwards develops into the acorn cup (cupule). The involucre is regarded as representing the four bracteoles of the lateral flowers. The ovary is trilocular, and each loculus has two pendulous anatropous ovules, but only one loculus and one ovule develop. The fruit (acorn) is a nut seated in a cupshaped cupule. *Quercus Ilex* is the Holly Oak; *Q. Suber*, the Cork Oak.

Beech (fig. 266):—The catkins are developed in spring in the axils of foliage leaves. The male is really a clustered dichasial cyme of about twelve flowers borne on an elongated pendulous peduncle. The flowers are shortly stalked and there are no bracts. The female catkin is stalked, erect, and consists of only two flowers (dichasial cluster) enclosed in a fleshy, four-lobed involucre (cupule), and a number of small outer scales. The homologies of these parts have not been clearly made out. The flowers themselves resemble those of the oak. The cupule is spiny and resembles a capsule. It contains two triangular nuts ("beech mast") and separates into four valves.

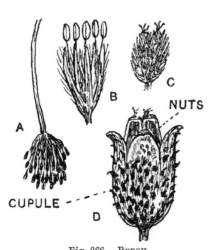

Fig. 266.—BEECH.
A, Male inflorescence; B, Male flower; C, Female inflorescence; D, Cupule with nuts.

Chestnut:—The catkins are axillary, and bract and bracteoles are all present. There are usually seven male flowers in the axil of the bract owing to the bracteoles of the lateral flower also having flowers. The female bracts bear three flowers and the cupule is formed by the four bracteoles of the lateral flowers. The mature cupule is spiny. It contains three nuts and separates into four valves.

The Cupuliferæ, Salicaceæ (p. 237), and other orders (e.g. Juglandaceæ, the walnut order) are combined under the cohort **Amentales** (= Amentaceæ or Amentiferæ), the group of catkin-bearing plants. The Cupuliferæ are now, however, usually divided into three orders—Betulaceæ (Birch and Alder), Corylaceæ (Hazel and Hornbeam), Fagaceæ (Beech, Oak, and Chestnut). The Betulaceæ are distinguished from Corylaceæ by the absence of a cupule, and both of these orders from Fagaceæ by the bicarpellary pistil and the ovule solitary in each loculus.

The Betulaceæ and Corylaceæ are chalazogamic (p. 199). This mode of fertilization is associated with certain peculiar developments in the ovule and embryo-sac which recall features found in Gymnosperms. Taken along with the other characters—the unisexual, anemophilous flowers, the absence or rudimentary character of the perianth, etc., it indicates that in these plants we have very primitive types, a conclusion which is supported by their early appearance in time as shown by fossils. They diverged from the main Angiosperm stem at an early period, and our modern types have survived because, having the habit of trees, they to a large extent escaped competition with more progressive types.* The Fagaceæ represent the most advanced section of the group as shown by their porogamic fertilization and their floral characters generally. In this connection the occurrence of rudimentary pistils in the male flowers is interesting.

Most, if not all of the Cupuliferæ, have mycorhizæ, which are ectotropic (ectophytic), *i.e.* the mycelial threads do not penetrate the cells of the root.

The roots of some Cupuliferæ, *e.g.* Hazel and Beech, are sometimes infested by a total parasite *Lathræa Squamaria*, the Tooth-wort, belonging to the order Orobanchaceæ. The parasite, which is of a white or purplish colour, feeds on the roots by means of suckers (haustoria). It has a rhizome with scaly leaves and sends up aerial shoots bearing racemes of purple flowers. It is possible that small cavities which are found in the scale leaves may function as insect traps.

* Progressive and up-to-date orders, *e.g.* Compositæ, are largely or entirely represented by herbaceous forms.

§ 17. Urticaceæ.

Distinguishing characters:—Flowers regular, usually unisexual; perianth of four or five segments; stamens four or five, opposite the perianth leaves; pistil monocarpellary, with one basal orthotropous ovule; fruit an achene.

This order consists chiefly of herbs, and is well represented in tropical and warm temperate climates. The only plants found wild in Britain are *Parietaria officinalis*, the Wall Pellitory, a perennial herb, common on walls and waste ground, and three species of *Urtica*, the Nettle. *U. urens*, the small nettle, and *U. pilulifera*, the Roman Nettle, are annuals; *U. dioica* is a perennial. These, however, illustrate very well the characters of the order. The plants are mostly herbs with opposite (*Urtica*) or alternate (*Parietaria*) stipulate leaves. Cystoliths are found in most of them, including *Parietaria* and species of *Urtica*. The order may be divided into two groups according as stinging hairs are present (*Urtica* *) or absent (*Parietaria*).

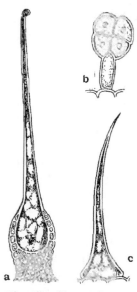

Fig. 267.—URTICA DIOICA (NETTLE).
a, Stinging hair; b, Multicellular glandular hair; c, Ordinary unicellular pointed hair. (Highly magnified.)

The **inflorescences** are usually axillary cymes which may be panicled in appearance or more or less clustered (glomerules). The **flowers** (figs. 268, 269) are regular, monochlamydeous, unisexual or, occasionally, hermaphrodite. *U. urens* and *U. pilulifera* are monœcious, the male and female flowers in the latter being borne in different panicles; *U. dioica* is diœcious; *Parietaria* is polygamous. **Perianth**

---

* The stinging hair of the nettle (fig. 267) has a siliceous point which is readily broken off. When a nettle is touched by the hand, the point breaks and a little wound is made in the skin, into which the acid sap is forced by the sudden contraction of the base of the hair.

APPENDIX.

of four or five leaves, poly- or gamo-phyllous (four and gamophyllous in British species), green, inferior, persistent.

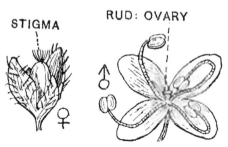

Fig. 268.—Male and Female Flowers of Nettle.

Stamens equal to the perianth segments and opposite to them. The stamens are at first folded inwards and downwards in the flower, but when ripe, or when moved, they spring up violently and give out a little cloud of pollen (an adaptation for pollen protection and wind pollination). **Pistil** monocarpellary; *ovary* superior, unilocular, with one basal, orthotropous ovule. Stigmas tufted and often sessile. The male flowers have a rudimentary pistil. The **fruit** is an

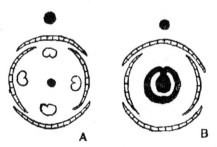

Fig. 269.—Floral Diagram of Nettle.
A, Male; B, Female flowers.

achene enclosed in the persistent perianth. **Seed** albuminous. The flowers are anemophilous.

The various kinds of "Artillery Plant" grown in our hot-houses are species of *Pilea* and are so called because of their explosive stamens.

The orders Ulmaceæ (*e.g.* Elm), Moraceæ (*e.g. Morus* the Mulberry, *Ficus Carica* the Fig-Tree, *Ficus elastica* the India-rubber Tree, *Ficus benghalensis* the Banyan, *Artocarpus incisa* the Bread-fruit Tree), Cannabinaceæ (*e.g. Cannabis sativa* the Hemp, and *Humulus Lupulus* the Hop) are sometimes included under Urticaceæ as sub-orders; but they differ in important respects. The following notes on some of the types mentioned may be of interest—

Ulmus (fig. 270):—*U. campestris* is the common elm; *U. montana* is the Scotch, Wych or Mountain Elm. The clustered dichasial cymes (glomerules) are produced in the axils of the leaves of the previous year. They have a few scale leaves at the base and come out before the leaves (an adaptation to wind pollination). The flowers are hermaphrodite, P (4-6) A 4-6 G (2). Only one loculus and one ovule develop. Fruit a samara (fig. 139, B). Seed exalbuminous.

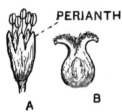

Fig, 270.—Elm.
A, Flower; B, Gynæceum of flower.

Moraceæ:—Trees and shrubs with latex. Perianth segments usually four; stamens as many and opposite them; pistil of two carpels, syncarpous. As a rule only one loculus and one ovule develop and form an achene or nutlet, but usually the axis or perianth becomes fleshy and a syncarp is formed. The Mulberry is monœcious and the flowers are in spikes. For fruits of Mulberry and Fig see p. 215. In *Artocarpus* the inflorescence becomes fleshy and has a texture resembling that of bread.

Hop:—The Hop is a twining plant. Its stem twines in a right-handed spiral (clockwise) like that of the Honeysuckle.* It is diœcious. The male flowers are small and produced on axillary panicled cymes. Each has a five-partite perianth and five stamens. The female inflorescence somewhat resembles a cone (fig. 271). It consists of a series of membranous bracts with two female flowers on the upper surface of each. Each flower has a tubular perianth and is invested by a scale (bracteole). The pistil is bicarpellary. The true fruits are achenes, but the whole inflorescence is regarded as a syncarp (p. 215). The flowers are anemophilous. The bracts, which are covered with glands serving to keep off insects and other intruders, finally separate and help to disperse the fruits.

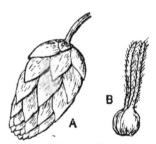

Fig. 271.—Hop.
A, Female inflorescence; B, Pistil of female flower.

* In most twiners the twining is in the opposite direction, *e.g. Calystegia sepium*, the large Hedge Convolvulus.

Cystoliths are found in various species of *Ficus* (p. 28), *Morus*, *Cannabis*, and *Humulus*.

§ 18. Chenopodiaceæ.

Distinguishing characters:—Flowers monochlamydeous, regular, hermaphrodite or unisexual, hypogynous, typically pentamerous. Characters of ovary and seed. An order of halophytes.

The plants belonging to this order are widely distributed in maritime regions, many of them (halophytes) growing in salt marshes or on muddy foreshores, and showing marked xerophytic characters (see p. 156). It is well represented in Britain by the Wild or Sea Beet (*Beta maritima* or *vulgaris*), the Glasswort or Marsh Samphire (*Salicornia herbacea*), the Saltwort (*Salsola Kali*), the Seablite (*Suæda maritima*), and various species of Goosefoot (*Chenopodium*), and Orache (*Atriplex*). *Chenopodium* is not so markedly halophytic as the other genera, various species (e.g. *C. album*) being commonly met with on waste or cultivated ground and presenting the ordinary herbaceous characters. The plants are mostly herbs, in which the stems and leaves are often succulent and fleshy and serve for the storage of water. The leaves are occasionally absent (e.g. *Salicornia*); when present they are exstipulate and alternate, or occasionally opposite (sp. of *Atriplex*). They often feel granular or mealy to the touch owing to the presence of small hairs; this is very noticeable in species of *Chenopodium*.

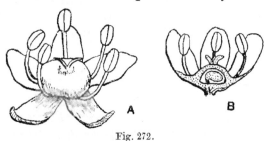

Fig. 272.

A, Flower of species of *Chenopodium*; B, Flower of Beet in section.

The **inflorescence** is frequently mixed; racemes, panicles, and spikes of small cymes are common. The **flowers** (fig. 272) are small and inconspicuous, regular, monochlamydeous, hypogynous, hermaphrodite, or occasionally (*Atri-*

plex) unisexual and either monœcious or diœcious. **Perianth** poly- or gamo-phyllous, small, sepaloid, and persistent. It usually consists of five leaves (the rule in *Chenopodium, Beta, Salsola,* and *Suæda*), sometimes of three or four (*Salicornia*); in the female flowers of *Atriplex* there are only two. **Stamens** usually as many as the leaves of the perianth and opposite them, hypogynous, sometimes perigynous (*Beta*); *Salicornia* has either one or two. **Gynæceum** of two, sometimes three, carpels, syncarpous; *ovary* unilocular, superior (half inferior in *Beta*) with one basal campylotropous ovule. **Fruit** a small nut enclosed in the persistent perianth. **Seed** albuminous or occasionally exalbuminous; the embryo is curved or spirally twisted round the endosperm.

Salicornia herbacea is a small leafless plant which is widely distributed in Britain and grows on muddy shores. It has succulent jointed stems. The flowers are placed two or three together in little cavities, two of which are found opposite to each other at each node. The flower has a fleshy perianth with three or four teeth, one or two stamens, and a pistil of two carpels.

There are some familiar cultivated forms. The Garden Beet, the Sugar Beet and the Mangold Wurzel are cultivated varieties of the Wild Beet. They are biennials and sugar is stored up in their roots. *Chenopodium Bonus-Henricus* (All-good or Good King Henry) is cultivated under the name of "Mercury." *Spinacia oleracea* is the Spinach; the flowers are in dichasia and are diœcious.

(b) **POLYPETALÆ**.

§ 19. **Polygonaceæ**.

Distinguishing characters:—Flowers polypetalous, hypogynous, usually hermaphrodite; trimerous or sometimes dimerous, but the number of parts often increased by duplication or diminished by suppression. Characters of ovary, ovule and fruit. The presence of an ochrea is characteristic of the order.

This order is represented in the British Flora by the three genera *Rumex* (Docks and Sorrels), *Polygonum* (e.g. *P. aviculare*, the Knotgrass) and *Oxyria* (the Mountain

Sorrel). Other genera are *Rheum* (e.g. *R. rhaponticum*, the Rhubarb, *R. officinale* the medicinal Rhubarb) and *Fagopyrum* the Buckwheat. They are mostly herbs. The leaves are simple and alternate, with ochreate stipules, and the stems are swollen at the nodes. The acid properties found in most of the plants are due to the presence of various oxalates (p. 36).

The **inflorescence** in most cases is mixed, commonly a raceme or panicle of cymes. The **flowers** are hypogynous and usually hermaphrodite. They are typically trimerous, sometimes dimerous, but the number of parts is often increased by duplication or diminished by suppression. Unisexual flowers occur in the Sorrels; *Rumex acetosa*, the Sorrel, is monœcious; *R. acetosella* the Sheep's Sorrel, is diœcious. The

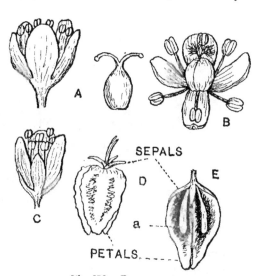

Fig. 273.—POLYGONACEÆ.
A, Flower and pistil of *Polygonum Persicaria* (2 styles); B, Flower of *Oxyria*; C, Flower of a species of *Rumex*; D, Fruits of species of *Rumex* showing persistent perianth (a = swollen midrib of petal).

perianth typically consists of three sepals and three petals resembling each other, and either sepaloid or petaloid (fig. 273, C, 274, A). It is polyphyllous, imbricate in aestivation, inferior and persistent. This typical condition is found in *Rumex* and *Rheum*, and in these genera the *inner* segments (petals) enlarge during the development of the fruit and invest it. In *Polygonum* (figs. 273, A, 274, B) the anterior segment of the inner series (petal) is suppressed so that the perianth consists of five leaves (P_{3+2} or K_3C_2); here the three *outer* segments become enlarged and invest the fruit. In *Oxyria* (fig. 273, B) there are two sepals and two petals (P_{2+2} or K_2C_2). The **Androecium** consists typically of six stamens (A_{3+3}), but this typical condition is seldom found. Usually there is chorisis (= duplication or dédoublement) of one or more of the stamens of the outer series, and this may be accompanied by suppression of one or more members of the inner series (see fig. 274). In

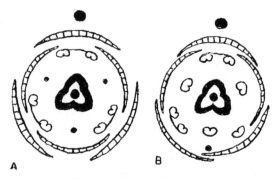

Fig. 274.—FLORAL DIAGRAMS.
A, *Rumex*; B, Species of *Polygonum*.

Rheum there are nine stamens, all the outer stamens being duplicated ($A_{3 \times 2 + 3}$). In *Rumex* the outer are all duplicated, but the inner suppressed ($A_{3 \times 2 + 0}$). In *Polygonum* there are five to eight stamens; usually two outer stamens are duplicated, and one or more of the inner ones suppressed. In *Oxyria*, where the arrangement is dimerous there are six stamens, the two outer ones being duplicated ($A_{2 \times 2 + 2}$). The **Gynæceum** is usually tri-carpellary and syncarpous; in *Oxyria* and some species of *Polygonum* (e.g. *P. amphibium*) it is bicarpellary (fig. 273, A). The ovary is unilocular, superior, with one basal orthotropous ovule (fig. 112, p. 180); *stigmas* two or three. The **fruit** (fig. 273, D) is ovoid when there are two carpels, triangular when there are three. The persistent membranous perianth provides for wind dispersal. The **seed** is albuminous.

An annular honey disc is present in *Polygonum* at the base of the stamens, and the flowers are entomophilous. Some species are marsh or water plants, e.g. *P. amphibium*, *P. Persicaria*, *P. Hydropiper* (water-pepper). *P. amphibium* also grows on dry ground where the long petioles of the aquatic form are lost. *P. viviparum* is an alpine form; it has small flowers, which rarely form fruit, and some of the lower ones are frequently replaced by little red bulbils. In *P. Convolvulus*, the Black Bindweed, the stem is twining. *P. aviculare* has cleistogamous flowers.

In *Rumex* there is no honey disc. The stigmas are long and feathery, and the flowers are wind-pollinated. *Rheum* is entomophilous. *Fagopyrum* resembles *Polygonum*, and is sometimes placed in that genus (*Polygonum Fagopyrum*).

The Polygonaceæ were formerly placed amongst the Incompletæ near Chenopodiaceæ, but the double perianth prohibits this arrangement. Its affinities amongst the Polypetalæ have not yet been clearly made out. It is not closely allied to any of the orders considered in this book.

§ 20. Violaceæ.

Distinguishing characters:—Flowers polypetalous, hypogynous, pentamerous; stamens five; pistil tricarpellary, syncarpous; ovary unilocular; placentation parietal; fruit a capsule. In European species the flowers are zygomorphic and the anterior petal is spurred.

The only European genus in this order is *Viola*, to which the various kinds of Violet and Pansy belong. *Viola odorata* is the Sweet Violet; *V. canina*, the Dog Violet; *V. sylvatica*, the Wood Violet; *V. palustris*, the Marsh Violet; and *V. tricolor*, the Pansy or Heart's-ease.

In this genus the plants are herbaceous, either annual, or perennating by means of rhizomes. Some species have runners (*V. palustris* and *V. odorata*). The leaves are alternate, and have large stipules, especially in the various kinds of Pansy, where the stipules are leafy and discharge the function of foliage leaves. The **flowers** (in British species) are solitary and axillary. They are polypetalous, hypogynous, pentamerous, irregular and zygomorphic, hermaphrodite, bracteate and bracteolate (figs. 118, 275). **Calyx,** of five sepals, polysepalous, inferior, persistent, imbricate; the sepals usually prolonged downwards below the point of attachment. **Corolla** of five petals, polypetalous, zygomorphic; the anterior petal has a spur serving as a receptacle

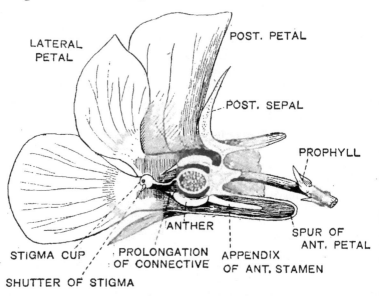

Fig. 275.—Pansy (Viola Tricolor).
Vertical Section of Flower.

for honey, and the lateral petals sometimes have a tuft of hairs; aestivation imbricate. **Androecium** of five free stamens with very short filaments, surrounding the ovary; anthers introrse and surmounted by membranous outgrowths of the connective; the two antero-lateral stamens bear greenish horn-like appendages projecting into the spur of the anterior petal and functioning as nectaries. **Gynæceum** of three carpels, syncarpous; *ovary* unilocular, superior; ovules ∞, anatropous, on three parietal placentas; *style* single; *stigma* usually pointed and oblique, sometimes capitate and hollow (Pansy). The **fruit** is a unilocular capsule dehiscing longitudinally into three concave valves, which by their contraction squeeze out the seeds, sometimes with considerable force. The **seed** is albuminous; in the Pansy a small aril is developed at the hilum.

Floral formula:—$K_5 C_5 A_5 G_{(3)}$.

The flowers are specially adapted to pollination by bees, although some of the smaller flowered species are also visited by flies and other short-tongued insects. In those species with pointed or hooked stigmas (without a valve or shutter), the insect may touch the stigma on withdrawing and self-pollinate the flower. In most of the British species, however, except *V. tricolor*, the flowers frequently fail to produce seed, and cleistogamous flowers are developed later (see p. 197). In these flowers the calyx is closed and the petals are rudimentary.

The other genera belonging to the order are tropical or subtropical. Most of them are distinguished from the genus *Viola* by their regular or nearly regular corollas, and the absence of nectar-spurs on the stamens. Many of them are shrubs and trees.

§ 21. Malvaceæ.

Distinguishing characters:—Flowers polypetalous, hypogynous, usually pentamerous. Stamens ∞ owing to branching, monadelphous, bearing half-anthers. Fruit usually a schizocarp.

Plants representing this order are widely distributed in the warmer regions of the globe. The Mallow (*Malva sylvestris, M. rotundifolia*) and the Marsh-Mallow (*Althæa officinalis*) are the commoner British forms. *Althæa rosea* is the Hollyhock. *Abutilon* and *Hibiscus* are frequently cultivated in hothouses. *Gossypium* is the Cotton plant.

The plants are herbs or shrubs with alternate, stipulate, multicostate leaves. The **flowers** (fig. 276) may be solitary or in cymose inflorescences. They are regular, hermaphrodite, hypogynous, and usually protandrous. **Calyx** usually gamosepalous and five-fid; valvate. An epicalyx is usually present, consisting of three or more leaves representing bracteoles and their stipules (see p. 168). **Corolla** regular, polypetalous, usually of five petals, which are adherent to the base of the staminal tube, twisted in aestivation. **Stamens** ∞ and monadelphous, bearing half-anthers. They are derived by the copious branching of five antipetalous

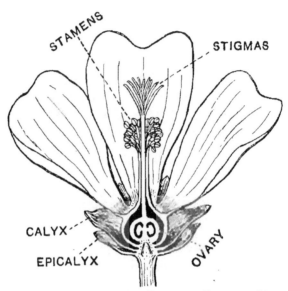

Fig. 276.—Vertical Section of Flower of Species of Mallow.

stamens; five antisepalous stamens, which have been lost, are still sometimes represented by staminodes (*Hibiscus*). The anthers have transverse dehiscence. **Gynæceum** polycarpellary (5—∞), syncarpous (occasionally almost apocarpous); *ovary* superior, multilocular; placentation axile; ovules 1—∞ in each loculus; styles united; stigmas free. **Fruit** (fig. 277) usually a carcerulus (see p. 211), splitting into as many mericarps as there are carpels, or a many-seeded capsule (*Hibiscus, Gossypium*). **Seed** with scanty endosperm, sometimes with hairs on the testa, *e.g.* in *Gossypium*, where the cotton consists of the hairs.

Fig. 277.—Carcerulus of Mallow.

Floral formula :— $K_{(5)} \overparen{C_5 A_\infty} G_{(\infty)}$.

In the different species of Mallow the column of stamens forms a convenient landing stage for a variety of insects, chiefly bees. Honey is secreted by the receptacle in five little pits lying between the bases of the petals and protected by hairs. Some species, however, e.g. *M. rotundifolia*, which is nearly homogamous, are often self-pollinated by the curling over of the stigmas.

§ 22. *Tilia* (the Lime), a genus placed, along with many tropical genera, in the order Tiliaceæ, is closely allied to the Malvaceæ. *Tilia Europæa*, the Common Lime, is the only British species.

The branching is sympodial (see p. 65). The leaves are asymmetrical (oblique), alternate and stipulate. The **inflorescence** is a small dichasial cyme (fig. 278). It arises in the axil of a

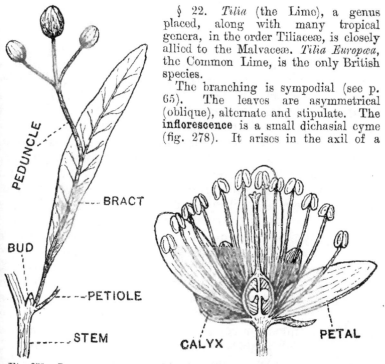

Fig. 278.—Inflorescence of Lime.

Fig. 279.—Vertical Section of Flower of Lime.

foliage leaf, and its peduncle is adherent (adnate) to a large lanceolate bract. This **bract** is developed from one of the scales of the bud which produces the inflorescence; in the axil of the opposite scale a resting bud is present which will develop in the following spring. The bract protects the flowers and later serves for dispersal of the fruits.

The **flower** (fig. 279) is pentamerous. The petals are imbricate, not twisted as in Malvaceæ. There are numerous stamens arranged in five antipetalous bundles. The anthers are normal. There are five carpels, and each cavity of the compound ovary has two anatropous ovules, but only one loculus, and as a rule only one ovule, develops. The **fruit** is a nut. The seed is albuminous. The flowers are protandrous, strongly scented and visited by numerous insects, chiefly bees. The **nectaries** are at the base of the sepals, and the honey is freely exposed.

§ 23. Geraniaceæ.

Distinguishing characters:—Flowers polypetalous, hypogynous, pentamerous, usually regular; stamens typically ten; fruit schizocarpic.

To this order, which is widely distributed, especially in temperate regions, belong the various species of Crane's-bill (*Geranium*) and Stork's-bill (*Erodium*), *Geranium pratense* is the Meadow Crane's-bill; *G. sylvaticum*, the Wood Geranium; *G. Robertianum*, Herb Robert. *Pelargonium* is a Cape genus to which our cultivated geraniums belong.

They are annual or perennial herbs with swollen, usually jointed nodes, and often covered with glandular hairs, The leaves are opposite or, occasionally, alternate, simple, palmately veined, deeply incised and stipulate. The **flowers** (fig. 280) may be solitary, but are usually in cymes, few or many flowered (see p. 190). They are usually regular and actinomorphic, pentamerous, and hypogynous. **Calyx** polysepalous, of five sepals, imbricate, inferior, persistent. **Corolla** of five regular, unguiculate petals, twisted in aestivation. **Stamens** typically ten, in two series, five long and opposite the sepals, five short and opposite the petals, obdiplostemonous, the bases of the filaments expanded and slightly connate. In *Erodium* there are only five, the five antipetalous ones being represented by scaly staminodes. In a few species of *Geranium* also (e.g. *G. pusillum*) these stamens have no anthers. **Gynæceum** syncarpous, of five carpels fused round a prolongation of the axis (carpophore); *ovary* superior, with five loculi; *ovules* one or two in each loculus, anatropous, pendulous; *styles* united to the carpophore; *stigmas* free

Fruit a schizocarp (see p. 210, fig. 144, D). The five carpels with their long persistent styles (*awns*) separate from the carpophore. In most species of *Geranium* the one-seeded portions (cocci) are dehiscent and roll up with

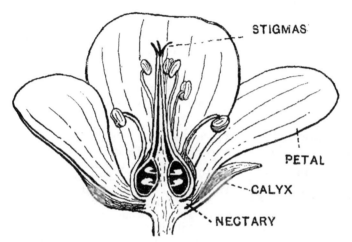

Fig. 280.—VERTICAL SECTION OF FLOWER OF SPECIES OF *Geranium*.

some force so that the seeds are shot out (explosive fruit). In *Erodium* the parts are indehiscent; the sharp-pointed awns are hygroscopic, curling up in a corkscrew fashion in dry weather, straightening out in damp weather. This tends to bury them in the soil. The **seed** has little or no endosperm.

Typical formula:—$K_5 C_5 A_{5+5} G_{(5)}$.

In *Geranium* and *Erodium* five nectar glands representing a disc are found as little cushions just outside the bases of the antisepalous stamens. The perennial species of *Geranium* (*G. sanguineum, G. sylvaticum, G. pratense, G. pyrenaicum*) are protandrous and entomophilous, but many of the annual species are only slightly protandrous (*G. Molle, G. Robertianum*), or protogynous (*G. pusillum*) and are frequently self-pollinated.

In *Pelargonium* the flower is zygomorphic. There are no glands at the base of the stamens, but the pedicel contains a glandular cavity which represents an adherent spur of the posterior sepal.

§ 24. The orders to which the Indian Cress or Garden Nasturtium (*Tropæolum*), the Wood Sorrel (*Oxalis acetosella*), and the Balsams (*Impatiens*, e.g. *I. noli-me-tangere*, Touch-me-not) belong, are sometimes regarded as sub-orders of Geraniaceæ, but are best kept distinct. The following notes indicate the chief points of difference:—

Tropæolum:—the leaves are exstipulate, the laminae peltate, and the petioles in some species function as tendrils. The flower is zygomorphic. The posterior sepal has a spur at the bottom of which honey is secreted (cf. *Pelargonium*). The three petals on the anterior side are unguiculate and fringed with hair-like processes. There are only eight stamens, the median one of each whorl being suppressed (fig. 281). The schizocarpic fruit (of three carpels) has no carpophore. There is no disc.

Fig. 281.—FLORAL DIAGRAM OF *Tropæolum*.

Oxalis acetosella has a monopodial rhizome (see p. 69), and radical trifoliate leaves. The flowers are solitary and closely resemble those of *Geranium*. The chief difference is in the gynæceum and fruit. The styles are free and the fruit is a loculicidal capsule with two seeds in each loculus. The seed is albuminous, and has a fleshy aril, which by its sudden inversion shoots out the seed (see p. 217). Cleistogamous flowers, developed underground, also occur. The leaves of *Oxalis* show sleep-movements (p. 153) and some species (*e.g. O. gracilis*) exhibit trimorphic heterostyly.

Balsams:—the flowers are very irregular. The calyx is petaloid. The posterior sepal is large and spurred; the two anterior ones may be minute or absent. The lateral petals on each side are fused. The stamens are syngenesious. The fruit is a five-valved explosive capsule with many seeds (see p. 217). Cleistogamous flowers are said to occur in *I. noli-me-tangere*.

§ 25. **Euphorbiaceæ**.

Distinguishing characters:—*Flowers hypogynous, unisexual, usually much reduced, occasionally dichlamydeous, but usually mono- or a-chlamydeous; stamens 1—∞; gynæceum usually of three (or two) carpels, syncarpous; fruit a schizocarp, often explosive.*

Nearly all the plants belonging to this large and interesting order have laticiferous "cells" or vessels.

Most of them live in warm climates, including the Castor Oil plant (*Ricinus communis*), and many species of *Euphorbia* and *Croton*. Shrubs and trees are common amongst these exotic forms, and many of them (*e.g.* species of *Euphorbia*) show very curious vegetative characters in adaptation to xerophytic conditions. In these xerophytic forms the leaves are frequently absent or reduced to spines. In such a case the stems take on the functions of assimilation and may form phylloclades (which, being vertically placed, transpire less than leaves), or may become swollen and more or less cactus-like and thus serve for the storage of water.

The only forms indigenous to Britain are a few small herbaceous species of *Euphorbia* (Spurge) and *Mercurialis*. *E. helioscopia*, the Sun-Spurge, is common on waste ground; *M. perennis*, the Dog's Mercury, in woods.

The **inflorescences** may be racemose or cymose and are often complex. The **cyathium** is the characteristic inflorescence in *Euphorbia* (figs. 130, 282). The **flowers** are unisexual, monœcious or diœcious, and often much reduced; in *Euphorbia*, for example, the male flower consists of a single stamen. Occasionally **calyx** and **corolla** are both present and are hypogynous (*e.g.* species of *Croton*), but frequently the corolla is wanting (fig. 283) and sometimes also the calyx. **Stamens** few or many, sometimes one (*Euphorbia*). The **gynæceum** is the part which shows the most constant characters. It is usually tricarpellary (or bicarpellary) and syncarpous; the *ovary* is trilocular (or bilocular) and superior: there are one or two pendulous, anatropous ovules in each loculus. The **fruit** is a schizocarp, breaking, often with violence, into dehiscent portions (cocci). The **seed** is albuminous and frequently has an aril developed from the micropyle (*e.g.* Castor Oil seed, fig. 138).

Fig. 282. Diagram of Cyathium of *Euphorbia*.

The British species of *Euphorbia* are either annuals (*E. helioscopia*) or perennials with rhizomes. Some of them grow on sandy shores (e.g. *E. Paralias*), and have rather thick coriaceous leaves. The cup-shaped or tubular involucre of the cyathium (figs. 130, 282) consists of five bracts with intervening glandular scales representing their fused stipules. Inside the involucre, the stamens (male flowers) are arranged in five groups, each group representing a scorpioid cyme, the oldest stamens towards the centre. Each stamen is articulated to a slender pedicel, at the base of which there is a scaly bract. In the centre of the involucre there is a single female flower borne on a stalk. It consists of the typical tricarpellary gynæceum with a single ovule in each loculus. The fruit is explosive.

Mercurialis perennis sends up numerous shoots with opposite leaves from a much branched rhizome. It is diœcious, the male plants being usually much more numerous than the female. The **flowers** (fig. 283) are borne in small clusters on axillary spikes, the female flowers few in number. Each male flower consists of a calyx of three small green sepals and from ten to twelve stamens. There is no corolla. The female flower has a similar calyx, and a bicarpellary pistil with large curved stigmas, and one pendulous ovule in each loculus. A few fine hairs are present representing rudimentary stamens. The fruit is similar to that of the spurge. The flowers are anemophilous.

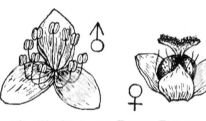

Fig. 283.—MALE AND FEMALE FLOWERS OF *MERCURIALIS PERENNIS*.

Ricinus communis is monœcious. Its flowers have a calyx, but no corolla, and the stamens are numerous and much branched. Castor oil is obtained from the seeds. The seeds of *Croton Tiglium* yield croton oil, and from *C. Cascarilla* cascarilla bark is obtained.

The Euphorbiaceæ were formerly placed in the Incompletae, but the occurrence of dichlamydeous forms, taken along with the extensive reduction of floral structure which is characteristic of the order, shows that they really belong to the Polypetalæ. The characters of the fruit indicate a relationship with Geraniaceæ.

§ 26. Crassulaceæ.

Distinguishing characters :—Flowers polypetalous, perigynous or subhypogynous, hermaphrodite, actinomorphic; usually pentamerous (or tetramerous), but the number of parts varies within wide limits; stamens usually twice as many as the petals; pistil apocarpous; fruit an etærio of follicles. *An order of succulent xerophytes.*

Common plants belonging to this order are the various species of Stonecrop (*Sedum*), the House-leek (*Sempervivum tectorum*), and the Pennywort (*Cotyledon Umbilicus*). Most of them are xerophytes growing on rocks, walls, housetops, or sometimes on sandy places near the sea, and have crowded fleshy leaves with waxy epidermis. Some are annuals or biennials, but the majority perennate by means of rhizomes. Propagation by offshoots is common (Fig. 41, p. 68).

The **inflorescences** are cymose (scorpioid), and the bracts are sometimes adherent to the axillary axes. The **flowers** (fig. 284), are regular, actinomorphic, hermaphrodite, subhypogynous or perigynous. *Sedum Rhodiola*, the Rose-root, is diœcious. The perigynous condition is not strongly marked. **Calyx** –poly- or gamosepalous, usually of five (or four) sepals, but occasionally of less or more, persistent, inferior. **Corolla** regular, polypetalous (exceptionally gamopetalous as in *Cotyledon* where it is tubular, fig. 284, B); petals as many as the sepals, imbricate in aestivation. **Stamens** usually in two whorls and twice as many as the petals, perigynous or subhypogynous. **Gynæceum** polycarpellary, apocarpous; carpels as many as the petals, sometimes slightly coherent; ovules ∞, anatropous, with marginal placentation. **Fruit** an etærio of follicles. **Seed** albuminous, but the endosperm small in amount.

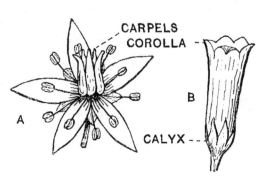

Fig. 284.
A, Flower of species of Sedum; B, Flower of *Cotyledon Umbilicus*.

Floral formula :—$K_n\ C_n\ A_{n+n}\ G_n$
where n is most frequently 5.

In the House-leek the parts are usually in sixes, and frequently the stamens are transformed into carpellary structures bearing rudimentary ovules.

In the majority of Crassulaceæ small nectariferous scales occur just outside the bases of the carpels. The flowers are usually protandrous, and as the honey is easily reached they are visited by many kinds of insects.

The Crassulaceæ are closely allied to Saxifragaceæ.

§ 27. Saxifragaceæ.

Distinguishing characters ;—Flowers polypetalous; subhypogynous, perigynous, or epigynous; pentamerous (or tetramerous). Stamens 10 (or 8). Carpels 2-5, frequently 2, connate, but often more or less free above; ovary unilocular or multilocular, superior, or more or less distinctly inferior. Placentation axile or parietal. Fruit a capsule or berry.

This order, which is closely allied to Rosaceæ and Crassulaceæ, has representatives in nearly all parts of the globe and includes plants of very diverse habit and appearance They are chiefly herbs or shrubs with radical or alternate occasionally opposite, exstipulate leaves. In a few exoti genera trees are found, and some of them have stipulat leaves. The order has been subdivided into a number c tribes or sub-orders, which some botanists, however, regar as distinct orders. Of these the most interesting is that c the Saxifrageæ, which consists largely of plants growin on rocks and mountains, often at high altitudes. The are only four British genera—*Saxifraga* and *Chrysospleniu* (sub-order Saxifrageæ), *Parnassia* (sub-order Parnassieæ and *Ribes* (sub-order Ribesieæ). Amongst exotic gene the better known are *Hydrangea, Deutzia, Escallonia* ar *Philadelphus;* the last named is commonly cultivated shrubberies under the name of *Syringa* or Mock Orange.

The **inflorescence** may be racemose or cymose; som times the flowers are solitary. The **flowers** (fig. 285) a hermaphrodite, and are interesting as showing various stag in the transition from hypogyny to perigyny and epigyn

The **calyx** consists of 5 (or 4) sepals and may be polysepalous or gamosepalous, superior or inferior. **Corolla** regular, polypetalous, usually of 5 (or 4) petals; aestivation imbricate. Occasionally the corolla is wanting (*Chrysosplenium*, fig. 285, B). **Androecium** usually of 10 (or 8) free stamens, obdiplostemonous; subhypogynous, perigynous or epigynous. **Gynæceum** of 2-5 carpels, frequently 2, connate, but usually more or less free above; *ovary* unilocular or multilocular, superior, or more or less distinctly inferior; *ovules* ∞, anatropous; *placentation* parietal or axile; *styles* free. **Fruit** a capsule or berry. **Seeds** albuminous.

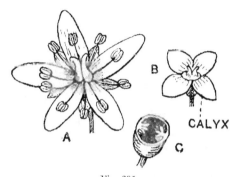

Fig. 285.
A, Flower of a Saxifrage. B, Flower of *Chrysosplenium*. C, Ovary

Saxifraga (Saxifrage):—most species of Saxifrage are alpine plants and show xerophytic adaptations. In some there are chalk glands

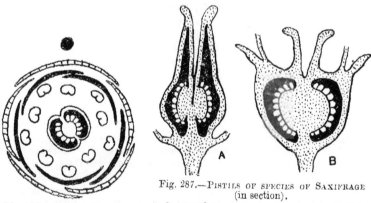

Fig. 286.—Floral Diagram of *Saxifraga*.

Fig. 287.—Pistils of species of Saxifrage (in section).
A, *S. granulata*, with superior ovary; B, *S. tridactylites*, with inferior ovary.

(p. 52) on the leaves, leaving in dry conditions a deposit of carbonate of lime which closes the pores. The flowers (fig. 285, A, 286) are usually in cymes, sometimes racemose or panicled. The pistil is bicarpellary, the two carpels usually diverging from each other above, and either free from the thalamus or more or less sunk in it (fig. 287). The ovary is bilocular; the placentation axile. Fruit a capsule more or less divided above. Honey is secreted round the ovary and is accessible to short-tongued insects of various kinds (flies, etc.). The flowers are usually protandrous. Typical alpine species are *S. oppositifolia* and *S. stellaris*. *S. granulata* (Meadow Saxifrage) and *S. tridactylites* are found in the low grounds. The former produces little pink or brownish bulbils in the axils of the lowest leaves. The latter grows on walls. *S. umbrosa* is London Pride.

Chrysosplenium oppositifolium is the Golden Saxifrage. It has small clustered cymes of tetramerous, apetalous flowers (fig. 285, B, C). The ovary is unilocular with two parietal placentas, and is deeply sunk in the thalamus. It is slightly protogynous and is visited by small insects. *C. alternifolium* is very similar, but has alternate, not opposite, leaves. Its flowers are homogamous. Both species grow in wet or marshy places, often together.

Parnassia palustris, the Grass-of-Parnassus, grows in bogs. It has radical leaves and solitary flowers. The flowers (fig. 288) are pentamerous, but the five stamens opposite the petals are transformed into

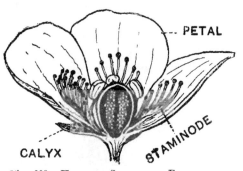

Fig. 288.—VERTICAL SECTION OF FLOWER OF
Parnassia palustris.

branched staminodes, the branches ending in glistening knobs which often deceive flies. The honey is secreted on these staminodes at their bases. The flowers are markedly protandrous and the stamens dehisce in succession above the ovary. To obtain the honey insects must reach the middle of the flower, either from above or by making their way over the barricade of staminodes. The unilocular ovary is slightly sunk in the thalamus and has four parietal placentas. Fruit a loculicidal capsule.

In *Ribes* (fig. 289), the ovary is inferior and has two (occasionally three) parietal placentas. There are only five stamens. The fruit is a berry (see Gooseberry, p. 213). *R. rubrum* is the Red Currant; *R. nigrum*, the Black Currant. In both species the flowers, which are in racemes, are homogamous and often self-pollinated. *R. Grossularia*, the Gooseberry, has spines (p. 124). The flowers are borne 1 or 2 together. They are protandrous and, owing to the elongation of the thalamus above the ovary, are adapted for pollination by small bees. *R. alpinum* is dioecious. *R. sanguineum* is the Flowering Currant of our shrubberies.

In the cultivated *Hydrangea* the flowers are in massive umbellate cymes; all the flowers are neuter and have large petaloid sepals. In *Philadelphus* (*Syringa*) the flowers have an inferior 4-locular ovary and numerous stamens.

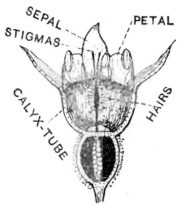

Fig. 289.—VERTICAL SECTION OF FLOWER OF GOOSEBERRY.
(In *Ribes nigrum* and *Ribes rubrum* the calyx-tube is not so deep).

§ 28. Onagraceæ.

Distinguishing characters:—flowers polypetalous, epigynous, usually regular, tetramerous or dimerous. Stamens 8, sometimes 4 or 2. Ovary multilocular. Placentation axile. Fruit usually a capsule or berry.

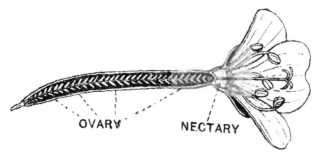

Fig. 290.—VERTICAL SECTION OF SPECIES OF *Epilobium*.

This order is confined for the most part to temperate regions, and is chiefly represented by herbaceous perennials. A few forms only are shrubby. Common British plants are the Willow-herbs (*Epilobium*), the

Evening Primrose (*Œnothera biennis*), the Enchanter's Nightshade (*Circæa*). The Fuchsia is a familiar exotic.

The leaves are alternate or opposite and exstipulate. The **flowers** (figs. 290-293) may be solitary, or borne in

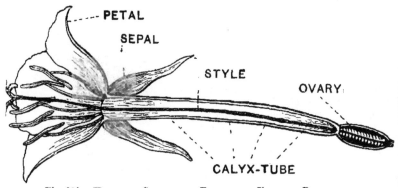

Fig. 291.—Vertical Section of Flower of Evening Primrose.

racemes, spikes, etc. They are regular or sometimes slightly zygomorphic, hermaphrodite, tetramerous, or (in *Circæa*) dimerous, epigynous. There is a honey disc on top of the ovary and the thalamus is frequently prolonged above the ovary into a tube bearing the sepals, petals, and stamens— markedly so in *Œnothera biennis* (fig. 291), which is pollinated in the evening by moths. **Calyx** of 2 (*Circæa*) or 4 sepals, polysepalous, superior, sometimes petaloid (*Fuchsia*); aestivation valvate. **Corolla** usually regular, polypetalous, of 2 (*Circæa*) or 4 petals; rarely absent as in *Ludwigia**; aestivation twisted. **Androecium**

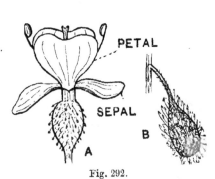

Fig. 292.
A, Flower; B, Fruit of Enchanter's Nightshade.

---

* *Ludwigia palustris* is the only British species. It grows in bogs in Sussex and Hampshire, but is very rare.

APPENDIX. 423

usually of 8 free stamens, sometimes 4 (*Ludwigia*) or 2 (*Circæa*). **Gynæceum** polycarpellary, syncarpous, usually of 4 carpels, or 2 (*Circæa*); *ovary* inferior, 4- or 2- locular; *ovules* one in each loculus (*Circæa*) or ∞, anatropous; *placentation* axile. **Fruit** a loculicidal capsule (*Epilobium* and *Œnothera*), a

Fig. 293.—FLORAL DIAGRAMS OF ONAGRACEÆ.
A, *Epilobium*; B, *Circæa*.

berry (*Fuchsia*) or a nut (*Circæa*, fig. 292, B). **Seed** exalbuminous, sometimes with a hairy aril (p. 204).

Floral formulae: $-Epilobium-\text{K}_4\text{C}_4\text{A}_{4+1}\text{G}_{\overline{(4)}}$,

$Circæa-\text{K}_2\text{C}_2\text{A}_2\text{C}_{\overline{(2)}}$.

The Willow-herbs (*Epilobium*) are plants of diverse habit. *E. angustifolium* is found in copses; *E. hirsutum* and *parviflorum* grow by the side of ditches; *E. palustre*, in bogs; some, e.g. *E. alpinum*, which is an alpine species, ascend to considerable heights. The flowers are adapted to pollination by bees and butterflies and are usually protandrous, but some of them, e.g. *E. hirsutum*, *E. parviflorum*, and *E. montanum* are nearly homogamous, and are often self-pollinated by the bending back of the lobes of the stigma.

In the Enchanter's Nightshade (*Circæa*) the fruit, a nut, has hooked bristles (fig. 292, B). *C. lutetiana* grows in damp woods; its fruit has two seeds. *C. alpina* is found in hilly districts; its fruit is one-seeded. In the process of pollination the visiting insect tucks the two stamens underneath its body and gets dusted with pollen. *C. alpina* may be self-pollinated by the stamens bending over to the stigmas.

(c) GAMOPETALÆ.

§ 29. Ericaceæ.

Distinguishing characters:—*Flowers gamopetalous, hypogynous, epigynous in Vaccinium; pentamerous or tetramerous; stamens usually twice as many as the sepals and petals, and not epipetalous; anthers often with appendages and opening by apical pores; pollen in tetrads; placentation axile. Shrubby plants growing usually on moors and hills.*

This is a widely distributed order consisting of woody shrubs, often of low growth, with alternate, opposite, or verticellate, simple, exstipulate leaves. Familiar examples are the whortleberry and cranberry, heaths, rhododendrons and azaleas. They are commonly alpine plants, or grow on moors and hills, generally in peaty soil. Many are evergreens, and most of them in adaptation to their environment are more or less markedly xerophytic. This is especially the case in the heaths where the leaves are narrowed and often rolled back, so as to cover over the under surface bearing the stomata, and thus prevent excessive evaporation. The roots of most Ericaceæ have mycorhizæ.

The **flowers** (figs. 294-297) are usually in racemes or racemose clusters. They are bracteate, hermaphrodite, regular and actinomorphic, or slightly zygomorphic (*Rhododendron*), hypogynous, or (in *Vaccinium*) epigynous. **Calyx** gamosepalous, 4- or 5- partite, persistent, inferior, or (in *Vaccinium*) superior. **Corolla** regular or slightly zygomorphic (*Rhododendron*), gamopetalous, 4- or 5-fid, usually globose, urceolate, or broadly campanulate, imbricate in aestivation, sometimes persistent (*Erica* and *Calluna*). **Stamens** eight or ten in number, rarely five (*Azalea*), obdiplostemonous,

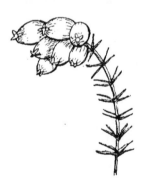

Fig. 294.—Sprig of *Erica Tetralix*.

hypogynous or (in *Vaccinium*) epigynous.* The anthers (figs. 295, 296) often have horn-like appendages (absent in *Rhododendron*), and open by apical slits or pores. The pollen is in tetrads (*i.e.* the four pollen grains formed in each mother cell remain united), and may be powdery (*Erica* and *Calluna*) or sticky (*Rhododendron*). **Gynæceum** of four or five carpels, syncarpous; *ovary* four-or five-celled, superior or (in *Vaccinium*) inferior; *ovules* one to ∞ in each loculus, anatropous; *placentation* axile; style simple; stigma capitate or four to five lobed. **Fruit** a septicidal or loculicidal capsule, or a berry. **Seed** albuminous.

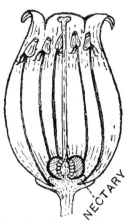

Fig. 295.—Vertical Section of Flower of *Erica*.

In the hypogynous forms there is a well-developed honey disc at the base of the ovary (fig. 295). The flowers are usually protandrous, and pollination is generally effected by bees, to whose visits the flowers, often pendulous, are adapted.

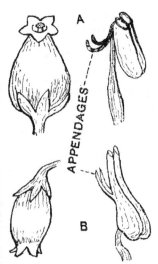

Fig. 296.

A, Flower and Stamen of *Arctostaphylos Uva-ursi*; B, Flower and Stamen of *Vaccinium uliginosum*.

Rhododendron (fig. 297) :—The flowers are pentamerous. They are placed more or less horizontally, and the stamens and style are bent slightly upwards so as to meet the body of the insect when it alights, the stigma being touched first. The anthers have no appendages. The fruit is a septicidal capsule. *Azalea* differs from *Rhododendron* chiefly in having five stamens.

Erica. The flowers (fig. 295) are tetramerous. The corolla is

* Ericaceæ and Campanulaceæ are the only gamopetalous orders in which the stamens are not epipetalous (see p. 391).

persistent. The anthers have appendages. The fruit is a loculicidal capsule. *E. Tetralix* (fig. 294) is the Cross-leaved Heath, *E. cinerea* is the Bell-Heather. These are the commonest British species. Bees visiting the pendulous flowers first touch the projecting stigma and then shake the anthers by pushing against their appendages.

Fig. 297.—FLORAL DIAGRAM OF RHODODENDRON.

Calluna vulgaris (sometimes called *Erica vulgaris*), the Ling or Heather, is very abundant and widely distributed. It resembles *Erica*, but it has a petaloid calyx and a small corolla cut almost to the base. It is often wind-pollinated as the pollen tetrads are dry and readily blown out of the anthers.

Andromeda (the wild Rosemary) is a plant with capsular fruit which grows in peat bogs. *Arbutus Unedo* (Strawberry Tree) with baccate fruit is found at Killarney. *Arctostaphylos Uva-ursi* (fig. 296, A) and *A. alpina* the Red and Black Bearberry, are alpine plants found in the Highlands. In all these the anthers have appendages and the ovary is superior.

Vaccinium is the only genus in which the ovary is inferior. For this reason it is sometimes placed by itself in an order Vacciniaceæ, but in all other essential respects it agrees with such forms as *Erica* and *Arctostaphylos*. The fruit is baccate. *V. Myrtillus* is the Whortle or Blae-berry; *V. Vitis-idæa*, the Cowberry; *V. Oxycoccos*, the Cranberry; *V. uliginosum*, the Bog Whortleberry (fig. 296B).

Two plants forming a small order (Pyrolaceæ) have also sometimes been included in the Ericaceæ—*Pyrola*, the Winter-green, and *Monotropa Hypopitys*, the Bird's Nest. The latter is a saprophyte with small yellow scaly leaves found in fir and beech woods. It obtains nourishment by means of a mycorhiza.

§ 30. Oleaceæ.

Distinguishing characters: Flowers gamopetalous, rarely apetalous, hypogynous, regular, usually 2- or 4- merous; stamens 2, epipetalous, inserted on the base of the corolla; pistil bicarpellary, syncarpous; ovary bilocular; seeds few.

This is an order of trees and shrubs widely distributed in temperate and tropical regions. They have opposite decussate, simple or compound, exstipulate leaves. The only native British representatives are the Ash (*Fraxinus excelsior*) and the Privet (*Ligustrum vulgare*). Other well-known forms are *Syringa vulgaris*, the Lilac, *Jasminum*, the Jasmine; *Olea Europæa* is the Olive.

The **inflorescence** may be racemose or cymose; it is frequently a racemose or thyrsoid cyme (*e.g.* Privet and Lilac). The **flowers** (figs. 298, 299) are regular, usually hermaphrodite (polygamous in the Ash), hypogynous. No honey disc is developed. The **calyx** is gamosepalous, usually 4-*fid* (5-10 in *Jasminum*), inferior; absent in common Ash; valvate in aestivation. **Corolla** regular, gamopetalous, usually tubular and 4-lobed (5-10 in *Jasminum*); absent in common Ash; valvate or (in *Jasminum*) imbricate in aestivation. **Androecium** of two free stamens usually placed in the lateral plane, epipetalous and inserted on the base of the corolla. **Gynæceum** bicarpellary, syncarpous; *ovary* bilocular, superior; *ovules* anatropous, one or two in each loculus, suspended (Ash, etc.) or ascending (Jasmine); *style*, single or absent; *stigma*, cleft. **Fruit** a loculicidal capsule (Lilac), a berry (Privet), a drupe (Olive) or a samara (Ash). **Seeds** 1-4, albuminous or (Jasmine) exalbuminous.

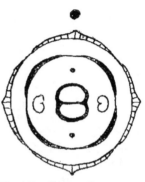

Fig. 298.—FLORAL DIAGRAM OF OLEACEÆ.

Common formula : $K_{(4)} \overgroup{C_{(4)} A_2} G_{(2)}$.

Ash. The inflorescence in the common Ash is racemose and clustered. The plant is polygamous. The flowers come out before the leaves; this, as in many other cases, is an adaptation to wind-pollination. The hermaphrodite flowers (fig. 299) are protogynous, and consist of two stamens inserted under the bicarpellary pistil, and alternating with the two carpels. Only one loculus and one seed come to maturity (see fig. 36, p. 59). It is interesting to notice that *Fraxinus Ornus*, the Manna Ash of S. Europe, has both calyx and corolla, the latter 4-partite.

Fig. 299. HERMAPHRODITE FLOWER OF ASH.

The Lilac is pollinated by bees and butterflies, and Jasmine in the evening by moths, to which the length of its tube is adapted.

Jasminum differs considerably in some of its characters from the other genera, as indicated above, and is sometimes placed in a distinct order.

§ 31. Boraginaceæ.

Distinguishing characters: Flowers gamopetalous, hypogynous, usually regular, pentamerous; stamens 5, epipetalous; structure of ovary and fruit. Mostly hairy herbs.

Common plants belonging to this order are the Borage (*Borago*), the Comfrey (*Symphytum*), Myosotis (*e.g. Myosotis palustris*, the Water Myosotis, or Forget-me-not, and *M. sylvatica*, the Wood Myosotis), the Viper's Bugloss (*Echium vulgare*), the Gromwell (*Lithospermum*), and the Hound's-tongue (*Cynoglossum*). *Pulmonaria*, the Lungwort, is a rare plant in Britain. They are herbs with very hairy, sometimes even bristly, stems and leaves. The stems are cylindrical and the leaves alternate, simple and often entire, exstipulate.

The **inflorescence** is apparently a scorpioid cyme, or a dichasium of scorpioid cymes; but some botanists believe that many of the so-called scorpioid cymes are really unilateral racemes. **Flowers** (figs. 300, 301) usually regular, sometimes slightly zygomorphic (*Echium*), pentamerous, hypogynous, occasionally heterostylic (e.g. *Pulmonaria* and species of *Myosotis*). Bracts, when present, may adhere to the axillary axes (cf. Solanaceæ, Crassulaceæ, Tilia, etc.) **Calyx** gamosepalous, 5-partite, inferior, persistent. **Corolla** usually regular (slightly zygomorphic in *Echium*), gamopetalous, 5-lobed: tubular, funnel-shaped or rotate; usually imbricate in aestivation. In some genera (*e.g. Borago, Symphytum, Cynoglossum*) the petals are ligulate, and the ligules,

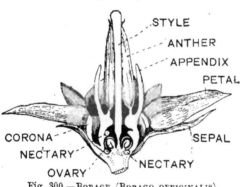

Fig. 300.—BORAGE (BORAGO OFFICINALIS). Vertical Section of Flower.

APPENDIX. 429

which are scaly, constitute a corona which closes in the throat of the corolla (fig. 300). **Stamens** 5, free, epipetalous. The anthers sometimes have appendages, *e.g.* in *Borago* (fig. 300). **Gynæceum and fruit** as in Labiatæ (q.v.) In some forms, chiefly tropical, the style is not gynobasic, but terminal, and the fruit is a drupe, *e.g.* Heliotrope. The seed is nearly exalbuminous.

Floral formula :—$K_{(5)} C_{(5)} A_5 G_{(2)}$.

Fig. 301.—FLORAL DIAGRAM OF BORAGINACEÆ.

The Viper's Bugloss is dimorphic. Some plants are smaller than the ordinary ones, and produce only female flowers. This is the gynodiœcious condition (cf. the gynomonœcious condition in Compositæ, p. 236).*

In correlation with the varying length of the corolla tube, the flowers of Boraginaceæ are adapted to a variety of insect visitors. The shorter-tubed forms, *e.g. Myosotis*, may be pollinated by flies in addition to bees, etc. The longer-tubed forms, especially those with pendulous flowers (*e.g.* Comfrey, Borage, etc.) are adapted to pollination by bees. The scales developed in the throat of the corolla keep off undesirable visitors, and cause the pollinating insect to keep to a definite path in seeking for the honey, which is secreted by a honey disc at the base of the ovary. In the Borage the insect touching the appendages of the stamens disturbs the anthers, and gets dusted with pollen.

The Boraginaceæ are closely related to Labiatæ, Convolvulaceæ, and Solanaceæ.

§ 32. Solanaceæ.

Distinguishing characters: Flowers gamopetalous, hypogynous, regular and pentamerous; stamens five, epipetalous, sometimes

---

* Gynodiœcism is found also in some Labiatæ, e.g. *Nepeta Glechoma*, the Ground Ivy, and species of *Plantago* (see under Plantaginaceæ).

syngenesious; pistil bicarpellary, syncarpous; a fruit capsule or berry.

This order is well represented in tropical countries, but only a few genera are found in Europe. It consists chiefly of herbs and shrubs with simple, more or less divided, exstipulate leaves, which are alternate in the vegetative region, but sometimes appear in pairs in the region of inflorescence. The following are some of the better-known members of the order:—*Solanum dulcamara*, the Nightshade or Bitter-sweet; *S. tuberosum*, the Potato; *S. Lycopersicum*, the Tomato; *Atropa Belladonna*, the Deadly Nightshade; *Hyoscyamus niger*, the Henbane; *Datura Stramonium*, the Thorn-apple; *Nicotiana Tabacum*, the Tobacco Plant; *Physalis Alkekengi*, the Winter Cherry; *Lycium barbarum*, the so-called "tea-plant" of English hedges and cottage gardens; and the various kinds of *Petunia*. *Mandragora officinalis*, the Mandrake, also belongs to this order. The Bitter-sweet, Henbane, and Deadly Night-shade grow wild in Britain. Many of these plants possess narcotic properties, and some are of medicinal importance, *e.g.*, *Atropa Belladonna*, from which the alkaloid atropin, the active principle in belladonna, is derived. Others have edible fruits (Tomato) or tubers (Potato).

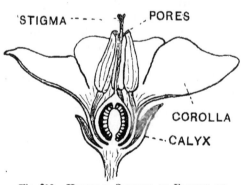

Fig. 302.—Vertical Section of Flower of *Solanum tuberosum*.

The **inflorescence** is usually a cyme, in which, sometimes, the bracts are adnate to the axillary axes (*e.g. Atropa*). The **flowers** (figs. 302, 303) are regular, or nearly regular, pentamerous and hermaphrodite. The **calyx** is gamosepalous, five-cleft, and persistent; in the Winter Cherry, for example, it forms the red bladder-like investment to the fruit. The **corolla** is usually rotate (Bitter-sweet), or campanulate (*Atropa Belladonna*). The **stamens** are five in number, epipetalous, and alternate with the lobes of the corolla. The anthers are sometimes connate (syngenesious, e.g. *Solanum*), and dehisce either by longitudinal slits (*Atropa Belladonna*) or by pores (*Solanum*). The **gynæceum** is bicarpellary and syncarpous; the *ovary* usually bilocular, but it sometimes becomes multi-locular owing to the formation of false septa (Datura). The two carpels are placed obliquely in the flower and not in the median plane (fig. 303). The **placentas** are axile, usually large and swollen, and bear numerous ovules. The *style* is single; the *stigma* simple or bilobed. The **fruit** is a capsule (*Hyoscyamus* and *Datura*) or berry (*Solanum* and *Atropa*). The **seed** is albuminous. The flowers are entomophilous. *Nicotiana* is pollinated in the evening by moths. *Solanum* is visited for pollen.

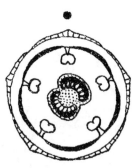

Fig. 303.—FLORAL DIAGRAM OF *Solanum*.

The Solanaceæ are closely allied to the Scrophulariaceæ. They are distinguished from the latter by their regular or nearly regular flowers, and by the oblique position of the carpels (a character, however, not easily recognised).

Floral formula :—$K_{(5)} \widehat{C_{(5)} A_5} G_{(2)}$.

§ 33. **Plantaginaceæ**.

Distinguishing characters: Flowers gamopetalous, hypogynous, regular and tetramerous, small and inconspicuous; stamens 4, with long filaments and versatile anthers; fruit a pyxidium (rarely a nut). Plants with ribbed and usually radical leaves.

This is a very small order though widely distributed in temperate regions. It contains only three genera, of which two are represented in Britain—*Plantago*, the Plantain, the only important genus, by five species, and *Littorella*, the Shore-weed, by one. The plants are annual or perennial herbs. The leaves are usually radical, without distinct petioles, exstipulate and strongly marked with parallel ribs. Some species of *Plantago* are xerophytic.

The **flowers** (figs. 304, 305) are borne in spikes on elongated axillary scapes. They are regular, isobilateral, usually hermaphrodite (unisexual in *Littorella*), hypogynous, small, inconspicuous, protogynous, and anemophilous. **Calyx** gamosepalous, deeply 4-partite, small, inferior, persistent. The sepals are diagonal. **Corolla** regular, gamopetalous, usually tubular and four-lobed (salver-shaped), thin and membranous, imbricate in aestivation, persistent. **Androecium** of 4 free stamens, epipetalous (hypogynous in *Littorella*) with long persistent filaments; anthers versatile with abundant pollen. **Gynæceum** bicarpellary, syncarpous; *ovary* superior, bilocular, sometimes 4-locular owing to formation of false septa (*Plantago Coronopus*) or unilocular (*Littorella*); *ovules* solitary and basal (*Littorella*) or, usually, two or more with axile placentation, anatropous or (in *Littorella*) campylotropous; *style* single; *stigma* simple or cleft. **Fruit** a pyxidium (*Plantago*) or nut (*Littorella*). **Seed** albuminous; in *Plantago* the testa is mucilaginous.

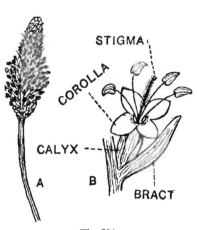

Fig. 304.
A, Spike; B, Flower of species of *Plantago*.

Plantago major is the Greater Plantain; *P. media*, the Hoary Plantain; *P. lanceolata*, the Ribwort Plantain. *Plantago Coronopus* and *P. maritima* (the Sea Plantain) are xerophytic forms with linear leaves. The former is found in sandy places usually near the sea. The occurrence of the latter at the sea-side, in salt marshes and on mountains is interesting as showing that these different habitats (sea-side and alpine) require somewhat similar xerophytic adaptation (see p. 157).

The protogynous condition is readily recognised in the spikes of *Plantago* (fig. 304, A). When the stamens of the lower flowers protrude after the withering of the stigmas, the stamens of the upper flowers are still in the bud and the stigmas ready for pollination. Some species may be visited by insects for pollen. *Plantago lanceolata* may be dimorphic, some plants having only pistillate flowers (the gynodioecious condition—see p. 429).

Littorella lacustris (Littorel or Shore-weed) is a small creeping aquatic growing on the margins or at the bottom of lakes and ponds. It has slender, radical, centric leaves with numerous cavities or spaces, and produces small spikes each of which bears two female flowers and one male. The flowers are anemophilous and the pistillate ones ripen before the male.

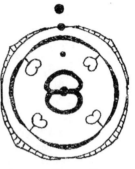

Fig. 305.—FLORAL DIAGRAM OF PLANTAGINACEÆ.

The Plantaginaceæ are regarded as an anomalous or degenerate order allied to Scrophulariaceæ. The tetramerous arrangement with diagonally placed sepals is generally supposed to have been derived from a pentamerous condition by suppression of the posterior sepal and the posterior stamen and the fusion of the two posterior petals (fig. 305, and cf. Speedwell, p. 234).

§ 34. Campanulaceæ.

Distinguishing characters:—Flowers gamopetalous, epigynous, pentamerous; stamens 5, usually not epipetalous; gynæceum of 2-5 carpels, syncarpous; placentation axile; fruit capsular and dehiscing by valves or pores. Mostly herbs with milky latex.

The plants referred to this order are usually herbs, often with milky juice, sometimes low shrubs. They are widely distributed in temperate regions. The leaves are simple, alternate, and exstipulate. Familiar examples are the Harebell (*Campanula rotundifolia*), *Lobelia*, and the Sheeps'-bit Scabious (*Jasione montana*), a plant common on hilly pasture ground.

The **inflorescence** may be racemose or cymose. In *Campanula* the apparent racemes are really racemose cymes as shown by the fact that there is a terminal flower which usually opens first. In *Jasione* and *Phyteuma*, the Rampion, the flowers are in capitula. **Flowers** (figs. 306, 307) regular, or zygomorphic and bilabiate (*Lobelia*), hermaphrodite, epigynous, markedly protandrous. In *Lobelia* the flowers are resupinate (see p. 240); the odd sepal is developed anteriorly (cf. Leguminosæ), but owing to the twisting of the pedicel it comes to occupy its usual posterior position. **Calyx** gamosepalous, 5-cleft, superior, persistent. **Corolla** of 5 petals, gamopetalous, regular (*Campanula*) or irregular (*Lobelia*); aestivation valvate. **Stamens** 5, free or syngenesious (*Lobelia*), usually epigynous* (epipetalous in *Lobelia*). The bases of the filaments are usually flattened and more or less triangular, occasionally united. **Gynæceum** of 2-5 carpels (often 3), syncarpous; *ovary* multilocular, inferior; *placentation* axile; *ovules* ∞, anatropous; *style* single; *stigmas* 2-5, reflexed when unfolded. **Fruit** (fig. 306, B) a capsule, dehiscing by pores at the base or apex, or by lateral valves. **Seeds** albuminous (fig. 306, C).

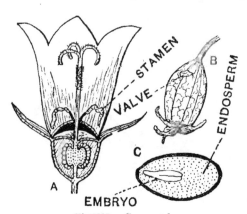

Fig. 306.—*Campanula*.
A, Vertical Section of Flower; B, Fruit; C, Longitudinal Section of Seed.

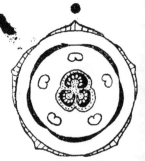

Fig. 307.—Floral Diagram of *Campanula*.

Floral formulæ:—*Campanula* : $K_{(5)} \ C_{(5)} \ A_5 \ G_{\overline{(3\text{-}5)}}$.

$\qquad$ *Jasione* : $K_{(5)} \ C_{(5)} \ A_{(5)} \ G_{\overline{(2\text{-}3)}}$.

$\qquad$ *Lobelia* : $K_{(5)} \ \overparen{C_{(5)} \ A_{(5)}} \ G_{\overline{(2)}}$.

* See footnote p. 425.

In *Campanula* (figs. 306, 307) the corolla is campanulate and 5-toothed, the gynæceum usually tricarpellary. The flowers are markedly protandrous, the stamens shrivelling up when the flowers open, except their triangular bases which persist and cover over the honey disc which is developed on top of the ovary. The pollen grains escape from the anthers when the flower is still in the bud condition and adhere to hairs on the style. The flowers are chiefly adapted to pollination by bees, but, failing cross-pollination, self-pollination may take place by the stigmas bending back and reaching the pollen on the style. The Canterbury Bell (*C. medium*) is a cultivated species with many varieties; it has five carpels.

In *Phyteuma* (the Rampion) the corolla is deeply cleft, but the tips of the petals are at first united and hold the stamens together so that the developing stigmas, still folded up, push through them and brush out the pollen, which, as in *Campanula*, adheres to the style. In *Jasione* also the stigmas act as a pollen brush, but here the anthers are slightly syngenesious. The same mechanism is recognised in a more perfect condition in *Lobelia* where the stamens are quite syngenesious (cf. Compositae). *Lobelia* is frequently placed in a separate order, Lobeliaceæ, on account of its irregular bilabiate flowers and syngenesious stamens; but this is scarcely justified as it is connected with the typical forms by *Jasione* and other genera and agrees with them in nearly all other important characters.

§ 35. Caprifoliaceæ.

Distinguishing characters:—Flowers gamopetalous, epigynous, regular or occasionally, zygomorphic, usually 5-merous, and hermaphrodite; stamens usually 5, epipetalous; gynæceum of 2-5 carpels, syncarpous; ovary multilocular; ovules pendulous or suspended; fruit baccate or drupaceous.

This is a small order of shrubs or, rarely, herbs, chiefly found in temperate regions. Common British plants are the Woodbine or Honeysuckle (*Lonicera Periclymenum* and other species), the Elder (*Sambucus nigra*), the Wayfaring Tree (*Viburnum lantana*), and the Guelder-rose (*Viburnum Opulus*). The leaves are opposite and decussate, usually simple and exstipulate; they are compound pinnate and stipulate in the Elder. Some species of *Lonicera* (*L. Periclymenum, L. Caprifolium*) are twiners. In *L. Caprifolium* the leaves are connate (p. 112).

The **inflorescence** is cymose, frequently a corymbose cyme (*e.g.* Elder, where the branching is multiparous), or cymose head (*e.g.* Guelder-rose and Honeysuckle). The

flowers (figs. 308, 309) are regular or occasionally zygomorphic, hermaphrodite, usually 5-merous, bracteate,

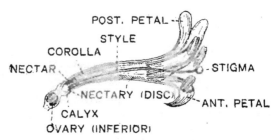

Fig. 308.—LONGITUDINAL SECTION OF FLOWER OF HONEYSUCKLE.

epigynous. **Calyx** gamosepalous, usually 5-fid or -toothed, superior. **Corolla** regular or zygomorphic, usually of 5 petals, gamopetalous, tubular, infundibuliform, campanulate, or rotate, sometimes more or less bilabiate (e.g. *Lonicera Caprifolium*); aestivation usually imbricate. **Androecium** usually of 5 free epipetalous stamens. **Gynæceum** of 2-5 carpels, syncarpous; *ovary* inferior, multilocular; *ovules* anatropous, $1-\infty$ in each loculus, usually pendulous or suspended; sometimes one or more of the loculi and ovules are abortive; *style* single, 3-5 cleft or absent; *stigma* capitate. **Fruit** a berry, or a drupe with 1-5 cavities and a cartilaginous endocarp. **Seed** albuminous.

Fig. 309.—FLORAL DIAGRAM OF *Lonicera*.

Floral formula :—$K_{(5)} \; \overline{C_{(5)} \; A_5 \; G_{\overline{(2-5)}}}$.

In *Lonicera* the inflorescence consists of small dichasia in which the middle flower is suppressed. There is a fleshy honey disc (fig. 308) at the base of the corolla tube, which is much longer here than in the other genera. The flowers of *L. Caprifolium* are protandrous and adapted to pollination by night-moths, though visited for pollen by other insects. Those of the other species have shorter tubes and are often visited by bees. The gynæceum consists of 2 or 3 carpels, and the fruit is a berry.

In *Viburnum* the ovary is trilocular with 3 pendulous ovules, but only one loculus with its ovule develops. The fruit is drupaceous, with one seed enclosed in a cartilaginous endocarp (it is often called a berry). In *V. Opulus* (Guelder-rose) the outer flowers of the inflorescence are neuter and have large corollas. This is the condition of all the flowers in the cultivated form. The flowers secrete honey, freely exposed.

In *Sambucus* the ovary has 3-5 loculi and the fruit is drupaceous (see p. 212). It is visited for pollen by various insects, but is often self-pollinated.

In *Linnæa borealis*, a small plant occasionally found in woods in hilly districts, the andrœcium consists of four stamens, and is didynamous, the posterior stamen being suppressed.

Symphoricarpus racemosus, the Snowberry, common in shrubberies, has a 4-5 celled ovary and a white berry. It is pollinated by wasps.

In *Adoxa moschatellina*, the Moschatel, a small graceful plant found in damp woods, the flowering shoot springs laterally from a rhizome with terminal growing point. It has a few radical leaves and a pair of compound leaves below the inflorescence which is cymose and consists of a cluster of 5 small greenish flowers. The terminal flower is tetramerous, the lateral ones pentamerous. The calyx is wanting, but in place of it there is an involucre of three leaves representing bract and bracteoles. The 4 or 5 stamens are branched so that there are really 8 or 10 stamens bearing half anthers. The fruit is 3-5 celled, and is drupaceous. The flowers secrete honey and have a musky smell. They are pollinated by flies. *Adoxa* is now usually placed in a separate order by itself (Adoxaceæ).

MONOCOTYLEDONS.

§ 36. Amaryllidaceæ.

Distinguishing characters :—Perianth petaloid; flowers epigynous; 6 stamens; inferior, trilocular ovary.

The plants of this order mostly perennate by means of bulbs. A few have rhizomes, e.g. *Agave*. They are found chiefly in hot, sunny, dry regions of the globe, their bulbs enabling them to tide over the dry, rainless season. They resemble Liliaceæ in most of their characters, but are readily distinguished by the inferior ovary. The only native British plants are the Snowdrop (*Galanthus nivalis*), the Snowflake (*Leucojum aestivum*), and two species of *Narcissus* (including *N. Pseudonarcissus*, the Daffodil).

Amongst the cultivated forms are numerous species and varieties of *Narcissus*, *Amaryllis*, and *Crinum*. *Narcissus Jonquilla* is the Jonquil. *Agave Americana* is the American Aloe or Century Plant*; it is a xerophyte with rhizome and massive rosettes of spiny leaves; after a storage period often of many years duration it produces a huge inflorescence and then dies down.

The **flowers** are produced on scapes, and are either solitary (Snowdrop, Daffodil), or two or more together in cymose inflorescences which are frequently umbellate (p. 190).

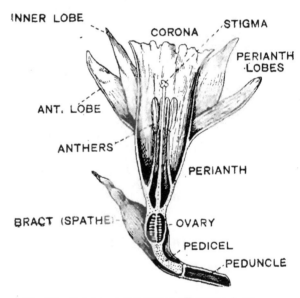

Fig. 310.—FLOWER OF DAFFODIL IN VERTICAL SECTION.

A spathe is present (fig. 310). The flowers are hermaphrodite, epigynous, usually regular and actinomorphic, occasionally zygomorphic. The **perianth** is superior and petaloid. It consists of six segments in two series (calyx and corolla), and may be polyphyllous (Snowdrop and Snowflake), or gamophyllous (Narcissus). In Narcissus (fig. 310) a cup-shaped or tubular corona is present (see

* The true aloes are liliaceous plants (see p. 239).

p. 170). **Stamens** 6, either epiphyllous (Narcissus) or epigynous (Snowdrop); anthers introrse or rarely extrorse. **Gynæceum** tricarpellary, syncarpous; *ovary* trilocular, inferior; *ovules* ∞, anatropous; *placentation* axile; *style* single; *stigma* simple or trilobed. **Fruit** a loculicidal capsule, occasionally a berry (*Agave*). **Seed** albuminous.

The Snowdrop produces only two foliage leaves and a lower sheathing leaf each year. The swollen bases of the foliage leaves form the fleshy scales of the new bulb. The flowering axis arises in the axil of the upper foliage leaf. The primary axis is monopodial, and continues to grow from year to year. A new bulb may arise as a bud in the axil of the sheathing leaf, sometimes also in the axil of the lower foliage leaf. The inner segments of the perianth (petals) differ from the outer (sepals). They are smaller and slightly cleft. Each has a green honey-secreting groove on its inner face. The anthers have pointed processes and closely surround the style. They dehisce by apical slits. A bee on entering the pendulous flower first touches the stigma and may effect cross-pollination; then it moves the anthers and gets dusted with pollen. If, however, cross-pollination is not effected self-pollination may take place. The anthers separate from each other and the pollen falls on the stigma, which, as the flowers are pendulous, lies below the anthers.

The Snowflake resembles the Snowdrop, but the leaves are more numerous, there are several flowers, and the inner and outer perianth segments are similar to each other.

In Narcissus honey is secreted by the base of the perianth-tube.

§ 37. Iridaceæ.

Distinguishing characters:—*Perianth petaloid; flowers epigynous; 3 stamens; inferior trilocular ovary.*

The Iridaceæ are largely represented in dry sunny countries (South Africa, etc.). Familiar plants are the Crocus, Iris or Flag, Gladiolus and Freesia. *Iris pseudacorus*, the Yellow Flag, *Crocus nudiflorous*, the Autumnal Crocus,[*] and *Gladiolus communis* are found in Britain, the last two probably naturalized. Most of them perennate by means of corms (*Crocus*, p. 70, fig. 44), or sympodial rhizomes (many species of *Iris*). The leaves are often equitant and isobilateral (fig. 85), e.g. *Iris*.

---

[*] Distinguish from *Colchicum autumnale*, the Meadow Saffron, which is also called the Autumn Crocus, see p. 238. In appearance the plants are very similar.

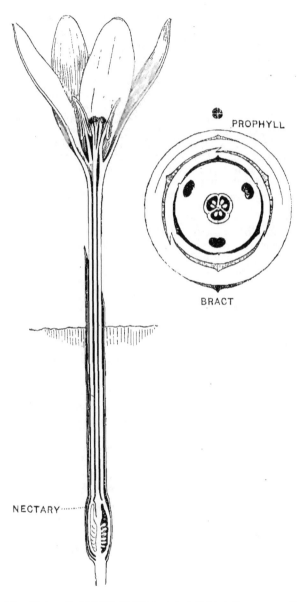

Fig. 311.—Vertical Section of Flower and Floral Diagram of Crocus.

The **inflorescences** are usually small cymes variously arranged. Thus in the Iris the flowering axis ends in a flower (which opens first), and has small lateral cymes each invested by a spathe. In *Gladiolus* and *Freesia* the lateral cymes are reduced to single flowers, each with a spathe (bract), so that the inflorescence is like a spike. In some species of *Crocus* the flowering axis ends in several flowers (cyme); but in the common species it is one-flowered (cf. Amaryllidaceæ). The **flowers** are hermaphrodite, regular (*Iris* and *Crocus*), or zygomorphic (*Gladiolus*, *Freesia*), epigynous. **Perianth** of six segments in two series, gamophyllous, petaloid, superior. **Andrœcium** of three epiphyllous stamens; they represent the outer whorl, the inner whorl being suppressed, and are situated between the carpels and the outer perianth segments. The anthers are extrose and

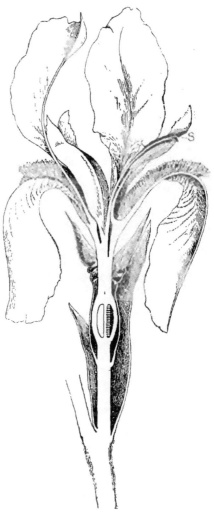

Fig. 312.—Vertical Section of Flower of Iris. (After Church.)

On the left are shown an outer perianth segment, a petaloid style entire, and an inner posterior perianth segment cut in half. On the right an outer anterior perianth segment and a style are cut in half showing a stamen lying between them; at the back is a lateral inner perianth segment (petal) s = stigma; bract and bracteoles are also shown.

lie on the outer side of the styles. **Gynæceum** tricarpellary, syncarpous; *ovary* inferior, trilocular, with ∞ anatropous ovules; *placentation* axile. The *styles* are united below, but free above, and sometimes expand into three large petaloid lobes (*Iris*). **Fruit** a loculicidal capsule. **Seed** albuminous.

Floral formula :—$P\overparen{_{(3+3)}}A_{3+0}\ G_{\overline{(3)}}$.

In the Crocus (fig. 311), honey is secreted by a nectary on top of the ovary (which is at first underground), and rises to the mouth of the long slender perianth tube. The flowers are protandrous, and are pollinated by bees or butterflies, which touch the stigmas before the anthers. Failing cross pollination, self-pollination may occur. In the Iris (see fig. 312), the anthers and pollen are protected by the petaloid styles. The stigmas are three thin membranes developed on the outer surfaces of the styles just above the anthers. Honey is secreted by the tissue of the basal portion of the perianth tube. A bee on entering the flower first pushes against the upper surface of the stigma which alone is receptive and then brushes the extrorse anthers. It is guided in many species by broad bands of hairs (known as the "beard") developed on the perianth segments.

INDEX.

(Generic names are given in italics.)

	PAGE
ABSCISS layer	128
Absorbed solutions, Course of	134
Abutilon	409
Acacia	227, 228
Accessory buds	63
Achene	206
Achenial fruits	205, 206
Achlamydeous	161
Aconitum	223
Acropetal succession	10
Actinomorphic	164
Acyclic flowers	162
Adaptation	5, 66, 96, 124, 156, 307, 342, 364
,, to aerial conditions	308, 350
Adaptive characters	307
Adelphous condition	172
Adhesion	182
Adnate	174, 411, 430
Adoxa	437
Adventitious buds	62
,, roots	95, 107
Aerobic	372
Æstivation	122, 170
Æthusa	230
Affinities of Fungi	349
Agamogenetic reproduction	15
Agaricus	4, 368
,, Development of	371
Agave	437, 438
Aggregate fruits	205, 214
Air-bladders	343
,, -chambers	314
Albumen	59
Albuminous seed	59, 201
Alburnum	87
Alchemilla	230
Alcoholic fermentation	358, 366
Alder	221, 392, 394
Aleurone grains	35, 145
,, layer	60
Algæ	4, 330
,, and higher plants	347
,, Gametophyte	330
Alkaloids	31, 146
Allogamy	193
Almond	212, 229
Alnus (see Alder)	392

	PAGE
Aloe	239, 438
Alopecurus	243
Alpine Plants	157, 420, 424
Alsinoideæ	225
Alternation of generations	265, 304
,, parts	163
Althæa	409
Amaryllidaceæ	221, 437
Amaryllis	438
Amentaceæ	221, 399
Amentiferæ	221, 399
Amentum	187
Amides	30, 142
Amphithecium	318, 327
Amyloplast	29
Anabolism	13, 14
Anaerobic	372
Analogy	15, 307
Anatomy	2
Andrœcium	158, 171
,, Descriptive terms	172
Andromeda	426
Androphore	164, 225
Anemone	69, 161, 222, 224
Anemophilous	194, 195
Angiosperm and Gymnosperm	289, 388
Angiosperms	4, 219
,, Classification of	218
,, Embryo of	55, 199, 202, 203
,, Evolution in	388
,, Fruit of	201, 203
,, Gametophyte of	302
,, Leaf of	109
,, Life-history of	192
,, Reproduction of	192
,, Root of	95
,, Seed of	55, 201, 203, 304
,, Sporangia and spores of	299
,, Sporophyte of	299
,, Stem of	61
Angle of divergence	114
Animal and Plant	11, 12, 13, 14
Annual rings	86, 150
Annuals	66, 197
Annular thickening	23
Annulus	257, 325

INDEX.

	PAGE
Anterior	160
Anther	171
,, -lobes	171, 300
Antheridial cell	292
Antheridiophore	312, 315
Antheridium	261, 277, 315, 320, 323, 345
,, Development of	262, 316, 324
Anthophore	164
Anthoxanthum	245
Antipetalous	163
Antipodal cells	180, 302
Antirrhinum	233
Apetalæ	220
Apheliotropism	153
Apical cell	254, 322, 323, 344
,, meristem	40, 78, 92, 98, 254, 282, 285
Apium	230
Aplanogametes	333
Apocarpous	176
Apogamy	267, 364
Apophysis of *Funaria*	325
,, ,, *Pinus*	289
Apospory	267
Apostrophe	153
Appendages (*see* Outgrowths)	8
Apple	213, 229
Apposition	26
Apricot	229
Aquilegia	222, 223
Araceæ	221
Arbutus	426
Archegoniophore	312, 316
Archegonium	261, 276, 295, 316, 320, 324
,, , Development of	262, 317, 324
Archesporium	258, 269, 274, 291, 300, 301, 318, 327
Archicarp	356
Arctostaphylos	426
Aril	204, 238, 409
Arrangement of leaves (*see* Phyllotaxis).	
"Artillery Plant"	401
Artocarpus	402
Arum	187, 195, 221
Ascocarp	357, 363
Ascogenous branches	356
Ascogonium	356
Ascomycetes	358, 363
Ascospore	356, 362, 366
Ascus	356, 362, 366
Asexual reproduction	15, 266, 331, 335, 339, 351, 355, 363, 374
Ash	58, 207, 426, 427
Ash-constituents	130
Asparagus	239
Aspen	237
Aspidium (*see* Fern)	248
,, , Leaf of	248, 249
,, , Rhizome of	248, 250

	PAGE
Aspidium, Root of	248, 249
Assimilation	11, 12
Assimilatory tissue	314, 323
Asymmetrical	164
Atriplex	403
Atropa	430
Atropin	430
Autogamy	193
Autumn wood	86
Avena	243, 244
Awn	243, 413
Axil of leaf	62
Azalea	424, 425
Azygospore	334
BACCATE fruits	205, 212
Bacillus	374
Bacillus subtilis	375
Bacteria	371, 372
Bacterium-cell, Structure of	373
Bacteroids	151, 373
Balm	232
Balsam	414
Bamboo	242
Banana	47, 213
Baneberry	224
Banyan	402
Barberry	124, 155
Bark	88, 106, 283
Barley	207, 244
Basal wall	264, 326
Basidiospore	364
Basidium	364
Basifixed	174
Bast (*see* Phloem)	45, 53
Bean	57, 227
Bearberry	426
Beech	207, 221, 392, 398
Beet	403
Belladonna	430
Bellis (*see* Daisy)	236
Berry	212
Beta	403
Betula (*see* Birch)	392
Betulaceæ	221, 399
Bi-collateral bundles	83
Biennials	66
Bifacial leaf	10, 51, 126, 156
Bilateral symmetry	9
Bindweed	407
Birch	221, 392, 394
Bisexual	162
Bittersweet	430
Blackberry	67, 214, 219, 229
Blaeberry	426
Blue-bell	239
Boraginaceæ	220, 428
Borago	427
Bordered pits	25
Botany	1
Bracken (*see Pteris*).	

INDEX. 445

	PAGE
Bract	110, 160
,, -scale	289
Bracteole	161
Branches, Development of	93
Branching	10, 64, 95, 115, 185
Brassica	226
Brazil-nut	35, 203
Bread-fruit	402
Brome-grass	245
Bromus	245
Broom	227
Bryophyta	4, 312
Buckwheat	405
Buds	62
Bulb	71, 238
Bulbil	68, 238, 407, 420
Bundle-sheath	43, 54, 76, 251
Burrs	217
Butcher's Broom	238
Buttercup	158, 222, 223
CABBAGE	226
Cæsalpineæ	228
Calceolaria	234
Calcium carbonate	27
,, oxalate	27, 143
Calluna	424
Callus	45
Caltha	224
Calyciflorœ	220
Calyptra	318, 326
Calyptrogen	100
Calyx	158, 161
,, -tube	165
,, , Descriptive terms	167
,, , Functions of	166
Cambial cells	19
,, , Division of	84
Cambiform cells	76, 91
Cambium	40, 76, 83, 88, 103
,, -ring	84
Campanula	433, 435
Campanulaceæ	47, 197, 221, 433
Campion	225
Cannabineæ	402
Cannabis	402
Canterbury Bell	435
Cap-cells	301
Capitulum	187, 236
Caprifoliaceæ	221, 435
Capsella	226
Capsular fruits	205, 207
Capsule of sporogonium	318, 321, 325
Capsules	209
Caraway	210, 230
Carbohydrates	19, 137
Carbon-assimilation	12, 137, 145
,, , Conditions of	139
Carcerulus	211, 232, 410
Carnation	225
Carnivorous Plants	125, 151
Carpel	158, 289

	PAGE
Carpinus (see Hornbeam)	392
Carpophore	210, 412
Carrot	230
Carum	230
Caruncle	204
Caryophyllaceæ	220, 224
Caryopsis	207, 244
Castanea (see Chestnut)	392
Castor-oil seed	35, 204, 415
Cataphyll	110, 282
Catch-fly	225
Catkin	187, 238, 392
Caudicle	241
Cauliflower	226
Cauline bundles	83, 93
Caulis	61
Celandine	47, 68, 222, 223
Celery	230
Cell, The	5, 11, 17, 19
,, -contents	28, 33
,, -division	36
,, -formation	36
,, -fusion	31
,, -sap	28, 33, 133
,, -wall	11, 19
,, ,, , Alteration of	26
,, ,, , Growth of	22
,, ,, , Thickening of	23
Cells, Changes in contents of	28
,, , Forms of	23
Cellular plants	18
,, structure	17
Cellulose	13, 19, 59, 145
Censor mechanism	216
Central cylinder	80
Centric leaf	9, 127
Centrifugal	185
Centripetal	185
Centrosome	21
Centrosphere	19, 21, 37
"Century Plant"	438
Chalaza	180
Chalazogams	199, 399
Chalk-glands	52
Cheiranthus	226
Chelidonium	47
Chemiotaxis	198, 263
Chenopodiaceæ	221, 403
Chenopodium	403
Cherry	212, 229
Chervil	230
Chestnut	207, 392, 398
,, , Horse	215
Chickweed	225
Chicory	235
Chlorophyceæ	330
Chlorophyll	5, 29, 139
,, -bands	332
,, -corpuscles	29
,, , Function of	12, 141
Chloroplasts	5, 29
,, , Functions of	30, 34

INDEX.

	PAGE
Chlorotic condition	141
Chorisis	182, 406
Christmas Rose	224
Chromatin network	21
Chromatophores	30, 332, 334, 346
Chromoplast	30
Chromosome	37
Chrysosplenium	418, 420
Cichorium	235
Cilia	39, 263
Circæa	422
Circinate	249
Circulation of protoplasm	148
Cladode	72, 74
Cladophyll	74
Cladoptosis	282
Class	219
Classification	2, 3, 218
Claviceps	358
,, Gametophyte of	363
Cleistocarp	363
Cleistogamous flowers	197, 407, 409, 414
Clematis	67, 161, 222, 224
Climbing plants	66
Clinostat	154
Closed bundles	92
Clover	227
Cocci	211, 413, 415
Coccus (micrococcus)	374
Coco-nut	212
Cœnocyte	39, 46, 339
Cœnocytic structure	339, 348, 355, 369
Cohesion	182
Cohort	219
Colchicum	70, 238
Collateral bundles	76
Collenchyma	42
Colouring matters	30, 33, 146
Colt's-foot	236
Columbine	222, 223
Columella	325
Comfrey	428
Comma	374
Common bundles	78, 92
Companion cells	76, 91
Compositæ	47, 161, 197, 221, 234, 391
Composite fruits	205, 215
Compound leaf	116, 120
Concentric bundles	83
Conceptacle	343, 345
Cone	288, 293
Conidiophore	355, 360, 368
Conidium	350, 351, 355, 360, 370
Conifer	280
Conjoint bundles	53
Conjugation	37, 39, 333, 337
,, -tube	333
Conjunctive ground-tissue	80, 100
Connective	171
Conopodium	230
Contact	155
Continuity of protoplasm	17

	PAGE
Contractile vacuole	336
Convallaria	239
Convolvulaceæ	220
Cork	43, 88, 94
,, -cambium	88, 105
,, -formation in leaf	126
,, -formation in wounds	94
,, -layer of leaf	128
Corm	70, 238
Cormophyte	7
Corncockle	225
Corolla	158, 161
,, Descriptive terms	168
,, Functions of	168
,, Origin of	310, 389, 391
Corona	170, 429, 438
Corpusculum	295
Cortex	54, 75, 82
Corylaceæ	221, 399
Corylus (see Hazel)	392
Corymb	186
Costa	230
Cotton-plant	409
Cotyledon	417
Cotyledons	55, 57, 110, 202, 264, 296, 298
Cover-scale	289
Cowberry	426
Cowslip	231
Cow-wheat	233
Cranberry	424, 426
Crassulaceæ	220, 416
Cratægus	229
Creeping stem	70
Cremocarp	210, 230
Cress	226, 414
Crinum	438
Crocus	70, 439, 442
Cross-fertilization	193
,, -pollination	193, 194
Croton	415
Cruciferæ	220, 226
Cryptogam	4
Culm (= Haulm)	62
Cupule	207, 392
,, of Marchantia	313
Cupuliferæ	221, 392, 394
Currant	421
Cuticle	26, 44
Cuticular absorption theory	137
Cuticularization	26
Cuticularized tissue	43
Cutin	26
Cyanophyceæ	372
Cyanophyll	140
Cyathium	191
Cycad	280
Cyclamen	231
Cyclic flowers	162
Cymes, Forms of	187, 189, 190
Cymose branching	11, 64, 95, 185
,, inflorescence	185, 188
Cynarrhodium	214

INDEX. 447

	PAGE
Cynoglossum	428
Cypripedium	241
Cypsela	206
Cystolith	27, 400, 403
Cytisus	227
Cytoplasm	19, 21
Cytoplastin	21
DAFFODIL	437
Daisy	188
"Damping-off"	350
Dandelion	47, 66, 235
Date	213
Datura	430
Daucus	230
Dead-nettle	232
Dédoublement	406
Deferred shoot	62
Definite branching	11, 64, 185
,, nucleus	180
Dehiscence of anther	174
,, ,, capsules	209
Delphinium	223
Dermatogen	79, 82, 99
Deutzia	418
Development	2
Diaheliotropism	154
Diandræ	241
Dianthus	225
Dichasium	188
Dichlamydeous	161
Dichogamy	194
Dichotomous branching	10, 344
Diclinous	162
Dicotyledon	4, 57, 219, 220, 222, 389
,, , Apical merstem	78
,, , Embryo	55
,, , Exceptional structure	83
,, , Leaf	115
,, , Root	95, 102
,, , Secondary growth	83, 103
,, , Seed	55
,, , Structure of stem	75, 82
Didynamous stamens	173
Differentiation	5, 18
,, of sex	278, 347
,, of tissue	78, 92
,, , Processes of	22
Digestive sac	107
Digitalis	234
Dimorphism	196
Diœcious	162
Disc-florets	237
Disciflorae	220
Dispersal of seeds and fruits	216
Displacement	182
Dissimilar members	8, 10, 16
Distribution	193, 216
Division	219
,, of labour	6
Dock	36, 404

	PAGE
Dog's Mercury	415
Dormant buds	62
Dorsifixed	174
Dorsiventral	10
Double Fertilization	304
Double Samara	211
Dracæna	93, 238
"Drawn plants"	140
Drupaceous fruits	205, 212
Drupe	212
Duplication	405, 406
Duramen	87
Dwarf-shoot	66, 282
ECHIUM	428
Egg-apparatus	180
Elaborated compounds	142
Elaboration of food-material	134
Elaters	319
Elder	85, 212, 435
"Elements"	18
Elm	207, 221
Embryo (see under Types)	2, 22, 55
,, -sac	180, 291, 299, 301
Embryonal cell	199, 200, 296
,, mass	200, 297
Emergence	52
Enchanter's Nightshade	422, 423
Endive	118, 235
Endocarp	212
Endodermis	43, 54, 76, 82, 90, 92, 100, 125, 252, 287
Endogenous development	106
Endopleura	57, 201
Endosmosis	132
Endosperm	59, 200, 297
Endosporium	257
Endothecium	318, 327
Energid	18
Energy	12, 13, 147
Entomophilous	194, 195
Environment	15, 144, 152
Epibasal cell	200, 264, 277, 318, 326
Epiblema	101
Epicalyx	168, 228, 409
Epicarp	212
Epicotyl	57
Epidermal cells	49
,, outgrowths	52
,, system	49, 82
Epidermis	43, 49
Epigeal	57
Epigyny	165, 390
Epilobium	421
Epipetalous stamens	173
Epiphyte	157, 239
Epipodium	109
Epistrophe	153
Epithelial layer	60
,, ,, of resin passage	48
Epithem tissue	52, 128

INDEX.

	PAGE
Equisetum	268
,, , Gametophyte of	269
,, , Life-history of	268, 269
,, , Sporangia and spores of	269
,, , Sporophyte of	268
Erect stems	66
Ergot	358, 359, 361
Ergotin	362
Erica	424, 425
Ericaceæ	220, 424
Erodium	412
Erythrin	330
Escallonia	418
Essential elements of food	131
,, organs	161
Etærio	214
Ethereal oil	31, 36, 146
Etiolated plants	140
Etiolin	140
Euphorbia	47, 191, 415, 416
Euphorbiaceæ	220, 414
Euphrasia	233
Eurotium	4, 354
,, , Gametophyte of	358
,, , Sporophyte of	358
Evening Primrose	422
Evolution in Angiosperms	388
,, of Flower	389
,, ,, Fungi	349
,, ,, higher plants	307, 347
,, ,, Muscineæ	328
,, ,, sexuality	338, 346
,, theory	305
Exalbuminous seed	59, 201
Exceptional dicotyledonous structure	83
,, secondary growth	93, 107
Excretions	13, 31, 146
Exine	174
,, of *Fucus*	346
Exodermis	102
Exogenous development	93, 127
Exosmosis	133
Exosporium	257
External morphology	1, 5
Extra-stelar ground-tissue	80, 82, 91, 92, 100
Extrorse	174
Eyebright	233

FAGACEÆ	221, 399
Fagopyrum	405
Fagus (see Beech)	392
False fruit	205
,, tissue	348, 369
Fascicle	191
Fascicular cambium	76, 81
Fats	36
"Feeder"	278
Fermentation	366, 372

	PAGE
Ferments, Organized	367
,, , Unorganized	31, 56, 146
Fern	4, 248
,, , Apical meristem	254
,, , Embryo of	264
,, , Gametophyte of	265
,, , Leaf of	255
,, , Life-history of	266
,, , Root	253
,, , Sexual organs	261
,, , Sporangia and spores	255, 257
,, , Sporophyte	265
,, , Steles	205
(*See also Pteris* and *Aspidium*.)	
Fertilization	40, 198, 263, 277, 295
	318, 325, 341, 346
,, -tube	353
Fibres	43
,, of Pericycle	86
Fibrous drupe	212
,, layer	174, 300
Fibro-vascular bundle	53
Ficus	402
Fig	215, 402
Filament	171
Fission	374
Flag	439
Flagella	374, 375
Floral contrivances	194
,, diagram	182
,, envelopes	161
,, formulæ	184
,, leaves	110
,, mechanisms	194, 196, 232, 236
Flower	7, 158, 280, 299, 302
Flowering Plants	4
,, and V. Cryptogams	305
Foliage-leaf	110
,, ,, , Functions of	156
Foliar gap	250
Follicle	208
Food-materials of Plants	11, 130
Fool's Parsley	230
Foot	264, 318, 319, 321, 325
Forget-me-not	428
Formic aldehyde	137
Foxglove	234
Fragaria	229
Fragmentation	21
Fraxinus (see Ash)	426
Free cell-formation	37, 38
Free water	132
Freesia	439
French Bean	58, 227
Fruit	201, 205
,, of Gymnosperms	297
,, *Pinus*	297
,, -mechanisms	217
Fuchsia	422
Fucus	4, 341
,, , Embryo of	346
,, , Gametophyte of	344

INDEX.

	PAGE
Fucus, Sexual organs of	344
Funaria	4, 312, 320
,, , Capsule of	321, 325
,, , Gametophyte of	320
,, , Life-history of	320
,, , Spore of	327
,, , Sporogonium of	321, 325, 326
,, . Sporophyte of	321
,, , Stem of	322
Fungi	4, 348, 349
,, , Affinities of	349
,, , Evolution of	349
,, , Life-history of	349
,, , Nutrition of	349
Fungus-cellulose	348
Funicle	179

GALANTHUS	437
Gametangia	330
Gametes	15, 39, 330
,, , Male	295, 353
,, , Origin of	338
Gametophyte (*see under* Types)	265, 308, 330
Gamogenetic reproduction	15
Gamopetalæ	220, 391, 424
Gamopetalous	168
Gamosepalous	166
Garden Nasturtium	414
Gemma	265, 313, 315, 322
Gemmation	365
Generative cell	197, 295
Genetic spiral	114
Genista	227
Genus	219, 306
Geotropism	154
Geraniaceæ	220, 412
Geranium	412
Germinal disc	320
Germination of pollen-grain	197
,, ,, seed	56, 157
,, ,, spores	259, 275, 294, 320, 328, 333
"Germs"	372
Germ-tube	320, 327, 359, 402
Gladiolus	439
Glands	48
Glandula	241
Glandular hairs	48
,, tissue	48
Glans	207
Glass-wort	403, 404
Glaux	231
Globe-flower	224
Globoid	35
Glomerule	400, 402
Glumes	243
Glumifloræ	220, 221
Gnetaceæ	280
Gongrosira	340
Gonidangium	336, 339, 352

	PAGE
Gonidiophore	365
Gonidium	335, 350, 365
Gonophore	164
Gooseberry	68, 421
Goosefoot	403
Gorse	227
Gossypium	409
Graminaceæ	221, 242
Grasses	242
Grass-of-Parnassus	420
Gravity	154
Gromwell	428
Ground Ivy	232, 429
Groundsel	197, 236
Ground-tissue	75, 80, 81, 82, 90, 100
,, system	49, 54, 82
Group	219
Growing-points	19, 22
,, , Properties of	148
Growth	11, 130, 144, 148
,, -forms	374
Guard-cells	49, 50, 136
Guelder Rose	435
Gymnosperms	4, 219, 280
,, and Angiosperms 289, 388
,, ,, V. Cryptogams	290
,, , Flowers of 280, 288, 289
,, , Fruits of	297
Gynæceum	158, 175
Gynandrous	173, 240
Gynobasic style	232, 429
Gynodiœcious	429
Gynomonœcious	236
Gynophore	164
Gynostemium	240

HAIRS	8, 48, 52, 118
Halophyte	157, 403
Hard bast	53, 76, 81
Harebell	47, 433
Haulm (*see* Culm).	
Haustoria	151, 349
Hawkweed	236
Hawthorn	229
Hay-bacillus	375
Hazel	207, 221, 392, 395
Heat	155
Heath	424
Heather	426
Helianthus	236
Helicoid branching	64
,, cyme	188
Heliotrope	429
Heliotropism	153
Helleborus	224
Help-cells	180
Hemicyclic flowers	162
Hemp	402
Henbane	430
Hepaticæ	312
Herb Robert	412

INDEX.

	PAGE
Heredity	194, 306
Hermaphrodite	162
Heteroblastic development	320
Heterogamous	331
Heterosporous	272
Heterostyly	196
Hibiscus	409
Hieracium	235
Hilum of ovule	180
,, ,, seed	57
,, ,, starch-grain	34
Histological differentiation	18, 22
Histology	2, 17
Holcus	245
Holly berry	212
Holoblastic	296
Homogamous	197
Homologies 268, 279, 289, 290, 307, 328
,, in Angiosperms	299, 302
Homology	15, 307
,, based on relationship	305
Homosporous	269
Honesty	226
Honey	195
,, , Concealment of	390, 391
"Honey-dew"	359, 360
Honeysuckle	435
Hop	67, 215, 402
Hordeum	244
Horehound	232
Hornbeam	392, 397
Horse-tail	4, 268
Hound's-tongue	428
House-leek	68, 417
Humulus	402
Hyacinth	239
Hydrangea	418, 421
Hydrophilous	194
Hydrotropism	155
Hygroscopic water	132
Hymenial layer (hymenium)	370
Hyoscyamus	430
Hyphæ	348, 349
Hypobasal cell	199, 264, 277, 318, 326
Hypocotyl	56, 108, 277
Hypodermis	54, 76, 82
Hypogeal	58
Hypogyny	164, 390
Hypophysis cell	201
Hypopodium	109
Hypsophyll	160

	PAGE
IDIOBLAST	48
Impatiens	414
Incision of lamina	119
Incompletæ	220, 221, 390, 392
Indefinite branching	10, 64, 185
Indian Cress	414
Indiarubber-plant	27, 49, 402
Indirect development	320, 322
Indusium	256

	PAGE
Indusium, False	258
Inferior calyx	167
,, ovary	179
Inflorescence	159, 185
,, , Compound	189
,, , Mixed	188
Innate	174
Inorganic compounds	131
Insect-visitation	195, 389
Insertion of anther	174
,, ,, floral leaves	164
,, ,, leaf	113
Integument	179, 301
Intercalary meristem	40
Intercellular spaces	32, 150
Interfascicular cambium	84
Internal morphology	2
Internode	61
Intine	174
,, of Fucus	346
Intra-stelar ground-tissue	80, 81, 82, 91, 100
Introrse	174
Intussusception	26
Inulin	33
Involucel	189
Involucre	188, 189
Iridaceæ	221, 439
Iris	439
Irregularity	163
Irritability	152
Isobilateral leaf	9, 51, 127
,, symmetry	9, 164
Isogamous	330
Ivy berry	215

	PAGE
JASIONE	433, 435
Jasminum	427
Jonquil	438
Juglandaceæ	399
Juniper	280

	PAGE
KARYOKINESIS	21, 37
Katabolism	13, 14, 143
Knotgrass	404

	PAGE
LABELLUM	240
Labiatæ	221, 232
Laburnum	227
Lactuca	235
Lady's Mantle	229
Lamellæ	369
Lamina	109
,, , Incision of	119
,, , Outline of	117
Lamium	232
Larkspur	223
Lateral branching	10
,, buds	62
,, rootlets	106, 286

INDEX. 451

	PAGE
Latex	46, 47
Lathræa	399
Laticiferous "cells"	39, 46
,, tissue	46
,, vessels	32, 46
Lavender	232
Leaf	7, 8, 57, 109
,, , Adaptation of	124
,, , Apex of	118, 128
,, , Branching of	115
,, , Definition of	129
,, , Descriptive terms	111, 117, 119
,, , Development of	127, 255
,, , Duration of	122
,, , Fall of	128
,, , Functions of	14
,, , Insertion of	113
,, , Margin of	118
,, , Structure of	125
,, , Texture of	122
,, , Venation of	115
Leaflets	116
Leaf-structure, Types of	110
Leaf-trace	78
Legume	208
Leguminosæ	220, 227
Lenticel	89, 105
Lepidium	226
Lettuce	235
Leucojum	437
Leucoplasts	29
,, , Functions of	29, 34
Lid-cells	317
Life-cycle	266
Light	139, 152
,, as source of energy	12
Lignification	27
Lignin	27
Ligule	52, 112, 270
Liguliflora	235
Ligustrum	426
Lilac	427
Liliaceæ	221, 238
Lilium	238
Lily of the Valley	239
Lime	87, 411
Linaria	234
Ling	426
Linnæa	437
Lipoxeny	364
Lithophyte	157
Lithospermum	428
Littorella	431, 433
Liverworts	4, 312
Lobelia	433, 435
Loculicidal	209
Lodicule	244
Lolium	243, 245, 359
Lomentum	210, 228
London Pride	420
Lonicera	435, 436
Lousewort	233

	PAGE
Ludwigia	422
Lunaria	226
Lungwort	428
Lychnis	225
Lycium	430
Lysigenous cavities	32, 48

MAIZE ... 60, 242
Male Shield Fern (see *Aspidium*).

Mallow	409
Malva	409
Malvaceæ	220, 409
Mandragora	430
Mandrake	430
Mangold Wurzel	404
Marchantia	4, 312
,, , Embryo of	318
,, , Sexual organs of	312
,, , Sporogonium of	313
,, , Sporophyte of	313, 319
Marsh Mallow	409
,, Marigold	224
,, Samphire	403, 404
Matthiola	226
Meadow Rue	224
,, Saffron	238
,, -sweet	229
Mechanical contrivances, Origin of	316
Medlar	214
Medulla	75, 81
Medullary rays	54, 76, 81, 82, 86, 87, 105, 283, 284
Megasporangium	271, 290, 299
Megaspore	271, 279, 290, 291, 299
Megasporophyll	290
Megazoospore	335
Melampyrum	233
Member	1, 8
Mentha	232
Mercurialis	415, 416
"Mercury"	404
Mericarp	210
Meristele	126, 287
Meristem	40, 92
Meristematic cells	19, 22
Meroblastic	296
Mesocarp	212
Mesophyll	126, 156
Mesopodium	109
Metabolism	13
Metallic elements	143
Microbe	372
Micrococcus	373
Micropyle of seed	57
,, ovule	179
Microsporangium	271, 290, 299
Microspore	272, 279, 290, 299
Microsporophyll	290, 299
Microzoospore	337

	PAGE
Middle lamella	24
Milfoil	236
Milk-tubes	46
Milkwort (Sea-)	231
Mimosa	228
Mimoseæ	228
Mineral crystals	36
,, matters of cell-wall	27
Mint	232
Mitosis (*see* Karyokinesis).	
Mock Orange	418
Modes of nutrition	150
Modifications	8
,, of floral structure	182, 389
Monandræ	241
Monkshood	222, 223
Monocarpellary	175
Monochlamydeæ	220, 221
Monochlamydeous	161
Monoclinous	162
Monocotyledons 4, 60, 219, 221, 238, 389	
,, , Apical meristem of	92
,, , Embryo of	60, 204
,, , Exceptional secondary growth in	93
,, , Leaf of	115
,, , Root of	96, 102
,, , Seed of	60, 204
,, , Stem of	90
Monœcious	162
Monopodial branching	11
Monostelic	80, 91, 99, 282
Monostely	252
Monosymmetrical	9
Monotropa	426
Moraceæ	402
Morphia	31
Morphological differentiation	6
Morphology	1
Morus	402
Moschatel	437
Mosses	4, 312
Moss-plant	322, 328
Mother-axis	160
Mucilage	27
Mulberry	215, 402
Mullein	234
Multicellular plants	5
Multicostate venation	115
Muscineæ	4, 312
,, and V. Cryptogams	328
Mushroom	368
Mycelium 348, 349, 351, 354, 359, 366, 368	
Mycorhiza	151, 399, 424
Myosotis	428
"NAKED" cells	19
Narcissus	437, 439
Natural orders	218, 219, 307
Natural selection	306
Neck-canal-cells	262, 317, 324

	PAGE
Nectar-cups	170
,, -glands	48, 196, 310
Needles	282
Nepeta	429
Nettle	400
Neuter flowers	162, 237
Nicotiana	430
Nicotin	31
Nigella	224
Nightshade	430
Nitrification	373
Nitrogenous substance	141
Node	61
Non-essential organs	161
Normal roots	95
Nucellus	179
Nuclear disc	38
,, membrane	21
,, spindle	38
Nucleo-hyaloplasm	21
Nucleolus	21, 38
Nucleus	5, 19, 20, 29
,, , Division of	21
Nutation	149
Nutrition	11, 130
,, of Algæ	330
,, ,, Fungi	349
,, , Modes of	150
Nyctitropic movements	153
OAK	207, 221, 392, 397
Oat	207, 243, 244
Obdiplostemonous	163
Ochrea	113, 405
Octants	200, 264, 318
Œnothera	422
Offset	67, 68
Oil	30, 36, 145
"Old Man's Beard"	222
Olea	427
Oleaceæ	220, 426
Olive	427
Onagraceæ	220, 421
Oögonium	341, 345
Oöphyte	265
Oösphere 40, 180, 261, 295, 301, 317, 324, 331, 340, 346, 352	
Oöspore 40, 199, 263, 295, 331, 341, 346, 353	
Open bundles	76
Operculum	325
Orache	403
Orange	213
Orchid	241
Orchidaceæ	221, 239
Orchis	241
Organ	3
Organic acids	30
Orobanchaceæ	399
Orthostichies	114
Osier	237

INDEX. 453

	PAGE
Osmosis	132
Ostiole	315, 345, 362
Outgrowths	8, 52, 172
Ovary	159, 175
Ovule	175, 290, 299
,, , Development of	291, 301
,, , Forms of	181
,, , Structure of	179, 290
Ovuliferous scale	289
Ovum (see Oösphere)	40, 261
Oxalis	414
Oxyria	404
PÆONIA	195, 224
Paleaceous	235
Paleæ	243
Palisade parenchyma	41, 126, 156
Palm	93, 221
Palmaceæ	221
Palmella-stage	336
Panicle	189
Pansy	112, 196, 204, 407
Papaveraceæ	47, 220
Papilionaceæ	227
Pappus	167
Parachute mechanism	216
Parallel venation	115
Paraphysis	258, 345, 370
Parasite	14, 151, 349
Parasitic root	98, 151
Paratonic influence (of light)	153
Parenchyma, Forms of	41
Parenchymatous cells	23
Parietal layer	28
Parietaria	400
Parnassia	418
Parnassieæ	418
Parthenogenesis	267, 338
Pea	227
Pear	213, 229
Pearl-wort	225
Pedate leaf	12)
Pedicel	158
Pedicularis	233
Peduncle	160
Pelargonium	190, 412, 413
Penicillium	358
Pennywort	417
Pentstemon	234
Perennation	68
Perennial	66
Perianth	161, 170, 299, 383
Periblem	79, 82, 99
Pericambium	100
,, , Functions of	107
Pericarp	55, 205
Perichætia	316, 321
Pericycle	54, 81, 82, 90, 92, 100, 125, 251, 285, 287
Pericycle fibres	86
,, , Functions of	107
Periderm	88

	PAGE
Perigonia	321
Perygynium	318
Perigyny	164, 390
Periplasm	352
Perisperm	202, 297
Peristome	326, 328
Perithecium	362
Permanent tissues	40, 41
Petal	158
Petaloideæ	220, 221
Petiole	109
,, , Structure of	125
Phæophyceæ	330
Phanerogams	4, 219
Phelloderm	88
Phellogen	87, 105, 283
Philadelphus	418, 421
Phleum	243
Phloem	45, 53
,, -parenchyma	53
,, -sheath	252
,, , Primary	76, 81
,, , Secondary	81, 84, 104
Photosynthesis (see Carbon-Assimilation)	12, 137
Phycophæin	330
Phylloclade	74, 238, 415
Phyllode	124, 228
Phyllopodium	109
Phylloptosis	128
Phyllotaxis	113
,, , Floral	162
Physalis	430
Physiological differentiation	6
Physiology	1, 3, 5
Phyteuma	433, 435
Picea	282, 293, 296
Pigment spot	336
Pig-nut	230
Pilea	401
Pileus	369
Piliferous layer	101
Pimpernel	231
Pine-apple	215
Pink	225
Pinnule	121
Pinus	4, 280
,, , Apical meristem	283, 285
,, , Cones of	288, 289
,, , Embryo of	295
,, , Flowers of	287, 288, 289
,, , Fruit of	297
,, , Gametophyte of	292, 293
,, , Leaf of	286
,, , Life-history of	298
,, , Ovule of	290
,, , Pollination in	291
,, , Seed of	297
,, , Sporangia and spores of	290
,, , Sporophyte of	290
,, , Structure of root of	285
,, , ,, ,, stem of	282

INDEX.

Pistil 158, 175
Pistillate 162
Pitchers 124
Pith 54, 75, 81, 82, 100
Pits 24
,, , Bordered 25
Pitted thickening 24
Placenta 176, 256, 289
Placentation 176
,, , Forms of ... 176, 177, 178
Planogamete 337
Plant and Animal 11, 12, 13, 14
Plantaginaceæ 221, 431
Plantago 195, 429, 431
Plantain 431
Plants without chlorophyll 14
Plastic substances 13, 30, 145
Plastids 5, 19, 22, 29
Plerome 79, 82, 99
Pleurococcus 5
Plum 229
Plumule 56, 57, 296
Poa 245
Pod 208
Polar nuclei 302
Pollards 62
Pollen 174
,, -grain 171, 279, 288, 299
,, -sac 171, 288, 299
,, ,, , Development of ... 291, 299
,, -tetrads 425
,, -tube 198
,, , Protection of 390
Pollination 193, 292, 304, 309
Pollinium 174, 240
Pollinodium 352, 356
Polyandrous 172
Polycarpellary 175, 176
Polyembryony 296
Polygamous 162
Polygonaceæ 220, 404
Polygonatum 238
Polygonum 404
Polymerization 138
Polymorphism 358, 363, 374
Polypetalæ 220, 390, 404
Polypetalous 168
Polysepalous 166
Polystelic 83, 100, 250
Polystely 252
Pome 205, 213
Poplar 221, 238
Poppy 47
Populus 237
Pores of *Marchantia* 314
Porogams 199, 399
Posterior 160
Potato 72, 430
Poterium 230
Prefloration 122, 170
Prefoliation 122
Pressure 149

Prickles 8, 52
Primary endosperm 294
,, meristem 40, 81
,, root 264, 297
,, stem 264
Primordial cell 19
,, utricle 29
Primula 231
Primulaceæ 220, 231
Privet 426
Procambial strand 80, 99
Pro-embryo 320
Promycelium 353
Prosenchyma 42
Prosenchymatous cells 23
Protandry 195
Proteid 29, 142
,, crystalloids 35
,, grains 30, 35
Prothallus (prothallium) 260, 269, 270, 275
,, , Female... 270, 276, 295, 302
,, , Male 270, 277, 292, 302
,, , Reduction of 276, 279, 304
Protogyny 195
Protonema 320, 321, 328
Protophloem 81
Protoplasm 5, 11, 12, 20
,, , Continuity of 17
Protoplasmic strand 29
Protoplast 18
Protoxylem 76, 81, 100
Prunus 229
Pseudobulb 239
Pseudocarp 205
Pteridophyta 4
Pteris 248
,, , Rhizome of 249
,, , Steles of 253
Pyxis 122, 170, 249
Pullulation 365
Pulmonaria 428
Pulvinus 110, 155
Putrefaction 373
Pyrenoid 332, 334
Pyrola 426
Pyrolaceæ 426
Pyrus 229
Pythium 4, 350
,, , Sporophyte of 353
Pyxidium 209, 231

Q *UERCUS* (see Oak) 392
 Quince 214

R ACEME 185
 Racemose branching 10, 64, 65, 185
,, inflorescence 185
Radial symmetry 9
,, vascular bundle 102

INDEX.

Radicle 55, 57
Ragged Robin 225
Ramenta 249
Rampion 433, 435
Ranunculaceæ 161, 220, 222
Ranunculus 223
Raphe 181
Raphides 36
Raspberry 214, 219, 229
Ray-florets 237
Receptacle of capitulum 188
„ „ flower 158
„ „ Marchantia 313
Receptive spot 341
Reduced stems 66
Regma 211
Regularity 163
Rejuvenescence 37, 39, 336
Relation to environment 15
Relationship amongst plants 305, 347
Replum 209
Reproduction by seed 193
„ , Methods of 15, 266
Reproductive shoot 7, 259, 312
Resin 31, 36, 146
„ -passage 33, 48, 283
Respiration 13, 144
Resupinate 240, 434
Reticulate thickening 24
„ venation 114
Rheum 405
Rhinanthus 233
Rhizogenic cell 254
Rhizoid 260, 312, 320, 328
Rhizome 68, 248, 249, 250
Rhizophore 271
Rhododendron 424, 425
Rhodophyceæ 330
Rhubarb 405
Ribes 418, 420
Ribesieæ 418
Ricinus 415, 416
Robinia 227, 228
Rock-plants 157
Root 7, 8, 57, 95, 156
„ -absorption 131, 132
„ -cap 95, 99
„ -cell 334
„ -hairs 52, 98, 133, 260, 312
„ -pressure 135, 137
„ -process 338, 342
„ -stock 68
„ -tuber 97, 222
„ , Adaptation of 96
„ , Branching of 95
„ , Development of 105, 254
„ , Forms of 95, 96
„ , Functions of 14, 96
„ , Structure of 99, 253
Rosaceæ 220, 228
Rose 229
Rosemary 232

Rose-root 417
Rostellum 241
Rotation of protoplasm 148
Rubus 219, 229
Rumex 404
Runner 67
Ruscus 74, 238
Rye-grass 243, 244

SACCHAROMYCES 4, 364
Sacs 48
Sago 172, 232
Salicaceæ 221, 237, 399
Salicornia 403, 404
Salix 237
Salsola 403
Salts 131
„ , Insoluble 133
Salt-wort 403
Salvia 232
Samara 207
Sambucus 435, 437
Sand-wort 226
Saprophyte 14, 151, 239, 349
Saxifraga 418, 419
Saxifragaceæ 220, 418
Saxifrageæ 418
Scabious 433
Scalariform thickening 26
Scale-leaves 63, 110
Scape 160
Schizocarpic fruits 205, 210
Schizogenous formation 33
Schizomycetes 371
Schizophyta 372
Schizostele 126
Schulze's solution 20
Scilla 239
Sclerenchyma 43
Sclerenchymatous fibres 43
Sclerotic cells 43
Sclerotium 359, 360
Scorpioid branching 64
„ cyme 188
Scots Fir 4, 281
Scrophularia 234
Scrophularineæ 221, 233
Scutellum 60
Sea-blite 403
Secale 359
Secondary cortex 88
„ growth 83, 93, 103, 283
„ meristem 40
„ nucleus 180, 200, 302
„ roots 95
„ tissue 83
Secretion 13, 30, 146
„ -reservoir 48
Sections 285
Sedum 417
Seed 55, 201, 304

	PAGE
Seed-coat	55, 57
Selaginella	4, 268, 270
,, , Embryo of	277
,, , Gametophyte of	275, 279
,, , Sexual organs of	275
,, , Sporangia and spores of	271, 274
,, , Sporophyte of	270
Self-fertilization	193
,, -pollination	193, 197
Seminiferous scale	289
Semi-parasites	233
Sempervivum	417
Senecio	236
Sensitive Plant	155, 228
Sepal	158
Septa, true and false	179
Septicidal	209
Septifragal	209
Series	219
Seta	321, 325
Sexual organs	261, 269, 275, 312, 330, 344, 356
,, reproduction	15, 193, 261, 266, 330, 332, 337, 340, 344, 349, 352, 356
Sexuality, Origin of	338, 347
Shepherd's Purse	226
Shoot	7, 57
Shore-plants	157
Shore-weed	431, 433
Sieve-plate	45
,, -tissue	45
,, -tube	45, 143
Silene	225
Silenoideæ	225
Silica	27
Silicula	209, 226
Siliqua	208, 226
Similar members	8, 10, 16
Simple fruits	205
,, leaves	116
Sleep-movements	228, 414
Smilax	124, 238
Snap-dragon	233
Snowberry	437
Snowdrop	437, 439
Snowflake	437, 439
Soft bast	53, 76, 81
Solanaceæ	220, 429
Solanum	430
Solomon's Seal	69, 238
Sonchus	235
Sorosis	215
Sorrel	404
Sorus	256, 258
Sow-thistle	235
Spadicifloræ	220, 221
Spadix	186
Spathe	186, 190
Special mother-cells	259, 279, 291, 300
Specialization	6
Specialized characters	310

	PAGE
Species	219, 306
Speedwell	234
Spermaphyta	4
Spermatocytes	261, 316, 323
Spermatozoids	261, 263, 276, 309, 316, 324, 340, 346
,, in Gymnosperms	309
Sphacelia	359
Sphæraphides	36
Spike	186
Spikelet	243
Spinach	404
Spine	73, 113, 124
Spiræa	229
Spiral thickening	23
Spirillum	374
Spirogyra	4, 331
,, , Embryo of	334
,, , Gametophyte of	334
Spongy parenchyma	41, 126, 156
Sporangiferous spike	269
Sporangium	256, 257, 269, 271, 29), 299, 352
,, , Development of	258, 274
Spore	15, 256, 257, 266, 269, 271, 290, 299, 319, 327, 335, 374
,, -mother-cells	258, 274, 291, 300, 319
,, -reproduction	15, 266, 365, 374
,, -sac	325
Sporocarp	357, 363
Sporogonium	313, 318, 321, 325, 326
Sporophyll	259, 268, 290, 299
Sporophyte	265, 308
Spring wood	86
Spruce (see Picea)	282, 293, 296
Spurge	47, 415
Spurs	196
,, , Bifoliar	282
Stalk-cell	295
Stamen	158, 171, 288, 299
Stamens, Explosive	401
,, , Irritable	155
Staminate	162
Staminode	172
Starch	30, 138
,, -formation	145
,, -grain	34
,, -layer	76, 126
Stele	80, 82, 83, 102, 250
Stem	7, 8, 61, 155
,, , Adaptation of	66
,, , Descriptive terms for	61
,, , Forms of	65
,, , Functions of	65
,, , General characters of	74
Stereid bundle	43, 53
Stexigma	3c5, 360, 370
Stigma	159, 175
Stimulus	144, 152, 155
Stipe	369
Stipule	110, 112

INDEX. 457

	PAGE
Stitchwort	225
Stock	226
Stolon	67, 68
Stomata	50, 136, 315, 325
Stomatal cells	49, 50
Stomium	259
Stone-cells	43
Stonecrop	417
Storage-products	142, 145, 146
Stork's-bill	412
Strawberry	214, 229
Strengthening zone	90
Strobilus	215
Stromata	359
Struggle for existence	216, 305
Style	159, 175
Suæda	403
Sub-class	219
Suberin	26
Suberization	26
Suberized tissue	43
Sub-hymenial layer	370
„ -kingdom	219
Subsidiary cells	51
Succulent fruits	205
„ „ , Evolution of	310
Sucker	67, 68
Sugar	30, 33, 138
Sunflower	236
„ , Seed and fruit of	55
„ , Stem of	75
Superior calyx	167
„ ovary	179
Survival of the fittest	306
Suspensor	200, 202, 296
Sweet William	225
Syconus	215
Symbiosis	151
Symmetry of members	9
„ , Floral	164
Symphoricarpus	437
Symphytum	428
Sympodial cyme	181
Sympodium	64
Syncarp	205
Syncarpous	176
Synergidæ	180, 198, 301
Syngenesious stamens	173
Syringa	418, 421, 427
TANNIN	33
Tapetal cells (Tapetum)	258, 274, 291, 300
Tap-root	95, 96
Taraxacum	235
Tegmen	57
Tendrils	16, 67, 72, 98, 113, 124, 155
Terminal buds	62
Testa	55, 201
Tetradynamous	173, 226
Thalamifloræ	220

	PAGE
Thalamus	74, 158, 164
Thalictrum	224
Thallophyta	4, 6, 330
Thallus	6. 312. 314, 330, 331, 334, 338, 342, 343, 348
Theca	321
Thickening, Types of	23
Thistle	236
Thorn	73
„ -apple	430
Thylosis	150
Thyme	232
Thyrsus	189
Tilia	411
Tiliaceæ	411
Timothy Grass	243
Tissue-systems	49
Tissues	40
„ , Differentiation of	78, 92
„ , Meristematic	40
„ , Permanent	40
Toadflax	234
Tobacco Plant	430
Tomato	430
Torula	358
Torus of bordered-pit	25
„ „ flower	74, 158
Tracheal tissue	44
Tracheidal cells	287
Tracheides	44, 284
Trama	364
Transfusion tissue	287
Transpiration	14, 136, 137
„ -current	134
Trichomes	52
Trifolium	227
Triticum	243, 244
Trollius	224
Tropæolum	414
True fruits	205
Trunk	61
Tuber	71, 97
Tubulifloræ	235
Turgidity	133, 134, 149, 155
Turnip	97, 226
Tussilago	236
Twining plants	66, 227, 402
„ stems	149, 155
ULEX	227
Ulmaceæ	221, 402
Ulmus (see Elm)	402
Ulothrix	4, 334
„ , Gametophyte of	335
„ , Sporophyte of	335, 337
Umbel	187, 230
Umbelliferæ	220, 230
Underground stems	68
Unicellular plants	5
Unicostate venation	115
Unisexual	162

Bot. 30

INDEX.

Urtica 400
Urticaceæ 221, 400

VACCINIUM 424, 426
Vacuoles 28
Vagina 110
Vaginula 325
Variation 306
Variations 307
Varieties 307
Vascular bundle ... 53, 75, 76, 83, 91
„ „ , Development of 80
Vascular Cryptogam 4
„ „ and Flowering Plant 305
Vascular Cryptogam and Gymnosperm 290
Vascular plants 18
„ system 49, 53, 82
Vascular Systems:
  Angiospermous leaf 125
  „ root 100
  Aspidium 250
  Dicotyledonous stem 76
  Monocotyledonous stem ... 90
  Pinus 283
  Pteris 253
  Selaginella 272
Vaucheria 4, 338
„ , Gametophyte of 341
Vegetative cell 197, 292, 302
„ reproduction 15, 192, 259, 266, 322, 365
„ shoot 7, 259
Veins of leaf 156
„ , Ending of 128
Velamen 240
Velum 369
Venation of leaf 114
Venter of archegonium ... 261, 316
Ventral canal-cell ... 262, 295, 317, 324
„ scales 315
Verbascum 234
Vernal Grass 245
Vernation 122, 123
Veronica 234
Versatile 174
Vertical sections 182
Verticillaster 190, 232
Vessels 32
Vetch 227
Viburnum 435, 437
Vicia 227
Viola 196, 407
Violaceæ 220, 407

Violet 172, 197, 210, 407
Viper's Bugloss 428
Vitta 230

WALL-FLOWER 226
  Wall Pellitory 400
Walnut 212, 399
Water-culture 132
„ -plants 156
„ -pores 52, 128
„ -stomata 52
„ , Ascent of 137
„ , Importance of 20, 130
Wayfaring Tree 435
Weak stems 66
Wheat 207, 243, 244
Whorl 64, 113
Whortleberry 424, 426
Willow 221, 238
Willow-herb 421, 423
Winter Cherry 430
„ -green 426
Wood 44, 53
„ -fibres 53
„ -parenchyma 42, 53
„ -vessels 32, 44
Wood Sorrel 414
Woodbine 435

XANTHOPHYLL 140
  Xerophyte 156, 239, 417
Xerophytic characters ... 157, 298, 415, 419, 424, 432
Xylem 44, 53
„ , Functions of 134
„ , Primary 76, 81
„ , Secondary ... 81, 84, 104

YEAST (see *Saccharomyces*).
  Yellow Rattle 233
Yew 280, 288, 297
Yucca 93, 238

ZEA 242, 244
  Zoocœnocyte 340
Zooglœa 374
Zoogonidium 335, 339, 352
Zoophilous 194
Zoospore 335, 336, 352, 353
Zygomorphic symmetry 9, 164
Zygospore 40, 330, 335, 337
Zygote 15, 39, 330
Zymase 367

Lightning Source UK Ltd.
Milton Keynes UK
UKHW012019060622
404004UK00002B/642